T0252476

RESEARCH IN MARITIME HISTORY
NO. 55

ECONOMIC WARFARE
AND THE SEA

Grand Strategies for Maritime Powers,
1650–1945

DAVID MORGAN-OWEN
AND
LOUIS HALEWOOD

Recently Published Books in the Series

RESEARCH IN MARITIME HISTORY
NO. 55

Economic Warfare and the Sea

Grand Strategies for Maritime Powers,
1650–1945

DAVID MORGAN-OWEN
AND
LOUIS HALEWOOD

First published 2020 by
Liverpool University Press
4 Cambridge Street
Liverpool
L69 7ZU

This paperback edition published 2023

British Library Cataloguing-in-Publication data
A British Library CIP record is available

ISBN 978-1-78962-159-4 cased
ISBN 978-1-80207-826-8 paperback

Typeset by Carnegie Book Production, Lancaster
Printed and bound by CPI Group (UK) Ltd, Croydon CR0 4YY

Contents

Figures

Tables

Acknowledgements

In editing this volume we have incurred a great many debts. This project was conceived in the margins of a conference held at All Souls College, Oxford in the summer of 2017. As such, we would like to thank all of the participants in and organisers of that event for their part in a wide-ranging and thought-provoking discussion, which provided the initial inspiration for this volume.

As editors, we would also like to extend our thanks to all of our contributors themselves, for committing to the project and for working with us to bring the volume to publication successfully.

Beyond those who wrote chapters in this volume, many others contributed to it in other ways, and we would like to extend our appreciation to Nicholas Rodger, Richard Blakemore, Avram Lytton, Richard Unger, Sir Hew Strachan, Evan Wilson, and Peter Wilson for their advice and support. We would also like to thank All Souls College, Oxford; Merton College, Oxford; the Faculty of History, University of Oxford; and The Guy Hudson Trust. Countless others have provided valuable advice and support during the process of publication, but we are particularly grateful to the series editor and anonymous reviewers for their perceptive comments on the initial manuscript. Alison Welsby and the team at Liverpool have been exemplary throughout, and their care and hard work have added materially to the final version of the text.

The editors and authors would like to thank the help of a number of archives and archivists who have been of assistance in the production of this volume: the Archives départmentales des Bouches-du-Rhône, the Archives départmentales de la Gironde, the Archivo General de Indias, the British Library, the Bundesarchiv-Militaerarchiv, the Centre des Archives diplomatiques de Nantes, the William L. Clements Library, the Federal Reserve Bank of St Louis, the Historical Society of Pennsylvania, the Hull Record Office, the Nationaal Archief, the National Archives and Records Administration, the National Archives of the United Kingdom, the National Maritime Museum, the Naval Historical Branch, the New York Public Library, the Peabody Essex Museum, and the Rijksmuseum. Every effort has been made to secure the necessary permissions for copyrighted material. The editors extend their sincerest apologies for any lapses which may result in copyright having been infringed inadvertently.

Introduction:
Strategy, Economics, and the Sea

David Morgan-Owen and Louis Halewood

The sea has been the crucial medium across which global commerce has flowed for centuries. The renowned American maritime theorist Alfred Thayer Mahan may have exaggerated the influence of sea power when he wrote that 'control of the sea, by maritime commerce and naval supremacy, means predominant influence in the world'. Yet the justification he provided for this argument did strike to an essential truth: 'however great the wealth product of the land, nothing facilitates the necessary exchanges as does the sea'.[1] This aspect of Mahan's argument remains as relevant today as it did when he wrote it in the late nineteenth century: in 2018 some 90 per cent of world trade remains seaborne, and 99 per cent of global digital communications are carried through sub-ocean cables.[2]

Given the crucial importance of seaborne trade to economic development and the creation of wealth – particularly from the eighteenth century onwards as the volume of such trade increased ever more rapidly – it is unsurprising that strategic thinkers and statesmen have debated how controlling the seas could serve their military and political objectives for centuries.[3] Indeed, the complex interrelationship between these factors played a key role in driving the emergence of a distinct body of 'maritime' strategic thought between the late nineteenth century and 1945.[4] Since that point, however, a disjuncture has developed between these

---

[1]   A.T. Mahan, *Interest of America in Sea Power, Present and Future* (Port Washington, NY: Kennikat, 1897), p. 124.

[2]   United Nations, International Maritime Organization: https://business.un.org/en/entities/13; *Joint Doctrine Publication 0-10: UK Maritime Power* (Swindon: Ministry of Defence, 2017), p. 12.

[3]   R. Findlay and K.H. O'Rourke, *Power and Plenty: Trade, War, and the World Economy in the Second Millennium* (Princeton, NJ: Princeton University Press, 2007), chap. 5.

[4]   H. Strachan, 'Maritime Strategy and National Policy', in H. Strachan (ed.), *The Direction of War: Contemporary Strategy in Historical Perspective* (Cambridge: Cambridge University Press, 2013), pp. 153–8.

historical bodies of thought and practice and mainstream understandings both of warfare and of international relations.[5] As the discipline of history has become increasingly peripheral to debates about strategy since 1945, so the relationships between maritime power, strategy, and trade have become progressively less and less well understood.[6]

Efforts to bridge the divide have been made in recent decades, with varying degrees of success. Arguably the most significant have been the concepts of 'economic statecraft' and 'grand strategy'. The former aims to understand economics as an instrument of politics, the latter to 'bring together all of the elements, both military and non-military, for the preservation and enhancement of the nation's long term (that is, in wartime and peacetime) best interests'.[7] However, despite a number of significant interventions, each approach remains beset by conceptual difficulties. Literature on economic statecraft has, until recently, 'consist[ed] of a rather sterile debate about whether economic sanctions or incentives can achieve important foreign policy objectives, with less attention to the more policy-relevant issue of when and under what conditions economic statecraft can achieve these goals'.[8] It has also struggled to incorporate the role of commercial companies into its analysis, or to broaden its understanding beyond US-centric accounts.[9] The idea of grand strategy has similarly become embroiled in discussions over US foreign policy,[10] to the detriment of its aim of incorporating non-military means into a holistic interpretation of national power.[11] The result has been a persistent failure to link sea power, economics,

---

[5]   C. Nolan, *The Allure of Battle: A History of How Wars Have Been Won and Lost* (Oxford: Oxford University Press, 2017), *passim*; D.A. Baldwin, *Economic Statecraft* (Princeton, NJ: Princeton University Press, 1985), p. 4.

[6]   For an important recent intervention, see D. Moran and J.A. Russell (eds), *Maritime Strategy and Global Order: Markets, Resources, Security* (Washington, DC: Georgetown University Press, 2016).

[7]   P.M. Kennedy, 'Grand Strategy in War and Peace: Toward a Broader Definition', in P.M. Kennedy (ed.), *Grand Strategies in War and Peace* (New Haven, CT: Yale University Press, 1991), p. 5.

[8]   J.M.F. Blanchard and N.M. Ripsman, 'A Political Theory of Economic Statecraft', *Foreign Policy Analysis* 4 (2008), p. 371.

[9]   W.J. Norris, *Chinese Economic Statecraft: Commercial Actors, Grand Strategy, and State Control* (Ithaca, NY: Cornell University Press, 2016), pp. 14–21.

[10]   H. Brands, *What is Good Grand Strategy? Power and Purpose in American Statecraft from Harry S. Truman to George W. Bush* (Ithaca, NY: Cornell University Press, 2014); W.C. Martel, *Grand Strategy in Theory and Practice: The Needs for an Effective American Foreign Policy* (Cambridge: Cambridge University Press, 2015); T. Balzacq, P. Dombrowski, and S. Reich, 'Is Grand Strategy a Research Program? A Review Essay', *Security Studies* 40.1–2 (2017), pp. 295–324.

[11]   H. Strachan, 'The Meaning of Strategy: Historical Perspectives', in Strachan, *The Direction of War*, pp. 41–5.

and strategy together in a fashion which has proven convincing either to politicians or to the public.[12]

Now seems a particularly appropriate time to address this lacuna, as great power rivalry in the maritime domain has returned to the forefront of the global strategic agenda. Whether in the case of the United States and China in the Asia-Pacific, or NATO and Russia in the North Atlantic, Baltic, and Black Sea, confrontation in the maritime commons appears a more realistic prospect now than at any time since the end of the Cold War. Economic competition and confrontation have also become increasingly prevalent issues within the calculus of great power rivalry, particularly in the context of Chinese economic development. Despite a scholarly consensus that economic sanctions 'are of limited utility in achieving foreign policy goals', trade restrictions and protectionism form a prominent part of Western efforts to confront Chinese, Russian, and Iranian ambitions.[13] Nationalist and coercive trade policies, rhetoric about 'economic aggression', and efforts to ensure energy and resource independence also feature prominently in US discussions of Sino-American relations and rivalry.[14] Such competition often occurs below the threshold of conflict, blurring the distinction between peace and war and elevating the importance of non-state actors: the British armed forces anticipate future conflicts involving a 'close co-ordination of cyberspace, exploitation of information, and political and economic warfare, for which expertise and resources, particularly at scale, may primarily exist outside the state'.[15] The security of the maritime commons, and its potential as a means of exerting economic and diplomatic pressure upon rivals, are thus issues of increasing importance in both military and policymaking circles.[16] Against this backdrop, the ideas of 'classical' theorists of maritime

---

[12] H. Strachan, 'Strategy in Theory; Strategy in Practice', *Journal of Strategic Studies* 42.2 (2019), pp. 185–6.

[13] G.C. Hufbauer, J.J. Schott, and K.A. Elliot, *Economic Sanctions Reconsidered: History and Current Policy* (2nd ed.; Washington, DC: Institute for International Economics, 1990), p. 29.

[14] K. Johnson and E. Groll, 'It's No Longer Just a Trade War between the US and China', *Foreign Policy* 4 Oct. 2018: https://foreignpolicy.com/2018/10/04/its-no-longer-just-a-trade-war-with-china-pence-spying/ (accessed 26 Nov. 2018).

[15] Ministry of Defence, *Global Strategic Trends: The Future Starts Today* (6th ed.; London: Ministry of Defence, 2018), p. 136.

[16] S. Mirski, 'Stranglehold: The Context, Conduct and Consequences of an American Naval Blockade of China', *Journal of Strategic Studies* 36.3 (2013), pp. 385–421; Z.K. Goldman and E. Rosenberg, *American Economic Power and the New Face of Financial Warfare* (Washington, DC: Center for a New American Security, 2015): https://s3.amazonaws.com/files.cnas.org/documents/CNAS_Economic_Statecraft_061115_v02.pdf?mtime=20160906081249 (accessed 26 Nov. 2018); P. Navarro, 'Mearsheimer on Strangling China & the Inevitability of War', *Huffpost*, 3 Oct. 2016: www.huffingtonpost.com/peter-navarro-and-greg-autry/mearsheimer-on-strangling_b_9417476.html?guccounter=1 (accessed 26 Nov. 2018); C.J. McMahon, 'Maritime Trade Warfare:

strategy are coming to play a prominent role in framing debates about how to respond to such instability and competition – particularly in the Asia-Pacific.[17] As such, reinterpreting what these theorists said, and questioning whether their writings accorded with historical reality, appear urgent requirements.

This collection of essays aims to foster a debate over these issues by providing a series of historical perspectives on the practice of maritime 'economic warfare', and by relating those examples to the study of strategy. It is concerned less with how warfare has impacted on trade (although this is clearly an important issue) than with the role economic warfare has played within the strategies of state and non-state actors across time. It therefore sets out to examine some of the foundational claims of 'classical' maritime strategists such as Mahan and his contemporaries regarding the benevolent relationship between sea power, economics, and strategy. Before outlining the contributions made by individual writers, this introduction will therefore examine the current state of our understanding of strategy and of the role of maritime economic warfare within it. These are topics which have sat awkwardly between a number of academic disciplines, and within the study of history itself. This, along with broader trends within strategic thought, has resulted in a series of difficulties of interpretation which need to be explored and understood in order to frame what follows.

Strategy, Maritime Strategy, and Naval History

Writers on military affairs began to propound general theories of strategy in the late eighteenth and early nineteenth centuries. Works by writers such as Antoine-Henri Jomini and Carl von Clausewitz sought to apply the scientific rationalism of the European Enlightenment to the conduct of war. As the result of the circumstances in which they wrote and their own professional motivations for doing so, their theories were primarily concerned with land battles on the continent of Europe – the use of the engagement for the purpose of the war. Insofar as economic considerations featured in their analysis, they pertained

A Strategy for the Twenty-First Century?' *Naval War College Review* (2017), pp. 15–38; G. Collins, 'A Maritime Oil Blockade against China – Tactically Tempting but Strategically Flawed', *Naval War College Review* 71 (2018), pp. 49–78; H. Brands and Z. Cooper, 'Getting Serious about Strategy in the South China Sea', *Naval War College Review* 71 (2018), pp. 13–21.

[17]   J.R Holmes and T. Yoshihara, *Chinese Naval Strategy in the 21st Century: The Turn to Mahan* (London: Routledge, 2012); D.C. Gompert, *Sea Power and American Interests in the Western Pacific* (Santa Monica, CA: RAND Corporation, 2013); D. Connolly, 'The Rise of the Chinese Navy: A Tirpitzian Perspective of Sea Power and International Relations', *Pacific Focus* 37.2 (2017), pp. 182–207; G. Allison, *Destined for War: Can America and China Escape Thucydides's Trap?* (London: Scribe, 2017).

primarily to the sustainment and supply of armies in the field.[18] Ideas of maritime strategy were, by contrast, closely associated with theories of economics and trade from the outset. In Walter Raleigh's words: 'For whosoever commands the sea commands the trade; whosoever commands the trade of the world commands the riches of the world, and consequently the world itself'.[19] Yet it was not until the late nineteenth century that systematic attempts were made to capture the essence of maritime strategy in the manner of Jomini or of Clausewitz.

From the outset, writing on the topic was closely associated with the requirements of naval education and with debates about contemporary strategy. In this respect it was not fundamentally different from Clausewitz. Similarities between writing on land and naval warfare extended to the historical method scholars such as Philip and John Colomb, John Knox Laughton, Mahan, and Julian Corbett used to draw their conclusions. In other ways, however, thinking about maritime strategy represented a fundamentally distinct intellectual tradition. The maritime strategists sought to explain how a state could 'achieve its political, economic, and military ends on and through the oceans, by means such as maintaining access to markets and transporting troops'.[20] In other words, they sought to explore what today would be referred to as grand strategy, and to answer fundamentally different questions from their military predecessors in the process. The role of economics in war was a foremost consideration within this discourse, and how it was understood by protagonists in this debate has exercised a considerable influence on all subsequent writing on maritime economic warfare.

Writing about maritime strategy was catalysed by contemporary trends in trade, international affairs, technology, and diplomacy. Foremost amongst these was the explosive growth of global seaborne trade in the second half of the nineteenth century, which created new strategic imperatives. The cost of transporting goods by sea decreased by up to 70 per cent between 1840 and 1910. Combined with improvements in communication and land-based transport infrastructure, this fostered the emergence of a global market in bulk commodities, and the convergence of prices between key markets.[21] These changes had a pronounced effect upon Britain, enabling her to abandon agricultural protectionism and to focus her economy upon manufacturing and financing overseas investments. This produced unprecedented wealth, but also new strategic vulnerabilities. By 1914, some 60 per cent of the calorific value of food consumed in Britain arrived by

---

[18]   Strachan, 'The Limitations of Strategic Culture: The Case of the British Way in Warfare', in Strachan, *The Direction of War*, p. 136 and Strachan, *Clausewitz's On War: A Biography* (New York: Grove Press, 2008).

[19]   Quoted in M. Howard, 'The British Way in Warfare: A Reappraisal', in M. Howard (ed.), *The Causes of Wars and Other Essays* (Cambridge, MA: Harvard University Press, 1983), p. 169.

[20]   J.R. Ferris, 'Intelligence, Information, and the Leverage of Sea Power', in Moran and Russell, *Maritime Strategy and Global Order*, p. 282.

[21]   Findlay and O'Rourke, *Power and Plenty*, pp. 378–87.

sea.[22] Protecting seaborne trade was thus a central factor in shaping British war planning throughout the second half of the nineteenth century. It provided an important impetus for her to develop the institutional frameworks through which to coordinate defence planning,[23] and encouraged a series of writers to propound theoretical frameworks to explain the conduct of maritime warfare.

One of the earliest examples of such an approach was Captain John Colomb's 1867 essay, *The Protection of our Commerce and Distribution of our Naval Forces Considered*. Colomb sought to establish the relative roles of overseas bases, local defences, and a mobile battle fleet in a global system of trade protection, foreshadowing many of the contemporary debates about British defence policy in the 1870s and 1880s.[24] The vulnerability of British seaborne trade also prompted her international rivals to develop alternative theories, intended to exploit her reliance upon the sea to ensure her defeat. The most notable example of this approach was the French *Jeune École*, whose proponents questioned whether new technologies such as the mine and torpedo had overturned the strategic logic of earlier conflicts, and made commerce raiding a potentially decisive instrument of national strategy.[25] Whilst approaching the issue from different perspectives, Anglo-French discourses showed a general agreement that the increasing volume of transoceanic seaborne trade and its relative importance to a country's economy made it a crucial source of national strength, and thus a clear vulnerability in time of war.

This case was taken to a general audience by the writings of the American naval officer Alfred Thayer Mahan, whose work rose to prominence with the publication of *The Influence of Sea Power upon History* in 1890. Mahan argued that states which possessed command of the sea by virtue of a superior battle fleet need not fear *guerre de course*, for significant disruption to commerce was possible only through the control of major trade hubs, something which could not be achieved through the mere raiding of merchant shipping.[26] His influential

---

[22] A. Offer, *The First World War: An Agrarian Interpretation* (Oxford: Clarendon Press, 1989), pp. 218–19.

[23] B. Ranft, 'The Protection of British Seaborne Trade and the Development of Systematic Planning for War, 1860–1906', in B. Ranft (ed.), *Technical Change and British Naval Policy* (London: Hodder and Stoughton, 1977), pp. 1–22.

[24] D.M. Schurman, *Imperial Defence, 1868–1887*, ed. J. Beeler (London: Frank Cass, 2000).

[25] J.T. Sumida, *Inventing Grand Strategy and Teaching Command: The Classic Works of Alfred Thayer Mahan Reconsidered* (Baltimore, MD: Johns Hopkins University Press, 1999); A. Roksund, *The Jeune École: The Strategy of the Weak* (Leiden: Brill, 2007), pp. 1–7.

[26] Mahan, 'The Possibilities of an Anglo-American Reunion', *Interest of America in Sea Power*, p. 563; Dirk Bönker, *Militarism in a Global Age: Naval Ambitions in Germany and the United States before World War I* (Ithaca, NY: Cornell University Press, 2012), pp. 151–8.

thesis is undoubtedly the most commonly cited argument in favour of maritime economic warfare as a decisive weapon, and has retained an enduring significance within the canon of strategic studies.[27] Yet, many of Mahan's arguments were challenged, sophisticated, and nuanced by contemporaries such as Julian Corbett, and have received sustained criticism from subsequent historians.[28] What cannot be denied is the potency of their impact, which illustrated the broader significance sea power enjoyed within contemporary conceptions of international affairs. This impression survived the First World War and waxed large in the writings of polemicists and strategists alike in the interwar years. The power of maritime economic warfare was fundamental to Basil Liddell Hart's concept of a 'British way in warfare' and was spelled out in a succession of historically based works by other authorities such as Admirals Raoul Castex and Sir Herbert Richmond during the interwar period. Reflecting back on the experience of 1914–18 after the Second World War, Richmond maintained:

> It is not too much to say that, but for the consecutive surrenders of maritime rights at Paris and in London which gave sanction to that opposition … economic pressure which eventually contributed to the defeat and downfall of the German aggressor would have begun to take effect far sooner, that the lives of thousands of men would have been spared … and that the vast expenditure of wealth which added to the difficulties of reconstruction would have been avoided.[29]

Writers in this 'golden age' of maritime strategic thought were often more guarded in their claims about the influence of sea power than contemporary navalists – who often exaggerated their arguments for political effect.[30] Nevertheless, broadly speaking, maritime strategists tended to agree that sea power conferred three key benefits, each of which was related to economics. First, that naval and maritime power was a more economically productive investment than military force. This argument hinged upon shipbuilding, insurance, and the relatively low cost and multiple uses of warships when compared to costly standing armies.[31] Second,

---

[27]   The inclusion of a chapter on Mahan by Margaret Sprout in E.M. Earle's *Makers of Modern Strategy: Military Thought from Machiavelli to Hitler* (Princeton, NJ: Princeton University Press, 1948) and an updated entry on the admiral by P.A. Crowl in Peter Paret (ed.), *Makers of Modern Strategy from Machiavelli to the Nuclear Age* (Princeton, NJ: Princeton University Press, 1986) were influential in ensuring this. See also Kennedy, 'The Influence and the Limitations of Sea Power', *International History Review* 10.1 (1988), pp. 2–3.

[28]   Most comprehensively in P.M. Kennedy, *The Rise and Fall of British Naval Mastery* (London: Allen Lane, 1976).

[29]   H.W. Richmond, *Statesmen and Sea Power* (Oxford: Clarendon Press, 1946), p. 286.

[30]   Kennedy, 'The Influence and the Limitations of Sea Power', pp. 6–7.

[31]   Sumida, *Inventing Grand Strategy*, p. 32.

that sea power enabled maritime states to limit their military involvement in conflicts by preserving their trade and using economic blockades to wear an enemy down over time.[32] Finally, that the ability to move armies by sea offered a cheaper and more efficient alternative to the maintenance of permanent standing armies, or the imposition of systems of compulsory military service upon a population.[33] In the minds of some, sea power was thus economically benevolent, politically enlightened, and militarily restrained, making it the most effective form of power in the international system.[34]

This emphasis upon sea power in international affairs began to alter after 1945. The experience of the Second World War and the advent of atomic weaponry appeared to show that technological development had relegated sea power from the forefront of world affairs and curtailed the 'influence of sea power upon history'. Even scholars who had written extensively about sea power earlier in their careers began to eschew further research in the area, preferring instead to begin theorising about how the nuclear age had transformed strategy.[35] During the Cold War, scholarship on strategy therefore became focused upon nuclear deterrence, containment, and proxy warfare. Thereafter, 'small wars' and counterinsurgency, cyber warfare, and artificial intelligence have dominated modern security studies. In some respects this was a reflection of NATO's dominance of the world's oceans in the early decades of the Cold War – the absence of a credible competitor removing some of the impetus which had caused maritime strategy to flourish in an age of naval rivalry. Yet combined with growing scepticism amongst historians about much of the 'history' which underpinned earlier writing on maritime strategy, these shifts meant that the amount of attention devoted to the study of 'sea power' as a concept has declined precipitously. Today, naval history is peripheral to debates about strategy, and more frequently used to justify modern ideas than as a source of inspiration for them.[36] This has left those seeking to understand the role of maritime power in the international system

---

[32] A. Lambert, *Seapower States: Maritime Culture, Continental Empires, and the Conflict that Made the Modern World* (New Haven, CT: Yale University Press, 2018), pp. 1–16 and 'The Naval War Course, Some Principles of Maritime Strategy and the Origins of "The British Way in Warfare"', in K. Neilson and G. Kennedy (eds), *The British Way in Warfare: Power and the International System, 1856–1956* (Farnham: Ashgate, 2010), pp. 219–56; Strachan, 'The Limitations of Strategic Culture', pp. 143–4.

[33] H.J. Mackinder, 'Man-Power as a Measure of National and Imperial Strength', *National Review* (1905), p. 137.

[34] P.A. Crowl, 'Alfred Thayer Mahan: The Naval Historian', in Paret, *Makers of Modern Strategy*, pp. 444–80; Kennedy, 'The Influence and the Limitations of Sea Power', pp. 5–7; Lambert, *Seapower States, passim*.

[35] H. Strachan, 'Technology and Strategy', in Strachan, *The Direction of War*, pp. 184–6.

[36] J.J. Widen, 'Julian Corbett and the Current British Maritime Doctrine', *Comparative Strategy* 28.2 (2009), pp. 170–85.

today with significant conceptual challenges. These stem from the nature of the subject that maritime strategists wrote about, and from the contexts in which they lived and worked.

The majority of the 'classical' works on maritime strategy were produced over a century ago, or at least before 1945. On one level this should not matter. Clausewitz's work has an enduring relevance due to the continuity which characterises its subject – the fundamental nature of war – and to its dialectical method.[37] Mahan, Corbett, and their contemporaries can also lay claim to a significant and long-lasting contribution to the study of strategy. By applying the word 'strategy' to the level of the state, and speaking in terms of 'major' or 'grand' strategy, they prefigured how the word would be used and understood for much of the remainder of the twentieth century. Yet the relevance of other portions of their work has suffered for two reasons.

The first of these is historical. As numerous contributors to this volume point out, much of the history on which the classical works of maritime strategy were based has not survived a century of additional scholarship.[38] Many of the more assertive claims they made about the influence of sea power have thus proven to be false.[39] This is particularly true of some of the economic arguments present within the writings of classical sea power theorists, which were on occasion more powerfully asserted than they ever were proven – particularly in the writings of those whom they influenced, such as Liddell Hart. The notion that a 'maritime' approach to strategy represented a 'cost-effective if not downright cheap' means of waging war has endured well past the heyday of writing about maritime strategy.[40] Yet modern scholarship has made clear that the archetypal modern 'sea power state' – Britain – did not achieve victory in the global conflicts of the seventeenth and eighteenth centuries by avoiding military spending. The British state raised more taxes, spent more money, and mobilised more manpower than France throughout the Napoleonic Wars. Far from limited war based upon maritime power and an 'indirect approach', this period witnessed the creation of a 'fiscal-naval state'.[41] The economic capacity to sustain this degree of spending

---

[37]   H. Strachan and A. Herberg-Rothe (eds), *Clausewitz in the Twenty-First Century* (Oxford: Oxford University Press, 2007).

[38]   See also Crowl, 'Alfred Thayer Mahan: The Naval Historian' and Kennedy, 'The Influence and the Limitations of Sea Power', pp. 5–7.

[39]   Notable exceptions include J.S. Corbett, *England in the Seven Years War: A Study in Combined Strategy* (London: Longmans, Green and Company,1907); Richmond, *Statesmen and Sea Power*.

[40]   J. Brewer, *The Sinews of Power: War, Money and the English State, 1688–1783* (Cambridge, MA: Harvard University Press, 1990), p. 136.

[41]   J.M. Winter, 'Introduction', in J.M. Winter (ed.), *War and Economic Development: Essays in Memory of David Joslin* (Cambridge: Cambridge University Press, 1975), pp. 4–6; N.A.M. Rodger, 'War as an Economic Activity in the "Long" Eighteenth Century', *International Journal of Maritime History* 22.2 (2010), pp. 1–18.

was not, as Mahan argued,[42] the by-product of a virtuous circle of naval power and overseas trade: British economic growth in the eighteenth century was underpinned by agricultural productivity, and the economic impacts of her investment in *naval* rather than *military* power have not been systematically assessed.[43] The causative relationship Mahan drew between sea power and national economic strength is thus difficult to sustain. Nor were the effects of maritime economic warfare anything like as clear or definitive as they have occasionally been presented in the literature,[44] or in the political arguments of navalists and their supporters.[45] As Richmond himself argued in 1918, 'There are those who say it [sea power] is the principal part and that war should have been made by sea only. I know of no war in which sea-pressure so employed has, by itself, brought an enemy to his knees … the final overthrow has been on land and has been brought about by military action'.[46] If the history on which much classical strategic theory was based was inaccurate, and few modern attempts are being made to produce new general concepts about maritime strategy, then it is perhaps unsurprising that sea power has become increasingly divorced from debates about strategy in the twenty-first century.[47]

A second factor limiting the relevance of writing about sea power today is more conceptual. Mahan, Corbett, and their contemporaries helped to lay the foundations for subsequent debates about grand strategy by expanding the use of the term to include issues beyond solely military affairs. Yet in doing so they simultaneously rooted much of their own commentary in their own times. The very act of incorporating contemporary patterns of trade, finance, and diplomacy into their work meant that, as conditions altered, so their theories lost much of their immediate relevance to understandings of contemporary warfare. This vulnerability was a reality inherent in the methodology of using history to inform writing on strategy. As Corbett noted in 1913:

> A broad distinction will at once suggest itself between histories of bygone wars, waged with material that is now quite obsolete and under conditions that no longer exist, from which we can only derive the

---

[42]   Crowl, 'Alfred Thayer Mahan: The Naval Historian', pp. 449–51.

[43]   Rodger, 'War as an Economic Activity', pp. 9–14.

[44]   For such an over-optimistic interpretation, see N.A. Lambert, *Planning Armageddon: British Economic Warfare and the First World War* (Cambridge, MA: Harvard University Press, 2012).

[45]   D. Redford, 'Collective Security and Internal Dissent: The Navy League's Attempts to Develop a New Policy towards British Naval Power between 1919 and the 1922 Washington Naval Treaty', *History* 96.321 (2011), p. 52.

[46]   National Maritime Museum, RIC/13/3: 'The Functions of the Royal Navy', June 1918, p. 18. See also Karl O'Brien, 'Global Warfare and Long-Term Economic Development, 1789–1939', *War in History* 3.4 (1996), pp. 437–50.

[47]   Strachan, 'Strategy in Theory; Strategy in Practice', pp. 14–15.

broader doctrines; and histories that deal with the wars of yesterday, in which, in spite of the rapid development of material, we seek for closer and more direct light on the wars of tomorrow.[48]

For Corbett, the antidote was continually to write new histories of recent wars, and to use the experience gained therein to modify theories informed by the broader sweep of history. Yet this method has fallen from favour since 1945. Neither the United States Navy nor the Royal Navy produced an official history of the Cold War, the closest equivalent being the British account of the Falklands Campaign.[49] Whether official accounts – with all of their challenges and failings – remain the most suitable vehicle for linking contemporary history with ongoing debates about strategy is open to debate. Collaborative working between historians and students of strategy may well represent a more rigorous and impartial approach, the benefits of which can be seen in this volume and in similar recent works.[50] However, the fact remains that the structures of the modern academe often militate against such endeavours,[51] and the relationship between naval and maritime history and debates about strategy is now a distant one.[52] It is no accident that maritime strategic thought has atrophied as a result.

With the bridge between the past and the strategic issues of the present eroded, maritime strategy has become reliant upon a canon of works produced in the period between 1870 and 1945[53] – a situation which would have been anathema to many of the most influential writers on sea power at the time, who understood the vital importance of using contemporary history to continually update their work. This presents a major challenge to those seeking to adumbrate new visions of sea power today, and to navies seeking to develop new ways of thinking about maritime strategy in the modern world. Both are obliged to refer back to the 'classical' age of maritime strategy, and to rely upon the past only for the 'broader doctrines' which can be drawn from a different age. The result has been a tendency to quote from Mahan and Corbett without meaningfully

---

[48]   J.S. Corbett, 'Staff Histories', in J.S. Corbett (ed.), *Naval and Military Essays* (Cambridge: Cambridge University Press, 1914), p. 25.

[49]   L. Freedman, *The Official History of the Falklands Campaign*, 2 vols (London: Routledge, 2005).

[50]   J. Lacey (ed.), *Great Strategic Rivalries: From the Classical World to the Cold War* (Oxford: Oxford University Press, 2016); Moran and Russell, *Maritime Strategy and Global Order*.

[51]   R. Betts, 'Should Strategic Studies Survive?' *World Politics* 50.1 (1997), pp. 22–6.

[52]   On the need for the more 'traditional' aspects of naval history to be incorporated into the vibrant field of maritime history, see Q. Colville and J. Davey, 'Introduction', in Q. Colville and J. Davey (eds), *A New Naval History* (Manchester: Manchester University Press, 2019), pp. 14–19.

[53]   A rare Cold War exception being P. Gretton, *Maritime Strategy: A Study of British Defence Policy* (London: Cassell, 1965).

engaging with their work,[54] and a persistent failure to produce concepts of maritime strategy which convincingly address the challenges of the twenty-first century.[55] As a 2015 report by the US Naval Postgraduate School concluded, 'today's Navy suffers from a strategy deficit'.[56]

Taken together, these points highlight the benefit of revisiting the history of economic warfare, and of doing so from the perspective of strategy. Fresh history provides a new basis upon which to think about the 'broad doctrines' which warfare in the past can offer. Such history is urgently needed as, in their creditable quest to qualify the excessively optimistic assessments of 'the influence of sea power upon history' forwarded by Mahan and Liddell Hart, historians now risk underrating the influence sea power exerted on world affairs.[57] Moreover, historical writing which addresses the relationship between sea power and strategy in an explicit sense remains surprisingly sparse.[58] The works produced during the classic age of writing on sea power are finite in number and were written without the benefit of the wealth of historical understanding we now possess, and often without much of the primary material which we can now access. We therefore possess an inadequately small number of historically informed attempts to advance general conclusions about the practice of maritime warfare.[59] Scholarship of this nature is vital in order to inform our understanding of strategy in today's world, which approximates much more closely to traditional ideas of maritime strategy than to earlier military uses of the term. This is particularly so when it comes to discussions of 'grand strategy'. Such debates are increasingly fashionable, yet risk depicting their subject overwhelmingly in terms of historical continuity, and of assuming that current practice is justified by a long process of 'progress' in strategic thought.[60] The reality, as Corbett

---

[54]   Widen, 'Julian Corbett and the Current British Maritime Doctrine'; S. Reich and P. Dombrowski, *The End of Grand Strategy: US Maritime Operations in the 21st Century* (Ithaca, NY: Cornell University Press, 2017), p. 176.

[55]   Strachan, 'Strategy in Theory; Strategy in Practice', pp. 14–15.

[56]   J.A. Russell et al., *Navy Strategy Development: Strategy in the 21st Century*, Naval Postgraduate School Naval Research Program, June 2015, p. 4: www.hsdl. org/?view&did=768350 (accessed 10 Apr. 2019).

[57]   Kennedy, 'The Influence and the Limitations of Sea Power', p. 7; D.G. Morgan-Owen, 'War as it Might Have Been: British Sea Power and the First World War', *Journal of Military History* 83.4 (2019), pp. 1095–131.

[58]   Exceptions being Kennedy, *The Rise and Fall of British Naval Mastery*; D. Baugh, 'British Strategy during the First World War in the Context of Four Centuries', in D. Masterson (ed.), *Naval History: The Sixth Symposium of the US Naval Academy* (Wilmington, DE: Scholarly Resources, 1987), pp. 85–110; and D. Baugh, 'Great Britain's "Blue-Water" Policy, 1689–1815', *International History Review* 10.1 (1988), pp. 33–58.

[59]   For an exception, see C.S. Gray, *The Leverage of Sea Power: The Strategic Advantage of Navies in War* (New York: Free Press, 1992).

[60]   Strachan, 'Continuity and Change', in Strachan, *The Direction of War*, pp. 259–61.

argued, is that strategy is in continual flux, and that history is a vital means of appreciating how it changes.[61]

Chapter Summary

A series of important recent studies have illuminated various aspects of naval blockades, and their varied employment in modern history.[62] Our approach aims to set these issues within their broader diplomatic, international, and political contexts. In order to do so, the collection begins with two thematic chapters which address our core themes of the history of economic warfare at sea, and the place of maritime economic warfare in the development of strategic thought. In the first of these, Stephen Conway views economic warfare through the lens of the *longue durée*, covering the period 1650–1945. Following the rise and fall of British sea power over this period, Conway explores the evolution of the relationship between economics and naval warfare more broadly, analysing continuity and change. He explains how economic and industrial advancement affected the Royal Navy – and how the navy in turn stimulated and shaped Britain's economy and the system of global trade which was established over these centuries.

Thereafter, Bleddyn Bowen offers an incisive critique of economic warfare in naval theory, reassessing traditional interpretations of the potency of economic warfare at sea in the works of classical naval writers. Casting his net wider than the oft-cited works of Alfred Thayer Mahan and Julian Corbett, he explores lesser-known works on maritime strategy by Raoul Castex, Charles Callwell, Raja Menon, and others. He evaluates the difficulties they encountered in expressing the subtle and indirect application of sea power, with too many consequently overstating its capabilities. He argues that economic pressure alone is not decisive but is no mere distraction – often it has been a crucial contributor to victory. In doing so, he highlights the more useful elements of the canon of naval theory, which remain worthy of consideration by practitioners today.

Our subsequent contributors highlight four primary themes: the role of neutrals in shaping attempts at economic warfare; the ability of non-state actors to utilise and benefit from economic coercion; the significance of non-naval bureaucracies to the conduct of an economic warfare campaign; and broader questions of sea power and strategies of blockade or counter-blockade in grand strategy. We shall explore these in turn.

---

[61]   Strachan, 'The Limitations of Strategic Culture', in Strachan, *The Direction of War*, pp. 140–50.

[62]   See, for instance, Lance E. Davis and Stanley L. Engerman, *Naval Blockades in Peace and War: An Economic History since 1750* (Cambridge: Cambridge University Press, 2006); B.A. Elleman and S.C.M. Paine (eds), *Naval Blockades and Seapower: Strategies and Counter-Strategies, 1805–2005* (London: Routledge, 2006).

Neutrality

Economic warfare is global in its impacts and consequences, but also in its implementation. Attempts to manipulate the flow of global trade have implications for all states involved in a globalised trading system, whether they are neutral or belligerent.[63] This means that approaching the study of maritime economic warfare from the perspective of either the actor attempting to prosecute it, or the intended target, presents at best a partial picture. In particular, it risks dramatically underestimating the role neutral states play in moderating attempts to use economic warfare, and also the possibility of neutrality itself being used as a disguise for targeted measures of economic coercion below the threshold of formal hostilities. This is reflected by several of our contributors, each of whom highlights different ways through which neutral powers affect the strategic utility of maritime economic warfare.

Anna Brinkman examines the role neutrality played within British grand strategy during the American War of Independence. She shows how both Britain and Spain used official commitments to a policy of neutrality to avoid formal hostilities, whilst simultaneously pursuing efforts to disrupt each other's trade and commercial activities in the New World. Economic warfare was thus sustained at a level below that at which open hostilities might break out, but sufficient to achieve Britain's broader strategic aim of limiting supplies to the rebellious American colonists. This highlights a significant shortcoming of much of the existing literature on maritime economic warfare, which approaches the subject primarily as a wartime activity rather than as an instrument of statecraft which has and can be deployed on a continuous basis.

Leading on from this, Maartje Abbenhuis examines how neutrality remained a credible strategic choice for great powers for much of the century after the Napoleonic Wars. During this period neutrality became central to British grand strategy, as it offered a means of safeguarding British shipping in the event of war without requiring constant readiness for conflict against a myriad of potential foes. The might of the Royal Navy afforded Britain a powerful voice in the definition of neutral rights in international law, and it exploited this fact to foster international consensus around a series of principles which – broadly speaking – favoured her own strategic interests.

Non-state Actors

The second major theme of this volume is the role and experience of non-state actors and their function in economic competition and conflict in the maritime realm. Campaigns of economic warfare and blockade are normally associated with the apparatus of state and naval power; however, several chapters in this

---

[63]   L. Coppolaro and F. McKenzie (eds), *A Global History of Trade and Conflict since 1500* (Basingstoke: Palgrave, 2013); M. Lincoln, *Trading in War: London's Maritime World in the Age of Cook and Nelson* (New Haven, CT: Yale University Press, 2018).

collection uncover the ways in which non-state actors interacted with, and sought to benefit from, episodes of economic warfare in the international system. Erik Odegard explores the interaction between private organisations and formalised state navies with reference to the admiralty boards and chartered trading companies of the Dutch Republic. His analysis challenges the traditional view that relying upon non-state organisations to contribute to a national navy in time of crisis was an inherent weakness of Dutch naval policy. By highlighting that the decentralised Dutch system offered crucial commercial advantages in peacetime, enabling Dutch merchants to provide a higher degree of security for their trading activities in a contested and hostile maritime environment, he reminds us of the danger of assessing naval force structures solely by their performance in battle. As Corbett argued, naval battle was a means to a broader end – the ability to use the sea – and neglecting the forces required to exercise sea command is just as dangerous as failing to prepare for a major engagement.

Silvia Marzagalli examines the impact of the French Revolutionary and Napoleonic Wars on merchant fleets, focusing upon how the United States responded to the disruption of existing flows of global trade. She explains how, with France's merchant fleets impeded by the Royal Navy, merchants in belligerent states sought to use the cover of neutrality to conduct trade and evade the blockade. The greatest beneficiaries were American shipowners, who adapted to take advantage of opportunities presented by the changing circumstances of war in Europe – underlining the importance of non-state actors as well as neutrals in the global context of economic warfare at sea. Consequently, Britain moved to circumscribe neutral rights, a process which damaged its own economy and relations with non-belligerent powers. In the case of the United States, this spilled over into war in 1812, ending a prosperous era for American shipping.

Naval Blockades in Context

Whilst economic warfare remains closely tied to the exercise of state power, too few accounts address the scope of the organisational and bureaucratic challenges involved with waging economic warfare in the context of a modern state.[64] This presents a notable contrast with the studies the British government commissioned of its campaigns of economic warfare, and of the coordination of industry and shipping, in both world wars.[65] Roger Knight seeks to rectify this in the era of the

---

[64]   Exceptions include N. Cohen, 'The Ministry of Economic Warfare and Britain's Conduct of Economic Warfare, 1939–1945', PhD thesis, King's College London, 2001; G. Kennedy (ed.), *Britain's War at Sea, 1914–1918: The War They Thought and the War They Fought* (Abingdon: Routledge, 2016); P. Dehne, 'The Ministry of Blockade during the First World War and the Demise of Free Trade', *Twentieth Century British History* 27.3 (2016), pp. 333–56.

[65]   M.C. Siney, 'British Official Histories of the Blockade of the Central Powers during the First World War', *American Historical Review* 68.2 (1963), pp. 392–401; Rodger, 'War as an Economic Activity', p. 17 n. 61.

Napoleonic Wars by shifting the spotlight to merchant captains and the forgotten naval officers involved in the British system of convoy. These seafarers were charged not with defeating French and Spanish battle fleets but protecting British commerce – the lifeblood of her economy. Mahan's battle-centric analysis significantly under-rated the significance of these more mundane – but strategically vital – activities, and Knight's chapter points to an important area of future research.

John Ferris addresses a similar theme in his study of the bureaucratic machinery behind Britain's campaign of economic warfare in the First World War. He highlights the central importance of intelligence to the conduct of the economic warfare campaign, examining the interception of submarine cable communications and postal censorship as key sources of information to the British government. Ferris examines the bureaucracy in Britain which processed intercepted mail, comprised of organisations largely ignored by historians of the blockade. The postal censorship was composed of a higher ratio of women – deemed particularly adept at interpreting the intercepted letters – than other departments of state, revealing a new window into gender and economic warfare at sea.

Grand Strategy

Our final key theme is the place of campaigns of economic warfare within the grand strategies of the world's leading naval powers – Britain, Germany, Japan, and the United States – in the nineteenth and twentieth centuries. John Hattendorf reassesses the British blockade of the United States during the War of 1812. Exploring the gap between traditional Mahanian orthodoxy of the importance of blockade against the apparent impotence of commerce raiding, along with recent scholarship on the war fuelled by the 2012 bicentenary, he reinterprets the role of maritime power in the conflict. He argues that the primary aim of the United States was not to drive the British from North America but to obtain equal status and respect as a rising power in the eyes of Britons. Although the United States was soundly defeated, and blockade had caused enormous damage to the American economy, the negotiations at Ghent which concluded hostilities – settling on the status quo ante – enabled the United States to obtain recognition that it was now an important, independent state in international relations.

Matthew Seligmann addresses economic warfare in German naval planning and strategy in the early twentieth century. He exposes the myth that Germany's highly damaging submarine campaign during the First World War was an afterthought, and that all German strategic thinkers were consumed by the desire for a potent battle fleet at the expense of commerce warfare. Rather, he explains how German planners and theorists developed designs for employing economic pressure against Britain, its chief rival at sea. Even Alfred von Tirpitz, architect of the *Hochseeflotte*, wished to wage economic warfare against Britain in the event of an Anglo-German confrontation, seeking to target the City of London so as to cause domestic turmoil. Aspects of such designs produced in the *Admiralstab* were adopted upon the outbreak of hostilities, notably the use of surface raiders to destroy a not inconsequential amount of Entente shipping.

Though this small effort did not prove critical, the Admiralty was relieved that the German navy chose not to pursue these plans on a larger scale.

Greg Kennedy provides a new perspective on the dynamic between the Royal Navy, United States Navy, and Imperial Japanese Navy on the eve of the Pacific War. He explains how it was Japan – rather than Britain or the United States – which proved the master of economic warfare in the Far East during this period. The Japanese leadership proved adept at employing economic coercion and the flexibility of sea power against regional rivals, notably China. Meanwhile, naval deficiencies and a lack of willpower stymied London and Washington in their efforts to deter Japan. Instead, they achieved only the exacerbation of tensions with Tokyo, setting Japan on the path to its devastating attack on the American battle fleet at Pearl Harbor in 1941.

Finally, Daniel Moran concludes the volume with a reflection on maritime economic warfare across the period considered by this volume, and considers the use of economic coercion during the Cold War and after. He offers his thoughts on what this may mean for the present and future, warning that modern conditions are inimical to the application of economic warfare as seen in the past. Considerations of humanitarianism and the resilience of existing global trading networks make it harder for navies to wage economic warfare successfully in lengthy attritional conflicts, in the fashion of the first half of the twentieth century. Yet less overt forms of coercion – whether sanctions, economic statecraft, or trade policies – have shown the enduring significance of economic pressure in stand-offs short of war. As Moran shows, taken together, these chapters presage a number of shifts in our understanding of economic warfare and the sea and highlight the diversity of work ongoing in the area.

Towards a Study of Economic Warfare

Inevitably, no single study can do justice to the full breadth and depth of such a multifaceted topic. In aiming to encompass a broad chronology we have been unable to address in depth issues such as the expectations attached to economic warfare between the two world wars or its role in the Cold War (if sufficient documents are yet available to do so).[66] Rather than lament these omissions, we will conclude this introduction by proposing a series of areas for potential future research, based upon the arguments and discussions forwarded by our authors. We posit three areas which may offer useful avenues for further study: global and transnational approaches; exploring *perceptions* of economic warfare (rather than treating arguments for or against it in rational or economic terms); and analysing the impact of economic warfare on military operations.

---

[66]   G. Till, 'The Cold War at Sea', in Moran and Russell, *Maritime Strategy and Global Order*, p. 84.

A Global Approach

The recent turn towards global history has reinvigorated interest in a series of questions of vital importance to the study of maritime economic warfare, which, by their very nature, have widespread impacts upon patterns of worldwide trade.[67] Foremost amongst these is the issue of globalisation itself, and our understanding of the functioning of the global economy.[68] John Darwin has argued that tracing flows of commodities may represent the 'fourth kind of global history', and involve a focus on the 'dense networks of information and exchange, and a crowd of agents and brokers to populate them'.[69] As several of our contributors argue, and as assessments of wartime trade produced by economic historians have shown,[70] such a methodology offers the potential for a much more holistic appreciation of the impacts which attempts at economic coercion and competition have upon the global economy. Moreover, as John Ferris argues in his chapter, developing such an understanding of global trade is a crucial aspect of enacting a modern campaign of economic warfare.[71] Much greater attention to the gathering, interpretation, and exploitation of commercial intelligence clearly offers a profitable avenue for further study.

More broadly, a central challenge of existing approaches to economic warfare is to develop a realistic picture of action and reaction, and to chart how these developed over time.[72] This is a formidable methodological challenge for historians, owing to the difficulties involved in disentangling the effects of economic warfare from other wartime factors affecting the economy, and of incorporating an understanding of cause and effect into a single narrative. Taking the First World War as an example, in recent years novel attempts have been

---

[67]   D. Baugh, *The Global Seven Years War, 1754–1763: Britain and France in a Great Power Contest* (Abingdon: Routledge, 2011); Coppolaro and McKenzie, *A Global History of Trade and Conflict.*

[68]   J. Belich, J. Darwin, and C. Wickham, 'Introduction: The Prospect of Global History', in J. Belich, J. Darwin, M. Frenz, and C. Wickham (eds), *The Prospect of Global History* (Oxford: Oxford University Press, 2016), p. 3. One recent example of an attempt to tell a 'globalised' history of war at sea is C.L. Symonds, *World War II at Sea: A Global History* (New York: Oxford University Press, 2018).

[69]   J. Darwin, 'Afterword: History on a Global Scale', in Belich et al., *The Prospect of Global History*, p. 181.

[70]   Patrick K. O'Brien, *The Economic Effects of the American Civil War* (Basingstoke: Palgrave Macmillan, 1988).

[71]   See also N.A. Lambert, 'Strategic Command and Control for Maneuver Warfare: Creation of the Royal Navy's "War Room" System, 1905–1915', *Journal of Military History* 69.2 (2005), pp. 361–410.

[72]   A. Kramer, 'Blockade and Economic Warfare', in J.M. Winter (ed.), *The Cambridge History of the First World War*, vol. 2, *The State* (Cambridge: Cambridge University Press, 2014), p. 462.

made to understand the impact on land of economic warfare at sea, including measuring the long-term growth rates of German children who lived through the *Hungerblockade*.[73] Yet many histories have been artificially limited by framing the issue as an Anglo-German struggle in the North Sea. In fact, the Entente's economic warfare was a joint strategy executed by Britain, France, Russia, Italy, and – eventually – the United States against Germany, Austria–Hungary, Turkey, and Bulgaria.[74] Economic warfare as a multinational endeavour – building on the recent 'transnational turn' in diplomatic history – can reveal new perspectives which challenge standard interpretations. Enacting such an approach would require in-depth collaborative study, between different areas of history, and across varied geographic areas.[75] Yet it would appear to offer the prospect of significant new insights into a variety of areas – from social to economic history.

Perception, Reality, and Representation

Historians must rightly continue to debate what impact forms of economic warfare actually had upon patterns of trade and production. Yet of even greater significance to the study of strategy is establishing what people at the time believed the actions of economic warfare to be, and how they made those assessments. In other words, we must integrate our interpretations of blockade as an economic instrument with the study of intelligence to a much greater extent than is currently the case. Only by doing so can we reproduce the dynamic interplay between actions and perceived effects.[76] Moreover, if we accept that governments and organisations who undertake campaigns of economic coercion cannot know or anticipate the full repercussions of their actions, it follows that how their actions were understood and represented were almost as important as the actual effects they had.

Returning to the example of the First World War, there was a clear sense in most European capitals that Britain's entry into the conflict would have considerable economic effects. Paris, Berlin, and St Petersburg all viewed a British campaign of economic warfare as a potent – and potentially war-winning – intervention. Yet the British government had done precious little to prepare itself to enact a campaign of economic coercion, or to integrate it into a broader scheme of grand strategy.[77] After the outbreak of war, debates raged within

[73]   M. Cox, 'Hunger Games: Or How the Allied Blockade in the First World War Deprived German Children of Nutrition, and Allied Food Aid Subsequently Saved Them', *Economic History Review* 68.2 (2015), p. 602.

[74]   Kramer, 'Blockade and Economic Warfare', p. 462.

[75]   See, for example, S. Broadberry and M. Harrison (eds), *The Economics of World War I* (Cambridge: Cambridge University Press, 2009).

[76]   Ferris, 'Intelligence, Information, and the Leverage of Sea Power', pp. 283–92.

[77]   M.S. Seligmann, 'Failing to Prepare for the Great War? The Absence of Grand

the British government about what effect the Entente's sea power was exerting on the economies of the Central Powers, and when (or whether) it could be expected to produce clear effects upon the military situation.[78] Whether or not these perceptions were accurate, they still exerted an important influence upon decision-making both during the war and after and deserve attention in their own right.

Similarly, the political and rhetorical dimensions of economic warfare offer a rich source for further discussion. In the defeated Central Powers, the experience of the blockade became a powerful source of reference in political debate in the interwar years and was mobilised as part of nationalist rhetoric by many on the right.[79] Yet, simultaneously, economic blockade informed Britain's liberal self-image and became used as a tool for enforcing inchoate attempts at cooperative international policing under the aegis of the League of Nations.

It is also important to consider representation as a mode of coercion in itself, and to ask how actors have deployed the threat or impact of blockade in rhetoric. Has the threat of economic warfare proven as powerful as the reality? Has economic warfare itself been used to convey messages beyond its physical impact? Ronald Reagan's presidency is closely associated with economic leverage and the apparent overwhelming of the Russian economy via an arms race, centring on the Strategic Defense Initiative. Yet his strategy was subtler than all-out economic warfare. Reagan sought instead to use sanctions in a strategy of 'cold economic warfare' to send diplomatic messages to Moscow, so to deter Soviet aggression in eastern Europe. Ultimately, Reagan wanted to bring the Russians to the negotiating table, not to destroy the Soviet Union outright. That the latter disintegrated so shortly thereafter was a fortuitous by-product of his actual strategy.[80] Applying this nuanced approach to campaigns of economic warfare at sea – and reassessing the intended aim, if not simply to destroy the economy of a state in a Mahanian fashion – may reveal new interpretations of seemingly well-understood events.

Strategy in British War Planning before 1914', *War in History* 24.4 (2017), pp. 414–37.

[78]   R. Smith, 'Britain and the Strategy of the Economic Weapon in the War against Germany, 1914–1919', PhD thesis, Newcastle University, 2000.

[79]   A. Tooze, *The Wages of Destruction: The Making and Breaking of the Nazi Economy* (London: Allen Lane, 2006), pp. 167–8; Kennedy, *The War at Sea*, p. 343; A. Watson, *Ring of Steel: Germany and Austria–Hungary in World War I* (New York: Basic Books, 2014); Kramer, 'Blockade and Economic Warfare', p. 474; Cox, 'Hunger Games', p. 601.

[80]   A.P. Dobson, 'The Reagan Administration, Economic Warfare, and Starting to Close Down the Cold War', *Diplomatic History* 6.1 (2005), pp. 534–55.

Economics and Military Operations

Equally significant is to improve our understanding of the relationship between economic warfare and the conduct of military operations. As Phillips Payson O'Brien has recently argued, during the Second World War allied operations in the air and at sea destroyed a larger volume of Axis military *matériel* than the climactic land battles which feature so prominently in the mainstream historiography.[81] This perspective, which draws upon ideas from political economy, raises important questions about the ways in which historians think about what kind of activities exert the greatest impact upon the outcome of wars. O'Brien's work is focused on the mid-twentieth century, when air power created a powerful new means of attacking enemy production. Yet, as he shows, the conceptual framework which underpinned the allied air–sea campaign was that of the maritime thinkers of the nineteenth century. The same question might be posed of the German army in 1918. Although it is clear that the decisive blows of the war were struck on land, the German army was severely hampered in its mobility and effectiveness by a lack of key raw materials, chiefly rubber, petrol, and lubricants.[82] These materials could no longer be imported from its western neighbours, with whom Germany had gone to war, nor could replacements be found from overseas because of the Entente's blockade. Evidently, this relationship is of tremendous importance, yet historians have so far only scratched the surface of the dynamic between economic warfare at sea and operational-level fighting on land. Elements of the political economy approach may thus have significant implications for the study of other wars, and of earlier periods.

The work of the canon of sea power theory remains alive and well in modern debates about strategy. As Admiral John Richardson, the US Navy's Chief of Naval Operations, claimed in 2015, 'the essence of Mahan's vision still pertains'.[83] Modern scholarship on sea power and strategy is unlikely ever to achieve the historical resonance of Mahan. Yet, as the essays in this collection illustrate, a wealth of significant historical scholarship on the relationship between the sea and strategy is being written from a wide variety of perspectives. This collection aims to bring that work into conversation with debates about contemporary strategy, and to forward a new approach to assessing 'the influence of sea power upon history'.

---

[81] P.P. O'Brien, *How the War Was Won: Air–Sea Power and Allied Victory in World War II* (Cambridge: Cambridge University Press, 2015).
[82] Kramer, 'Blockade and Economic Warfare', p. 488.
[83] Quoted in Reich and Dombrowski, *The End of Grand Strategy*, p. 176.

CHAPTER ONE

Economics, Warfare, and the Sea, *c*.1650–1945

Stephen Conway

The essays in this volume, taken together, cover the period from the sixteenth to the twentieth centuries. Each contribution, however, focuses on a particular period, place, and theme. It seems appropriate, therefore, to start with a more general piece, which considers the relationship between national economic strength and maritime warfare over a long time span. As Britain was the main European and world naval power from the eighteenth century until the Second World War (1939–45), when the United States Navy overtook the Royal Navy in size, the British lens is used here as a way to explore this vast period. This chapter is not intended as a chronologically organised survey; rather, it looks at various dimensions of what was a complex and two-way relationship. Economic developments influenced the nature of sea warfare, while naval strategies affected the health of the economies of Britain and its enemies. The approach adopted is necessarily selective: to consider (even briefly) all aspects of the subject, and provide examples across the chronological range considered here, would be possible only in a book-length study. If many of the examples are from the eighteenth century that is because Britain's eighteenth-century wars are the conflicts with which I am most familiar; but I have sought to provide illustrations from other periods.

The chapter's opening section begins by exploring the role of the seventeenth-century Navigation Acts (designed to regulate the overseas trade of the colonies in North America and the Caribbean) in building up English maritime strength and naval power. It then considers the importance attached to the Navigation Acts by British politicians throughout the eighteenth century, and even up to their eventual repeal in 1849. The second part of the chapter turns to the navy's protection of British overseas trade, particularly through the development of the convoy system to shield merchant shipping from enemy attack, but also by facilitating access to markets that enemies tried to close to British products. The third part of the chapter looks at the navy's attempts to disrupt enemy overseas trade, examining different forms of blockade and assessing their effectiveness. The next section considers perhaps the most obvious – and potentially most important, when considering the period as a whole – connection between the economy and the navy: the stimulus provided to various sectors of British industry and agriculture by the demands of an expanded wartime Royal Navy. The conclusion reflects on what changed, and what stayed much the same, over the 300 years covered in this chapter.

I

The Navigation Acts constituted a vital element in the architecture of British naval power for nearly two centuries from 1650. Yet most historians see them principally as economic instruments, intended to create a protectionist trading system that would enable England and then (from 1707) Britain to secure the maximum benefit from the possession of overseas colonies in the Americas. The first efforts to regulate colonial overseas trade can be dated back to 1621, when an order of the English Privy Council (a prerogative body, acting in the king's name) sought to ensure that all tobacco exported from the infant colony of Virginia came first to England, whatever its ultimate destination.[1] Only with the Civil War and Interregnum did Parliament start to play a role. In 1650 and 1651, statutes passed by the Westminster legislature insisted that the seaborne carriage of goods between English possessions should be on English ships. Parliament continued to flex its muscles after the Stuarts regained the thrones of England and Scotland in 1660. That year, MPs approved the first of a series of Navigation Acts designed to create a legislative framework for the overseas trade of the colonies; amplifying, clarifying, and extending measures followed in 1663, 1673, and 1696.[2] The authors of the Navigation Acts intended them to boost national prosperity by giving the home economy first call on colonial products and by providing home manufacturers with a protected market in English overseas possessions. Unsurprisingly, those interested in the history of economic policy, or political economy, tend to focus on these commercial functions of the system created by the Navigation Acts. The burdens the acts inflicted on the colonies divide opinion: some economic historians suggest a significant diminution of potential income, others a marginal impact. The role of the Navigation Acts in the coming of the American Revolution, a cause célèbre in the middle of the twentieth century, remains a contentious issue.[3]

---

[1]   *Calendar of State Papers, Colonial Series, 1574–1660*, ed. W. Noël Sainsbury (London, 1860), p. 26.

[2]   *English Historical Documents*, ix; *American Colonial Documents to 1776*, ed. Merrill Jensen (London: Eyre & Spottiswoode, 1969), pp. 354–64 provides texts of the various Navigation Acts.

[3]   See, for example, L. Harper, 'The Effects of the Navigation Acts on the Thirteen Colonies', in R. Morris (ed.), *The Era of the American Revolution* (New York: Columbia University Press, 1939), pp. 3–39; O. Dickerson, *The Navigation Acts and the American Revolution* (Philadelphia: University of Pennsylvania Press, 1951); R.P. Thomas, 'A Quantitative Approach to the Study of the Effects of British Imperial Policy on

Maritime scholars, unsurprisingly, appreciate better than do economic historians the contribution of the Navigation Acts to British naval strength. As Daniel Baugh, a leading historian of the eighteenth-century Royal Navy, has memorably put it, the Navigation Acts formed a central feature of a regulatory system designed to generate resources in Britain's Atlantic 'back-yard' for deployment in its European 'front-yard'.[4] They were intended, in other words, to boost national power as much as national prosperity. The 1660 Navigation Act was in this sense the vital piece of legislation, referred to by successive generations of politicians and writers as *the* Navigation Act.[5] It limited overseas carriage of goods between the crown's dominions to English-owned (including colonial) ships, on which at least three-quarters of the crew had to be the crown's subjects.

The shipping clauses of the 1660 Act promoted the growth of the merchant fleet by effectively removing foreign competition in colonial trade. The Dutch, who had dominated Atlantic trade, as they did trade on the North Sea and the Baltic, were largely squeezed out. The effective removal of the Dutch rivals created room for the English shipbuilding and maritime service industries to develop under monopoly conditions. New England, with its plentiful supplies of timber, benefited particularly, with ship construction leading to the growth of highly profitable carrying charges and marine insurance. As two leading American historians have succinctly put it, 'The New Englanders became the Dutch of England's Empire'.[6] In the long term, a large British merchant marine enabled normal trading to continue in wartime even while a significant number of ships served the state as hired auxiliary support for the navy, transporting soldiers and supplies to overseas

---

Colonial Welfare: Some Preliminary Findings', *Journal of Economic History* 25 (1964), pp. 615–38; R.L. Ransom, 'British Policy and Colonial Growth: Some Implications of the Burdens of the Navigation Acts', *Journal of Economic History* 28.3 (1968), pp. 427–35; R.P. Thomas, 'British Imperial Policy and the Economic Interpretation of the American Revolution', *Journal of Economic History* 28 (1968), pp. 427–40; Larry Sawers, 'The Navigation Acts Revisited', *Economic History Review* 45 (1992), pp. 262–84; S. Lynd and D. Waldstreicher, 'Free Trade, Sovereignty, and Slavery: Toward an Economic Interpretation of American Independence', *William & Mary Quarterly*, 3rd series, 71 (2011), pp. 597–630; S. Conway, 'Another Look at the Navigation Acts and the Coming of the American Revolution', in J. McAleer and C. Petley (eds), *The Royal Navy and the British Atlantic World, c.1750–1820* (London: Springer, 2016), pp. 77–96.

[4]   D. Baugh, 'Maritime Strength and Atlantic Commerce: The Uses of "A Grand Maritime Empire"', in L. Stone (ed.), *An Imperial State at War: Britain from 1689 to 1815* (London: Psychology Press, 1994), pp. 185–223.

[5]   Edmund Burke, *Observations on a Late State of the Nation* (Dublin, 1769), p. 74 explained that the 1660 Act was 'commonly called the act of navigation'. See also [Anon.] *A Letter to a Member of Parliament on the Present Unhappy Dispute between Great-Britain and Her Colonies* (London, 1774), p. 18, which refers to 'the famous Navigation Act of 1660'.

[6]   John J. McCusker and Russell R. Menard, *The Economy of British America, 1607–1789* (Chapel Hill: University of North Carolina Press, 1985), p. 92.

theatres. In 1808, during the Napoleonic War, the Transport Board hired more than 1,000 private ships, an extent of hiring made possible because the total number of trading vessels at that time exceeded 22,000.[7]

More importantly, the manning clauses of the 1660 legislation meant that the number of foreigners who could serve on English (then British) merchant ships in peacetime was limited to a quarter of the total crew, in order to build up a supply of domestic mariners. These British mariners, trained in the hard school of long-distance Atlantic seamanship, were then available for the use of the navy in time of war. The notorious press gangs rounded up trained seafarers and forcibly conscripted them into state service, at least for the duration of armed conflicts, while many other merchant sailors, encouraged by enlistment bounties, volunteered to serve on the crown's ships.[8] To avoid a shortage of labour on merchant vessels when British mariners transferred to the navy, the manning clauses of the 1660 Navigation Act were relaxed in time of war, to allow the temporary recruitment into the merchant marine of foreign sailors.[9]

Contemporaries showed a keen awareness of the importance of the Navigation Acts to British sea power. The anonymous author of a *New History of England*, published in 1757, maintained that a great advantage that Britain derived from the 'Act of Navigation' was that it 'increases her Shipping, and breeds up her Seamen'.[10] Another historian, writing a few years later, was equally sure that the Navigation Act 'threw a great weight of naval force into the hands of the mother country'.[11] But perhaps the clearest expression of contemporary association of the 1660 Act with naval power came from Joshua Gee, who argued in 1729 that the 'principal object of the Navigation Act' was the creation of 'a perpetual army of Seamen kept in constant pay … ready to be diverted to public service, when wanted'.[12] Gee's assessment seems to have been

---

[7]   R.K. Sutcliffe, *British Expeditionary Warfare and the Defeat of Napoleon, 1793–1815* (Woodbridge: Boydell & Brewer, 2016), p. 89.

[8]   J.R. Dancy, *The Myth of the Press Gang: Volunteers, Impressment, and the Naval Manpower Problem in the Late Eighteenth Century* (Woodbridge: Boydell & Brewer, 2015) argues that most sailors on board British warships were volunteers. His claims are far from universally accepted; impressment remains as controversial in our own time as it was in the eighteenth century. See N. Rogers, 'British Impressment and its Discontents', *International Journal of Maritime History* 30 (2018), pp. 52–73.

[9]   See S. Conway, *Britannia's Auxiliaries: Continental Europeans and the British Empire, c.1740–1800* (Oxford: Oxford University Press, 2017), pp. 147–8.

[10]   [Anon.] *A New History of England, from the Time of its First Invasion by the Romans, Fifty-four Years Before the Birth of Christ, to the Present Time*, 4 vols (London, 1757), iii. 252.

[11]   Hugh Clarendon, *A New and Authentic History of England: From the Remotest Period of Intelligence to the Close of the Year 1767*, 2 vols (London, [1770?]), ii. 561.

[12]   Joshua Gee, *The Trade and Navigation of Great Britain Considered* (London, 1767), p. 195.

accepted even by those who were less than enthusiastic about other aspects of the Navigation Acts. Adam Smith, the most celebrated critic, acknowledged in 1776 that 'The defence of Great Britain ... depends very much upon the number of its sailors and shipping. The act of navigation, therefore, very properly endeavours to give the sailors and shipping of Great Britain the monopoly of the trade of their own country'.[13]

Politicians, perhaps even more than writers, defended the Navigation Acts as a source of naval strength. In the 1760s and early 1770s, parliamentarians' conviction that the strength of the Royal Navy was a product of the system created by the Navigation Acts led them to resist concessions to the American colonists on taxation. British politicians feared that if the Americans established that Parliament had no right to tax them, they would go on to challenge Parliament's right to regulate their trade. If the colonists broke out of the constraints of the navigation system, ministers and MPs seem to have believed, Britain's economic and naval power would be fatally undermined. It would be little exaggeration to say that British ministers went to war with the Americans in 1775 primarily to keep them in the system formed by the Navigation Acts, which they saw as a vital prop for the Royal Navy.[14] Their fears for the navy explain why in 1778, when the British government sought to reach a compromise peace with the colonists before France intervened in the War of Independence, Parliament's right to tax the colonies was given up but the British representatives received instructions to make no concessions on parliamentary trade regulation.[15]

In 1783, with the war over, Parliament rejected Lord Shelburne's attempt to win over the independent Americans (and reconstruct some form of political relationship) by allowing them to be regarded as British subjects for the purposes of the Navigation Acts. Shelburne's opponents believed that if the Americans could carry goods to and from the British Caribbean, then the British merchant marine would be diminished, which would lead, in the words of William Eden, a fierce critic of the government's proposal, to 'the absolute destruction of our navy'.[16] Shelburne's successors moved quickly to introduce orders in council denying American ships access to the remaining British colonies. William Knox, one of the authors of these orders, was so convinced of their importance that he wished to have a copy 'engraved on my tombstone, as having saved

---

[13]   Adam Smith, *An Inquiry into the Nature and Causes of the Wealth of Nations*, ed. R.H. Campbell, A.S. Skinner, and W.B. Todd, 2 vols (Oxford: Oxford University Press, 1976), i. 463.

[14]   See Conway, 'Another Look at the Navigation Acts', pp. 83–91.

[15]   See William L. Clements Library, University of Michigan, Wedderburn Papers: memorandum, probably drawn up by Alexander Wedderburn, the solicitor general, in January 1778, on the British position in any negotiations.

[16]   Speech of William Eden, opposing the American Intercourse Bill, 7 Mar. 1783, in *The Parliamentary History of England from the Earliest Period to the Year 1803*, ed. William Cobbett and John Wright, 36 vols (London, 1806–20), xxiii. cols 602–615.

the navigation of England'.[17] In 1786, the British Parliament approved a new Navigation Act, which specifically denied ships of the United States access to trade with the British colonies. A shipping shortage caused by the early stages of the conflict with revolutionary France led the British government reluctantly to abandon this absolute prohibition in 1794, but the link between the Navigation Acts and naval strength remained almost axiomatic to successive generations of British politicians. This continued to be the case until well into the nineteenth century, despite the growing influence of free-trade ideas, which seemed to make commercial regulations such as the Navigation Acts appear as relics of the protectionist past. But from the 1820s, free-trade theories started to influence government policy and regulations to control overseas trade began to be relaxed. In 1846, Parliament repealed the Corn Laws, which restricted the entry into Britain of foreign cereals. The Navigation Acts, the remaining parts of what had once been a substantial protectionist edifice, then became the target of free-traders. Even during the parliamentary debates on repeal in 1849, however, MPs expressed fears that removing the Navigation Acts from the statute book, however much it might be justified on commercial grounds, would undermine the power of the Royal Navy.[18]

II

The Royal Navy, for nearly all of the period we are considering, tended to deploy an overwhelming majority of its ships in European waters. In 1707, during the War of the Spanish Succession, nearly three times as many vessels of all types served at home or in the Mediterranean than in all other theatres combined. In 1757, in the early stages of the Seven Years War, 83 per cent of British warships were stationed in home waters or the Mediterranean. In the War of American Independence, the total fell to just under a half – the only occasion in the eighteenth century when concentration gave way to dispersal, despite the opposition of the Earl of Sandwich, the First Lord of the Admiralty. But the Napoleonic Wars saw a return to the established system: in 1804, 72 per cent of the Royal Navy's vessels operated in European waters; by 1812 the figure had risen to 75 per cent.[19] The primary reason for this European deployment was to

---

[17]   William Knox, *Extra-Official State Papers*, 2 vols (London, 1789), ii. 53.

[18]   See Hansard HC vol. 102, col. 725 (14 Feb. 1849) for the speech of John Lewis Ricardo, an enthusiast for repeal, who recognised that his opponents' chief concern was for the well-being of the navy. He had already tried to reassure those who were worried about the future of the navy in his *Anatomy of the Navigation Laws* (London, 1847), pp. 105–12, which suggested (correctly, as it transpired) that the navy would continue to thrive without the prop apparently provided by the Navigation Acts.

[19]   N.A.M. Rodger, *The Command of the Ocean: A Naval History of Britain*, vol. 2, *1649–1815* (London: Allen Lane with the National Maritime Museum, 2004), pp. 612–13,

ensure the availability of sufficient vessels to prevent an invasion of the home territories, regarded by First Lords of the Admiralty and government ministers in general as the Royal Navy's most important duty. British decision-makers realised that an enemy landing would not necessarily aim at occupation and conquest of the whole of the home islands and the installation of a client ruler who would do the bidding of the invader. A more likely outcome would be the destabilising of British public finances the moment enemy troops landed, damaging the wider economy and forcing the British government to open negotiations. British strategists recognised that countering the threat of invasion required more than the husbanding of naval strength in home waters. The Royal Navy carried out these defensive operations in the Mediterranean too; warships shadowed the French fleet based at Toulon, with the aim of preventing it from joining the French Atlantic fleet stationed at Brest in Britany.

The concentration of warships near to home had several other benefits, however. Perhaps most importantly, it facilitated the protection of homeward- and outward-bound British overseas trade. During hostilities, enemy privateers – authorised by their governments to attack and capture British merchant ships – could easily take lightly armed trading vessels. They operated far away from the British Isles; in 1748, at the very end of the War of Austrian Succession (1740–48), Spanish privateers off the Delaware Capes succeeded in stymying Philadelphia's overseas trade.[20] The most acute threat posed by enemy privateers, however, was in the waters nearer to home – in the Western Approaches to the Channel, in the Channel itself, and in the North Sea. Losses could be considerable in eighteenth-century wars. Seventeen Bristol ships, valued at more than £59,000, were lost between 1739 and 1741.[21] Bristol's trade, admittedly, was particularly vulnerable to attack. Other western ports with extensive Atlantic trade, such as Liverpool and Glasgow, could route their Atlantic commerce around the north of Ireland, well away from enemy naval vessels and privateers. For vessels sailing to or from Bristol, however, the northern route was too long a diversion and the Western Approaches and Bristol Channel, even though within easy range of enemy ships, the only practicable possibility.[22] But even if Bristol's geographical position made it particularly at risk of severe losses, the overall impact of enemy attacks on British trade was far from negligible. British merchant shipping losses

---

614–16 and Appendix III; The National Archives, Kew (TNA), Admiralty Papers, ADM 8/100.

[20]   Philip L. White (ed), *The Beekman Mercantile Papers, 1746–1799*, 3 vols (New York: New York Historical Society, 1956), i. 48.

[21]   W.E. Minchinton (ed.), *The Trade of Bristol in the Eighteenth Century* (Bristol: Bristol Record Society, 1957), p. 152.

[22]   K. Morgan, *Bristol and the Atlantic Trade in the Eighteenth Century* (Cambridge: Cambridge University Press, 1993), pp. 11–32, 220–1 argues that the city's strategically vulnerable position in the long runs helps to account for its decline, with Liverpool and Glasgow the main beneficiaries.

in the War of American Independence are reckoned to have amounted to as much as £18 million.[23]

The danger diminished if merchant ships sailed in convoys, with many other vessels, which were given naval support. Convoys might comprise more than a hundred trading vessels, escorted by only a small number of warships, but the protection they provided encouraged insurers to view merchant vessels in convoy as much less at risk compared with trading ships sailing on their own.[24] The admiralty recognised that trade protection was an important duty and responded positively to requests for naval escorts for merchant shipping convoys leaving British ports for overseas destinations. For instance, when France and Spain entered the War of American Independence on the side of the rebellious colonies, Hull's corporation revived its committee for convoys, which had been founded during the preceding Seven Years War (1756–63).[25] The corporation directly approached the admiralty for the allocation of naval vessels for trade protection duties, and the town's MPs, mindful of the importance of pleasing local interests, lobbied vigorously for the same purpose.[26] These efforts bore fruit: warships escorted convoys of Hull trading ships across the North Sea to the Danish Sound (the gateway to the Baltic) with consignments of British manufactures, and then back again with timber products, iron, hemp, and linen yarn.[27] The convoy system was a mixed blessing for merchant shippers and those who provided them with their cargoes: cheaper insurance was undoubtedly a benefit, but large numbers of vessels arriving in port at the same time inevitably lowered the price of the goods they carried and so reduced the profitability of voyages.

In the twentieth century, the Royal Navy provided warship escorts for merchant vessels sailing to France, the United States and Canada, and neutral Holland in the First World War (1914–18). The principal threat now came not from enemy surface naval craft and privateers, but from submarines. German U-Boats, as they were called in English, attacked British, allied, and neutral shipping, both in home waters and in the Atlantic. Their success was limited, however. After the sinking of the Cunard liner *Lusitania* in 1915, with the loss of 128 American lives, protests from the neutral United States led the Germans temporarily to scale back on their submarine attacks. Even so, during the course

---

[23]   McCusker and Menard, *Economy of British America, 1607–1789* (2nd ed., 1991), p. 362.

[24]   For an eye-witness account of the departure of one convoy during the War of American Independence, see *The Diary of Sylas Neville, 1767–1788*, ed. Basil Cozens-Hardy (London: Oxford University Press, 1950), p. 271.

[25]   Hull Record Office, WM/3, Committee for Convoys minute-book, 1757–82.

[26]   Hull Record Office, Borough Records, BRL 1295, 1311, 1315.

[27]   TNA, Admiralty Papers, ADM 1/2393, Capt. Stephen Rains to Philip Stephens, 25 April, 30 June, 8 July, 22 Sept., 4 Nov. 1780, 23 Apr. 1782; ADM 1/1446, petition of Hull merchants to Capt. Alexander Agnew, Oct. 1780.

of 1916, U-Boats sank 104,700 tons of merchant shipping.[28] That December, the German high command decided that starving Britain into submission was the best way to break the deadlock on the western front.[29] Implemented in early 1917, the German declaration of unrestricted submarine warfare proved less successful than its architects had hoped. British, allied, and neutral shipping losses, initially considerable, fell significantly during the course of 1917 and 1918. During the last year of the war, German U-Boats sank just 33,798 tons of merchant shipping.[30] Convoys seem to have provided effective protection; many of the losses occurred when merchant vessels fell behind and offered isolated targets. In part, the success of the convoy system was a result of another technological development absent in the days of sail: aircraft added to the defence of merchant shipping, especially in coastal waters near to the home islands.

In the Second World War, the convoy system was revived. The Royal Navy escorted merchant ships carrying war supplies and food to the Soviet Union, through Arctic waters, to the ports of Archangelsk and Murmansk. Weapons and other military equipment from Britain and the United States, along with North American food and raw materials, were used to sustain the Soviet campaigns against Nazi Germany. Foodstuffs landed in the Arctic ports, which were transported inland by rail, also helped ease starvation in the besieged city of Leningrad. Arms manufacturers in Britain and the United States, and American and Canadian food producers, may have derived some benefit from the successful shipping of their goods to Russia. However, in this instance political considerations were paramount. The Arctic convoys demonstrated the Western Allies' commitment to the Soviet Union at a time when the European 'second front' against Germany consisted only of amphibious operations in the Mediterranean, and the Combined Bomber Offensive.[31]

A more obvious advantage to the British economy in the Second World War came in the form of the naval protection provided for the Atlantic convoys that brought vital supplies – especially foodstuffs – to Britain from the United States, Canada, the West Indies, and South America. Despite concerted efforts to increase home production of foodstuffs and war supplies, imports remained vital to national survival. German surface vessels wrought considerable damage to merchant shipping, particularly in the early years of the war, despite the efforts of the Royal Navy. The high-speed German 'pocket battleship' *Admiral Graf Spee*

---

[28]   Lance E. Davis and Stanley L. Engerman, *Naval Blockades in Peace and War: An Economic History since 1750* (Cambridge: Cambridge University Press, 2006), table 5.12 (p. 183).

[29]   D. Steffen, 'The Holtzendorff Memorandum of 22 December 1916 and Germany's Declaration of Unrestricted U-Boat Warfare', *Journal of Military History* 68 (2004), pp. 215–24.

[30]   Davis and Engerman, *Naval Blockades*, table 5.12 (p. 183).

[31]   For the Arctic convoys, see Richard Woodman, *Arctic Convoys, 1941–1945* (London: John Murray, 2004).

sank more than 50,000 tons of merchant shipping in the South Atlantic in the first three months of the war.[32] German aircraft, especially the light and medium bombers in which the *Luftwaffe* specialised, inflicted damage in the seaways within range of German airbases. But the greatest threat came not from surface warships or aircraft, but from beneath the waves. German submarines inflicted devastating losses – much more so than in the First World War – especially following the fall of France in the summer of 1940. The capture of the French oceanic ports of Saint-Nazaire and Lorient in Britany enabled U-Boats to extend their reach westwards across the Atlantic. German submarines were capable of attacking British-bound merchant ships assembling in the St Lawrence River in Canada, as well as ships within sight of the east coast of the United States.[33] The latter proved a particularly bountiful hunting ground, for American cities did not operate a blackout, ensuring that floating targets were silhouetted for U-boat captains facing the western horizon. In December 1942, submarines accounted for a staggering 94.8 per cent of allied shipping losses.[34] Only with the operational employment of sophisticated detection technology (sonar and radar), which enabled the more targeted use of high-explosive depth charges, was it possible to neutralise the danger to Britain's Atlantic lifeline posed by the German U-Boat fleet.

Trade protection was not simply a matter of escorting convoys, however. It also involved keeping open markets that enemies wished to close to British commerce. In the Berlin and Milan decrees of 1806 and 1807, Napoleon sought to exclude British goods from Europe and to deny Britain access to continental products. Napoleon intended his Continental System, as it was called, to use economic pressure to drive Britain – his most resilient enemy – to seek terms. In theory, the prohibition should have worked; nearly 70 per cent of British exports (including re-exports) went to mainland Europe. Napoleon's own forces controlled large areas of the Continent and most of the rest comprised territories of his allies or client states.[35] Yet the Continental System failed. In part, the explanation for its failure lies with the ability of British manufactures to find new markets beyond Europe. British exports to South America and the West Indies (British and foreign) rose from £4.2 million in value in 1802 to £10.7 million in 1810. But extra-European compensations are not the whole story; indeed, if we compare figures for the value of British exports to Asia and Africa in the same

---

[32]   The captain of the *Graf Spee* scuttled the ship after it was trapped in the River Plate estuary that December. See D. Pope, *The Battle of the River Plate: The Hunt for the German Pocket Battleship Graf Spee* (Ithaca, NY: McBooks Press, 2005).

[33]   See N.M. Greenfield, *The Battle of the St Lawrence: The Second World War in Canada* (Toronto: HarperCollins Canada, 2004).

[34]   Davis and Engerman, *Naval Blockades*, table 6.1 (p. 269).

[35]   Davis and Engerman, *Naval Blockades*, table 2.4 (p. 41), for the importance of the Continent for British exports. This table is the source of the figures in this and the following paragraph.

period, we see declines not increases – from £2.9 million to £1.7 million for Asia and from £1.1 million to just under £500,000 for Africa. Exports to the United States, valued at £7.8 million in 1810, unsurprisingly dropped dramatically with the coming of tension and then war between the two countries in 1812; by 1814, British exports to the United States were officially valued at a mere £7,000.

To a remarkable extent, trade with the European mainland continued despite the Continental System. In 1802, British exports to the Continent were worth £22.7 million. In 1810, with most of Europe supposedly closed to British trade, they stood at £19.6 million. That the export figures remained relatively strong owed much to the Royal Navy. Britain's Baltic trade continued to be buoyant, thanks largely to a significant British naval deployment in the region.[36] Vital Baltic naval stores – especially timber, tar, pitch, and hemp – still arrived in British ports. Just as importantly, British manufactured goods flowed into eastern and northern Europe through the Baltic doorway. Further holes were punched in the Continental System by the establishment, under naval protection, of trading footholds, on the islands of Lissa, in the Adriatic, and Heligoland, off the north German coast. The implications of the navy's undermining of the Continental System, for the British economy and for the course of the Napoleonic War, were considerable. The ability of Britain to keep trading overseas helped to generate the wealth that made possible (through taxation and borrowing) the generous British subsidies to Prussia, Austria, and Russia that eventually led to the defeat of Napoleon in 1814.[37]

III

The principal objective behind the policy of concentrating naval resources in European waters, as we have seen, was to safeguard the home territories from invasion. In the eighteenth and early nineteenth centuries, British ships based at home or in the Mediterranean could blockade enemy ports, preventing French or Spanish warships from coming out to sea. Blockades could be distant, with a few frigates watching a port while the bulk of the British fleet remained at its own bases, or close, with larger numbers of warships anchored within view of the enemy coastline. Long-distance blockades of the French Atlantic fleet became possible once the Royal Navy created a 'western strategy' (from the time of the War of the Austrian Succession), establishing its principal bases at Plymouth and

---

[36] A.N. Ryan, 'The Defence of British Trade with the Baltic, 1808–1813', *English Historical Review* 74 (1959), pp. 443–66; J. Davey, *The Transformation of British Naval Strategy: Seapower and Supply in Northern Europe, 1808–1812* (Woodbridge: Boydell Press, 2012).

[37] For more on this subject, see S. Conway, 'Sea Power: The Royal Navy', in Alan Forrest (ed.), *The Cambridge History of the Napoleonic Wars*, 3 vols (Cambridge: Cambridge University Press, 2019).

Portsmouth, which were much closer to Brest than were the Medway and Thames Estuary stations at Chatham and Sheerness. The Royal Navy's close blockade of Brest in 1803–5 effectively prevented invasion at the beginning of the War of the Third Coalition against Napoleonic France.[38] French troops has been massed at Boulogne for a descent on southern England, but with the Brest fleet unable to offer protection to the invading army the plans for a landing had to be abandoned even before Nelson's great victory over the French Mediterranean and Spanish fleets at Trafalgar in October 1805.

Confining enemy navies to port had the collateral benefit of exposing unprotected enemy merchant shipping to attack by the Royal Navy or British privateers. The navy seems to have been the most successful: it was responsible for more than half (52 per cent) of the captures recorded at the London prize courts between 1739 and 1748.[39] Even if enemy merchant vessels eluded capture, the risk often acted as a deterrent and reduced the number of successful trading voyages. In successive wars against France in the eighteenth and early nineteenth centuries, French trade with extra-European territories suffered. French western ports, such as La Rochelle, experienced the most serious commercial blows.[40] But the deprivation of access to colonial markets, and colonial and other extra-European products, did not bring the French economy to its knees. The disruption of French overseas trade and loss of its extra-European possessions proved far from fatal to the French economy in the Seven Years War, when the Royal Navy established total maritime dominance. Much of France's seaborne trade transferred onto Dutch and other neutral vessels, and landward trade with Germany and Spain increased to make up for a decline in colonial commerce.[41] The same happened in the long wars against revolutionary and Napoleonic France. The value of French exports in 1812 was considerably higher than in 1802 – 418.6 million francs compared with 339.1 million. Compensation for the loss of colonial markets came through increased exports to other European countries, which could be accessed overland and were therefore unaffected by a naval blockade. French exports to the Italian states rose exponentially between 1802 and 1812, from 3.6 million francs worth to 85.5 million.[42]

The Royal Navy's earlier attempts to institute a blockade of the coast of an extra-European territory proved similarly ineffective. At the start of the War of American Independence, the British government considered relying wholly on

---

[38]   See *Despatches and Letters Relating to the Blockade of Brest, 1803–1805*, ed. John Leyland, 2 vols (London: Navy Records Society, 1899–1902).

[39]   D.J. Starkey, *British Privateering Enterprise in the Eighteenth Century* (Exeter: University of Exeter Press, 1990), table 10 (p. 137).

[40]   J.G. Clark, La *Rochelle and the Atlantic Economy during the Eighteenth Century* (Baltimore, MD: Johns Hopkins University Press, 1981), esp. pp. 150–1.

[41]   J.C. Riley, *The Seven Years War and the Old Regime in France: The Economic and Financial Toll* (Princeton, NJ: Princeton University Press, 1986), pp. 224–5.

[42]   Davis and Engerman, *Naval Blockades*, tables 2.7 and 2.8 (pp. 45–6).

a naval blockade to destroy the resistance of the rebel colonies. The secretary at war, the minister responsible for the army, argued that a land war would be impossible to win, given the scale of the logistical challenges it would pose.[43] But other ministers – and perhaps most importantly the king – rejected a purely naval war on colonial trade, on the grounds that it would take too long and would leave Americans who remained loyal to the crown dangerously exposed and unprotected. For these reasons, the government placed its faith in deploying a large army in the rebel colonies for the 1776 campaign, which ministers hoped would deliver a knock-out blow to the rebellion. Even so, a blockade came into force alongside land operations, formally from December 1775, when the American Prohibitory Act authorised the King's navy to seize and confiscate American merchant ships and their cargoes. In the long run, the blockade helped disrupt and damage the American economy, leading to acute shortages of some products and fuelling inflationary pressures created by excessive printing of paper currency; but it did not bring the rebels to seek terms. They transferred much of their long-distance overseas trade onto neutral vessels (particularly Dutch ships) and received large quantities of imports via neighbouring North American and Caribbean colonies of the Spanish, French, Dutch, and even British. British manufactures even seeped out into the American-occupied hinterland of the British army's headquarters at New York City and other British-held enclaves in the territory of the United States.[44] That Americans received British goods at all reminds us of a fundamental problem that every attempted blockade faced: to deny the enemy imports no doubt seemed, from a government perspective, a legitimate weapon of war, but implementing that objective – on this occasion and more generally – was very difficult when one's own merchants wanted to carry on trading with the enemy. Private profit and public interest might work harmoniously in the case of contracting for supplies, but when it came to blockades, they nearly always clashed.

The transfer of much enemy trade onto neutral carriers, in both the War of Independence and earlier and later conflicts, led the Royal Navy to stop and search neutral ships on the high seas. But the so-called Rule of 1756, promulgated by British Admiralty Courts to justify the navy's seizure of neutral vessels trading with the enemy in ways that they would not have been permitted to do in peacetime, was never formally recognised by other states. Indeed, neutral powers consistently contested the British conviction that the law of nations (what we would now call international law) provided sanction for their navy's actions.[45]

---

[43]   Shute Barrington, *The Political Life of William Wildman, Viscount Barrington* (London, 1814), pp. 158–9.

[44]   See Richard Buel Jr, 'Time: Friend or Foe of the Revolution?' in Don Higginbotham (ed.), *Reconsiderations on the Revolutionary War: Selected Essays* (Westport, CN: Greenwood, 1978), pp. 124–43, which argues that British goods were draining the rebel states of money and undermining their ability to continue the war.

[45]   G. Best, *Humanity in Warfare: The Modern History of the International Law of*

Tensions over the issue of neutral rights almost brought the British and Dutch to blows in the Seven Years War. It contributed greatly to conflict between the two states in the later stages of the War of American Independence. In December 1780, the British government declared war on the Dutch republic; the fourth Anglo-Dutch War did not formally end until 1784. British attempts to stop and search neutral merchant vessels also enraged other powers. Earlier in 1780, Catherine the Great of Russia organised a League of Armed Neutrality, which included many European states, to resist British pretentions. Russia inspired another League of Armed Neutrality in 1800, with the same aim.[46] British naval power, regarded by most continental commentators as dangerously excessive from the time of the Seven Years War, threatened to unite much of Europe against Britain, much as in the late seventeenth and early eighteenth centuries Louis XIV's military power had united much of Europe against France. In the mid- and late eighteenth century, then, British blockades not only failed to bring enemies to their knees; they also risked creating new enemies.

Economic blockades, designed to deprive the enemy of imported goods, may have been more successful in the twentieth century. By this time, industrialisation, urbanisation, population growth, and the increasing integration of the world economy had left many European states much more dependent upon in-flows of raw materials and food from beyond the Continent. In the First World War, the Royal Navy and its allies attempted to cut Germany and Austria–Hungary off from external sources of supply by closing both the North Sea and Mediterranean coasts of the Central Powers and searching neutral vessels on more distant oceans to ensure that they were not carrying 'contraband' (including foodstuffs) to the enemy. Total German imports dropped from 10.8 billion marks by value in 1913 to 7.1 billion in 1917.[47] There can be no doubt that the civilian population of the Central Powers, and especially Germany, experienced considerable hardships as food supplies diminished. Before 1914, the average German consumed 342 grams of bread a week and 950 grams of meat. By the last year of the war, rationing had reduced this to 160 grams and 135 grams respectively.[48] Historians continue to ponder the impact of the allied blockade and particularly whether it was responsible for the collapse of Germany and Austria–Hungary.[49]

Armed Conflicts (London: Methuen, 1983), pp. 67–74. For fuller coverage, see Richard Pares's old but still important study, *Colonial Blockade and Neutral Rights, 1739–1763* (Oxford: Clarendon Press, 1938), esp. chap. 3.

[46]   Pares, *Colonial Blockade and Neutral Rights*, pp. 242–79; H.M. Scott, *British Foreign Policy in the Age of the American Revolution* (Oxford: Clarendon Press, 1990), chap. 11.

[47]   Davis and Engerman, *Naval Blockades*, table 5.1 (p. 164).

[48]   Davis and Engerman, *Naval Blockades*, p. 203.

[49]   See, for example, C.P. Vincent, *The Politics of Hunger: The Allied Blockade of Germany, 1915–1919* (Athens: Ohio University Press, 1985); M. Cox, 'Hunger Games: Or How the Allied Blockade in the First World War Deprived German Children of Nutrition,

Most, however, now gather around a middle-ground position: the blockade contributed to allied victory but was not in itself sufficient to end the conflict.[50]

In the Second World War, the Royal Navy was again used to cut off Germany's supplies of food and other vital resources. The German economy experienced reduced access to imported raw materials and the German people suffered food shortages as increasingly stringent rationing took effect. Initially, the blow was softened by German access to (and exploitation of) the resources of much of the Continent, either through conquest and the establishment of puppet regimes, or trade with sympathetic neutrals, such as Spain, Switzerland, and Sweden.[51] Only when German armies began to retreat in the East, and the territory controlled by the Third Reich contracted, did the civilian population begin to experience real hardship. Even then, the blockade's impact must be put in context. The intensive bombing of German cities by the British and American air forces played at least as big a part in undermining the enemy economy; historians tend to accord the allied bombing offensive a significant role in the explanation of German defeat.[52] The blockade, furthermore, does not seem to have seriously eroded the enemy's will to keep fighting. Despite severe food shortages, Germany's resistance, unlike in 1918, continued until allied armies occupied much of its territory.

IV

We turn now to our last link between the economy and naval warfare. Throughout the period considered here, the Royal Navy's demand for new ships, weapons, dockyard and maintenance facilities, and food for its personnel, provided a stimulus to the development of the domestic economy. We should note that the British economy could benefit from orders for warships placed from abroad as well as at home. In the American Civil War (1861–65), the Confederate government paid for armed ships to be constructed in British yards to help break the Union naval blockade of the South and to attack northern shipping. The most celebrated (or notorious) of these British-built Confederate warships was the *Alabama*, constructed at Laird's yard at Birkenhead in 1862, which, even after it had been sunk, went on to be the cause of strained diplomatic relations between Britain and the United States.[53] We should also note that the Royal Navy's demands were not limited to wartime, because the British fleet remained in being (even if on a

---

and Allied Food Aid Subsequently Saved Them', *Economic History Review* 68.2 (2015), pp. 600–31.
[50]   Davis and Engerman, *Naval Blockades*, p. 214 argue that there is now a consensus that the blockade 'was not a weapon that by itself could have brought the war to an end'.
[51]   See Eric B. Golson, 'The Economics of Neutrality: Spain, Sweden, and Switzerland in the Second World War', PhD thesis, London School of Economics, 2011.
[52]   See, for example, R. Overy, *Why the Allies Won* (London: Pimlico, 2006).
[53]   See A. Cook, *The Alabama Claims* (Ithaca, NY: Cornell University Press, 1975).

smaller scale) even when the country was at peace. The navy kept the sea lanes open for trade, and provided protection from pirates, even when Britain was not engaged in international conflict. In some years of peace, such as 1765, 1784, and 1785, expenditure on the navy, and the dockyard facilities it needed, was as high as in most years of war.[54] In general, however, the requirements of the Royal Navy increased enormously once hostilities commenced, or even in anticipation of their beginning. Some examples, largely from the second half of the eighteenth century, and the beginning of the nineteenth, the period usually associated with Britain's Industrial Revolution, give us a better sense of what was involved.

In peacetime, the royal dockyards handled most of the work of building and repairing ships for the navy. In war, when the navy expanded greatly, the state yards lacked the capacity to cope. Rather than use large sums of money to build new royal facilities, which would be underused and expensive to maintain when peace returned, governments in the eighteenth century (and later) entered into contractual relationships with private shipbuilders to meet at least part of the wartime demand. Private yards built more than 130 warships between the beginning of the War of the Austrian Succession in December 1740 and October 1746.[55] Most of these vessels were smaller craft; the largest of ten ships build in Hull yards between 1739 and 1747 was the 50-gun *Tavistock*; the smallest the 16-gun *Raven*.[56] The War of American Independence provided a similar opportunity for private yards. Buckler's Hard, the Duke of Montagu's Hampshire shipyard, launched 11 naval vessels in the years preceding the conflict (1771–74), but six in 1779 alone, by which time French and then Spanish intervention had increased the demand for more British warships.[57] The naval estimates presented to the House of Commons at the end of 1774 envisaged spending just £17,574 in merchant yards over the following 12 months. By 1783, the naval estimates pointed to planned expenditure in private yards over the next year of £770,100. In all, Parliament agreed to the spending of in excess of £2.6 million in private yards during the war, and navy bills (credit instruments to cover unanticipated expenditure) may well have added substantially to this total.[58]

Private shipyards were not the only beneficiaries of wartime naval expansion. Warships needed weapons. In the eighteenth century, the Ordnance Board placed

[54]   Rodger, *Command of the Ocean*, Appendix VII (p. 644). C. Wilkinson, *The British Navy and the State in the Eighteenth Century* (Woodbridge: Boydell Press, 2004), makes the case for the importance of consistent investment – in peace as well as in war.

[55]   B. Pool, *Navy Board Contracts, 1660–1832: Contract Administration under the Navy Board* (London: Archon Books, 1966), pp. 83–5.

[56]   James Joseph Sheahan, *History of the Town and Port of Kingston-upon-Hull* (2nd ed., Beverley: J. Green, [1866]), pp. 367–8.

[57]   Beaulieu, Montagu Estate Papers, 'Ships Built for Government at Buckler Hard from Septr. 1743 to January 1791'.

[58]   Figures derived from *Journal of the House of Commons*, vol. 35, pp. 56–7, 434–5; vol. 36, pp. 38–9, 590; vol 37, pp. 34–5, 550–1; vol. 38, pp. 88–9, 634–5; vol. 39, pp. 167–8.

orders for guns for both the army and the navy. The army's demand for muskets easily outstripped the navy's, but the reverse was true when it came to cannons, most of which were purchased for the Royal Navy's warships. John Fuller, an iron manufacturer in the Sussex Weald, found his works stretched to breaking point in trying to meet orders placed by the Ordnance Board in the War of the Austrian Succession; in 1745, he made 136 cannons, ranging from nine to 32 pounders. Fuller's son provided the Ordnance Board with even larger quantities in the Seven Years War, when the demands of the armed forces – and particularly the navy – created great opportunities for profit.[59] In 1759, in response to the stimulus of wartime demand for cannon, the planning for the opening of the Carron works in Scotland began; while in South Wales Lewis and Company constructed the Dowlais furnace to meet Ordnance Board orders. By the end of the war, according to a leading historian of Britain's Industrial Revolution, 11 new smelting works had been opened.[60] The next conflict – the War of American Independence – had a similar effect. Anthony Bacon, a wealthy merchant and MP, secured contracts to supply just short of 1,000 cannons to the Ordnance Board between 1775 and 1779.[61] Samuel Walker's Rotherham ironworks, which reportedly employed 1,000 hands in 1777, doubled its capital between 1775 and 1782, in large part due to orders for the manufacture of large guns for use on Royal Navy warships.[62]

More ships necessitated more men to crew them. In the Seven Years War, according to the naval estimates, the number of mariners on board the king's ships rose steadily, peaking in 1760–62 at 70,000, compared with a mere 10,000 in 1752–54. The actual number of men borne may well have been larger: in 1759, when the House of Commons expected to pay for 60,000 sailors, the navy mustered more than 77,000.[63] In the War of American Independence, the navy appears to have grown even more spectacularly. At the beginning of the rebellion, the government asked the House of Commons to cover the cost of employing

---

[59]   *The Fuller Letters, 1728–1755: Guns, Slaves, and Finance*, ed. David Crossley and Richard Saville (Lewes: Sussex Record Society, 1991), pp. 130, 211; Somerset Record Office, Taunton, Dickinson Papers, DD/DN 498, esp. John Ward to James Cooper, 2 Nov., 9 Dec. 1757, William Sanson to Cooper, 21 Nov., 6 Dec. 1757, and Fuller to Ordnance Board, 9 Dec. 1757 (draft).

[60]   T.S. Ashton, *Iron and Steel in the Industrial Revolution* (2nd ed.; Manchester: Manchester University Press, 1951), pp. 128–33.

[61]   Sir Lewis Namier, 'Anthony Bacon MP, an Eighteenth-Century Merchant', in W.E. Minchinton (ed.), *Industrial South Wales, 1750–1914* (London: Frank Cass, 1969), pp. 83–4.

[62]   British Library, London, Egerton MS 2673, p. 400; *The Walker Family: Ironfounders and Lead Manufacturers, 1741–1893*, ed. A.H. John (London: Council for the Preservation of Business Archives, 1951), pp. 11–19.

[63]   N.A.M. Rodger, *The Wooden World: An Anatomy of the Georgian Navy* (London: Collins, 1988), Appendix XI (p. 369).

28,000 seamen and marines. By 1778, the figure had risen to 60,000; by 1782, to 100,000.[64] These men had to be fed – a squadron of ships, away from port for many weeks or even months, was like a town or small city; it produced no food of its own (apart from the fish its crews caught) but consumed vast quantities of foodstuffs. We can get a sense of the scale of the consumption if we consider the Victualling Board's purchases for the navy in 1760. These included 33,600 quarters of wheat, 34,100 hundredweight of flour, 41,400 hundredweight of biscuits, 96,700 hundredweight of beef oxen, 32,400 hundredweight of pigs, and 22,200 hundredweight of cheese.[65] Missing from this list is beer. The special naval breweries established to service the fleet in peacetime, much like the royal shipyards, could not cope with the increased demands associated with war. Contract brewers in London, Chatham, Portsmouth, and Plymouth met the shortfall. In London, Whitbreads supplied the navy with 3,000 barrels of porter in 1778, while Calverts provided 3,300 barrels.[66]

Money, then, flowed into the domestic economy as a result of the navy's wartime requirements for more ships and guns, and more food and drink. The chief beneficiaries must have been the major contractors, some of whom made a fortune and often faced bitter public criticism for their apparently bloated profits.[67] But a host of smaller winners can be identified, ranging from farmers and maltsters to shipwrights, sailmakers, and gunsmiths. The benefit to the wider economy is difficult to assess. The eighteenth-century Victualling Board's purchases of foodstuffs – mainly though not exclusively for the navy – may have helped to develop the infrastructure of supply that facilitated the feeding of the new urban centres that were growing rapidly at the end of that century and into the next. At the least we can say that the Victualling Board was almost certainly the biggest single buyer of foodstuffs in the British Isles, and its demands must have helped bring many producers into an emerging national market.[68]

Yet if the wartime demands of the Royal Navy brought profits to some sectors of the economy, war itself brought losses – often severe – to others. The

---

[64]   S. Conway, *The British Isles and the War of American Independence* (Oxford: Oxford University Press, 2000), table 1.1 (pp. 17–18).

[65]   Rodger, *Wooden World*, p. 84.

[66]   Peter Mathias, *The Brewing Industry in England, 1700–1830* (Cambridge: Cambridge University Press, 1959), p. 199.

[67]   See, for example, Tobias Smollett's novel *Humphrey Clinker* (1771), which denounced the 'commissaries and contractors, who have fattened, in two successive wars, on the blood of the nation'. *The Expedition of Humphrey Clinker*, ed. Angus Ross (Harmondsworth: Penguin, 1967), p. 65.

[68]   This is the argument of Christian Buchet, *Marine, économie et société. Un exemple d'interaction: l'avitaillement de la Royal Navy durant la guerre de Sept Ans* (Paris: Champion, 1999). For a contrary view, see Janet Macdonald, 'A New Myth of Naval History? Confusing Magnitude with Significance in British Naval Victualling Purchases, 1750–1815', *International Journal of Maritime History* 21 (2009), pp. 159–88.

state expenditure that paid for the navy's needs was often raised by borrowing. As more and more money was needed to pay for long and costly wars, investment that may have benefited the long-term development of the wider economy migrated into the public funds, attracted by the higher rates of interest that the government was able to offer. Domestic infrastructure projects, such as road making and canal construction, suffered especially from this transfer of resources from private to public during the War of American Independence.[69] In that conflict, perhaps the best we can say is that vastly increased state spending, a significant part of it on the navy, compensated for the fall in profits generated by overseas trade. The overall impact of war on the national economy, at least in this particular case, was broadly neutral.[70]

In the longer wars against revolutionary and Napoleonic France, some historians argue, the economy was a net gainer from government spending, with heavy industries, such as iron and steel making and shipbuilding, benefiting from the stimulus.[71] If there was indeed a generally beneficial effect – and other historians are deeply sceptical[72] – the demands of the Royal Navy certainly played a part. We should note, however, that the army had more government-raised money spent on it than did the navy. In 1804–5, the army received £15.75 million and the navy £11.75 million; in 1811, with a large British force fighting in Spain and Portugal, the split was £23.8 million for the army and £19.5 million for the navy.[73] We should also note that the spectacular growth in the importance of the cotton textiles in the overall economy cannot be attributed to the demands of the war. The boost to textile production caused by the demand for uniforms and seamen's clothing (bulk purchased from 1756) benefited the woollen industry more than cotton manufacturing. If the Industrial Revolution owed something to the wars against France between 1793 and 1815, the navy's contribution, then,

---

[69]   For canals, see J.R. Ward, *The Finance of Canal Building in Eighteenth-Century England* (Oxford: Oxford University Press, 1974), pp. 31–3, 102, 168. The *Minutes of the Chesterfield Canal Company, 1771–1780*, ed. Christine Richardson and Philip Riden (Chesterfield: Derbyshire Record Society, 1996), pp. 207, 216 tell the same story. For roads, see William Albert, *The Turnpike Road System in England, 1663–1840* (Cambridge: Cambridge University Press, 1972), pp. 129, 131. The number of parliamentary bills to enclose land also fell: see Michael Turner, *Enclosures in Britain, 1750–1830* (London: Macmillan, 1984), chap. 3.

[70]   Conway, *The British Isles and the War of American Independence*, p. 84.

[71]   See, for example, J.L. Anderson, 'Aspects of the Effect on the British Economy of the Wars against France, 1793–1815', *Australian Economic History Review* 12 (1972), pp. 1–20.

[72]   T.S. Ashton, despite recognising the stimulating effect on the iron and steel industry, argues that 'the Industrial Revolution might have come earlier' if Britain had not engaged in any wars in the eighteenth century: *Economic Fluctuations in England, 1700–1800* (Oxford: Clarendon Press, 1959), p. 83.

[73]   See C.D. Hall, *British Strategy in the Napoleonic War, 1803–1815* (Manchester: Manchester University Press, 1999), p. 16.

needs to be put in proper perspective. It may have helped the development of certain sectors of the economy, but it would surely be an exaggeration to say that it was vital to the whole complex process of industrialisation.

V

One of the advantages of taking the long view is that it enables changes, and continuities, to emerge with greater clarity. As always, change is the more eye-catching. In this case, developments unconnected with the navy itself altered the character of trade protection and the effectiveness of blockades. From the seventeenth century to the end of the nineteenth, trade protection was understood to be about the deployment of naval warships to shield merchant convoys from attack by enemy surface vessels – in their navy or privateers. Submarines, though making their debut as early as 1776, when an American underwater vessel tried to sink Admiral Lord Howe's flagship in New York harbour, only became a major weapon of war in the twentieth century. In the First World War they contributed greatly to the loss of more than 7.6 million tons of merchant shipping in the British Empire.[74] By the time of the Second World War, they posed a serious challenge to the carrying out of the Royal Navy's convoying duties, as did – to a lesser extent – the use of aircraft. Aerial bombing by the *Luftwaffe* accounted for nearly a quarter of allied shipping losses in 1941.[75] Indeed, the advent of air power made a difference in many ways. The Royal Air Force (RAF) created new demands for the products of British industry and so diminished, if only in relative terms, the role of the Royal Navy as a stimulus to manufacturing. More strikingly, the RAF – and still more the United States Air Force – through the use of mass-produced heavy bombers, inflicted more immediate and dramatic damage on German industry – by physically destroying plant and equipment and railway networks – than even the tightest naval blockade could hope to achieve. Blockades still had the capacity to inflict harm on the enemy, but much of this harm derived from the reduction in food available to sustain the civilian population. In this respect, the navy benefited from significant structural changes over which it had no control. Demographic growth in Europe, combined with urbanisation and industrialisation and the increased integration of the world economy (what we now call 'globalisation'), made the populations of many European states much more dependent upon food imports than they had been in the past, and therefore much more exposed to privation if a sea blockade cut off external sources of supply.

Continuities are observable, too. The movement of goods by sea, inward and outward, remained a crucial feature of the mid-twentieth-century British

---

[74]   Davis and Engerman, *Naval Blockades*, table 5.3 (p. 169).

[75]   Davis and Engerman, *Naval Blockades*, table 6.3 (p. 288).

economy as much as it had been an important element of national prosperity in the seventeenth, eighteenth, and nineteenth centuries. Britain was still an island, heavily reliant on its overseas trade. Long-distance seaborne commerce required protection during the Second World War as much as it had done in any earlier international conflict. That protection, in the British case, was provided principally by the Royal Navy. Its ships were now equipped to fend off both submarine and aerial attack (by using depth charges and anti-aircraft guns) but its duties remained the same as they had always been – to escort merchant ships safely to their destination. Sea blockades, as we have just seen, continued to play a part in economic warfare in the twentieth century, even if they now operated alongside aerial bombardment of industrial infrastructure as a way of degrading the enemy's capacity to resist. And, most importantly, the Royal Navy, though it now competed with the RAF as well as the army for government spending, continued to require ships and weapons that provided a stimulus for British industry, particularly the shipyards of the Clyde Valley, Belfast, and the English North-east and Merseyside.

CHAPTER TWO

Neither a Silver Bullet Nor a Distraction: Economic Warfare in Sea-power Theory

Bleddyn E. Bowen

The noiseless, steady, exhausting pressure with which sea power acts, cutting off the resources of the enemy while maintaining its own, supporting war in scenes where it does not appear itself, or appears only in the background, and striking open blows at rare intervals, though lost to most, is emphasized to the careful reader.

Alfred Thayer Mahan, *The Influence of Sea Power upon History, 1660–1783*, p. 209.

Introduction

It is something of a truism that the international system of the early twenty-first century resembles that of the early twentieth century: maritime commerce and supply routes dominate the world's most productive and consumptive economies in a highly globalised trading system. Indeed, with China and the United States – whose economies depend heavily upon seaborne trade – set to dominate the international system as we move away from the United States's 'unipolar moment', sea-power theory ought to have a bright future as more scholars and analysts seek to examine past instances of great power rivalries at sea.[1] Maritime economic warfare is a topic particularly open to reconsideration

---

[1]   On the rise of China in the international system and sea-power, see, for example, A.S. Erickson, Lyle J. Goldstein, and Carnes Lord (eds), *China Goes to Sea: Maritime Transformation in Comparative Historical Perspective* (Annapolis, MD: Naval Institute Press, 2009); Eric Heginbotham et al., *The U.S.–China Military Scorecard: Forces, Geography, and the Evolving Balance of Power, 1996–2017* (Santa Monica, CA: RAND, 2015); Stephen G. Brooks and William C. Wohlforth, 'The Rise and Fall of the Great Powers

as part of this process, owing to the paucity of theoretical work done in this area. Possibly owing to the inherently indecisive nature of economic warfare and the consequent methodological challenges of assessing it, most literature on sea power and maritime strategy has tended to focus on the more dramatic struggles to impose and win battles on enemy fleets. As a result, sea-power theorists have struggled to theorise the more subtle and indirect economic pressure of sea power in warfare.

This chapter sets out an informal framework of how to think about the vicissitudes of economic warfare at sea by considering how past generations of sea-power theorists attempted to distil insights from past experiences and history from the indecisive yet not insignificant influence of economic warfare. It argues that economic warfare features in sea-power theory as an instrument that is often important, particularly in retrospect, but which has featured as something of an afterthought or something taken for granted, particularly because decisive results have eluded practitioners in the past. Alone, the economic pressure of sea power is indecisive, but that does not mean it is a mere distraction or that it cannot be effective as part of a broader grand strategy in war as well as peacetime coercion. With this grey area of theory in mind, the chapter outlines the points of theory that are useful for historical and contemporary studies of economic warfare. It demonstrates how the utility of economic warfare fluctuates with the target, capabilities available, and the length of a conflict. It is also affected by grand strategic considerations such as the escalation and spread of conflict, its impact upon neutrals, and the ethical and moral implications stemming from its effect upon civilians and societies at large. These theoretical observations are taken from a variety of theorists who examine a range of historical episodes in their works, and form a general consensus on the potential and pitfalls of maritime economic warfare in grand strategy.

Strategic Truth and the Constants of Sea-power Theory

The relationship between history and strategy is reciprocal and intimate. Just as history plays a vital role in shaping theory, so observations from theory offer an important tool for historians studying war and the practice of strategy.[2] Historians often lament the unwillingness of scholars of strategic studies to embrace the specificity and empirical detail of their own field. Yet the reverse is equally true: most historians have proven reluctant, not to say determined, to

in the Twenty-First Century: China's Rise and the Fate of America's Global Position', *International Security* 40.3 (2016), pp. 7–53; Tudor A. Onea, 'Between Dominance and Decline: Status Anxiety and Great Power Rivalry', *Review of International Studies* 40 (2014), pp. 125–52.

[2]   A case persuasively made in H. Strachan (ed.), *The Direction of War: Contemporary Strategy in Historical Perspective* (Cambridge: Cambridge University Press, 2013).

avoid the use of theory in their own work.[3] This is particularly unfortunate for the historical study of strategy, owing to the universal logic of the subject at hand. Strategic theory places general historical observations on a more defensible footing by attempting to merge the continuing changes of technology, economics, and politics with universal aspects of conducting war and strategy. Michael Handel observed that numerous theorists over the centuries have 'independently and ineluctably arrived at similar conclusions' regarding the nature of war.[4] It is in this spirit that this chapter outlines universal strategic truths relevant to the practice of maritime economic warfare. It focuses upon the contradictions and tensions *within* the works of key sea-power theorists, rather than *between* them, in order to illustrate the fundamental consensus on key aspects of the subject.[5]

Each theorist highlights different aspects of the same fundamental tensions of theory and practice, of chaos and reason, of passion and calculation. The pillars of maritime strategic thought agree on the need to seek strategic truth in the face of technological, political, and economic changes.[6] Strategic truth provides insight into some practical problems of strategy as the actual influence of economic warfare cannot be predicted in advance. This uncertainty shapes debate and decision. Such is the purpose of strategic theory, of which sea-power theory is a type. In other words, for those seeking to understand the essence of economic warfare and its principles, a framework of theory remains an indispensable tool.

It is important at the outset to consider how such concepts and principles should be used, so that changes in the character or application of theory are not mistaken as fundamentally new principles of war itself. This difference can be seen in the terms 'maritime strategy' and 'sea-power theory' themselves, which are different intellectual exercises and claims to knowledge. A maritime strategy is a strategy for a particular scenario and actor at a particular time, an actual plan for action;[7] a sea-power theory is a collection of timeless principles that should guide – but not dictate – the formation and the study of maritime strategies. The confusion of these terms is partly the result of one of the core texts of both disciplines: Julian Corbett's *Principles of Maritime Strategy*, which has led many to believe the book presents a maritime strategy.[8] Whilst Corbett made notable

---

[3]   C. Elman and M.F. Elman (eds), *Bridges and Boundaries: Historians, Political Scientists, and the Study of International Relations* (Cambridge, MA: MIT Press, 2001).

[4]   M.I. Handel, *Masters of War: Classical Strategic Thought* (Abingdon: Routledge, 2001), p. xvii.

[5]   Handel, *Masters of War*, p. 7.

[6]   A.T. Mahan, *The Influence of Sea Power upon History, 1660–1783* (London: Marston & Co., 1890), pp. 1, 88; Raoul Castex, *Strategic Theories*, trans. and ed. Eugenai Kiesling (Annapolis, MD: Naval Institute Press, 1994), pp. 17, 21; Julian S. Corbett, *Principles of Maritime Strategy* (Mineola, NY: Dover, 2004), pp. 6, 204.

[7]   This does not detract from the use of the word 'maritime strategy' in a looser and broader sense to mean a field of study within military history and strategic studies.

[8]   Corbett, *Principles of Maritime Strategy*.

contributions to contemporary British maritime strategy in other ways,[9] this text was an attempt to delineate the principles of creating a maritime strategy. It must therefore be understood as a text in the same vein as Clausewitz's *On War* – and as 'a seapower theory that is meant to help planners and strategists develop a maritime strategy for whatever contingencies they face'.[10] The basis of sea-power theory – and all strategic theory – is to explore, explain, and educate the universal elements of practice in war, which Clausewitz portrayed through the universal elements of passion, reason, and chance. Echoing that view, Corbett himself stressed that theory is a matter of education and deliberation, but not execution.[11] The central principle of sea-power theory for the development of any maritime strategy is not how to win battles but knowing how and when it is desirable to try to impose battle on the enemy and then how to exploit one's command of the sea towards the land.[12] Thus, by moving away from a focus upon battle and thinking in terms of sea-power theory, rather than maritime strategy, the discussion turns more naturally to the subject of this chapter: the role of maritime power and the economy within the broader picture of grand strategy. The grand strategic aspects of maritime economic warfare are sufficiently universal that it also informs sanctions and economic pressure in general, beyond the capabilities of navies alone and into other economic tools of statecraft and coercion.

The economic use of the sea and its potential military implications are constants of sea-power theory, incorporated under the idea of sea lines of communication (SLOC). Be it for mercantilist empires in the seventeenth century or the trade of bulk goods and raw resources in the twenty-first century, the sea remains the crucial global medium for the transport of military forces and commerce. Given this universality, many of the questions which have animated sea-power theorists remain pertinent today. Focusing on economic warfare in particular, sea-power theory proposes two enduring questions for the study and conduct of maritime economic warfare:

- When is maritime economic warfare decisive and feasible?

- Does maritime economic warfare help achieve the political objectives sought?

---

[9]   J. Goldrick and J.B. Hattendorf (eds), *Mahan is Not Enough: The Proceedings of a Conference on the Works of Sir Julian Corbett and Admiral Sir Herbert Richmond* (Newport, RI: Naval War College Press, 1993).

[10]   Bleddyn E. Bowen, 'From the Sea to Outer Space: The Command of Space as the Foundation of Spacepower Theory', *Journal of Strategic Studies* 42.2–3 (2019), pp. 532–56.

[11]   Corbett, *Principles of Maritime Strategy*, p. 3.

[12]   Bowen, 'From the Sea to Outer Space', pp. 6–9. On the pedagogy of strategic theory, see J.T. Sumida, *Decoding Clausewitz: A New Approach to* On War (Lawrence: University Press of Kansas, 2008); Harold R. Winton, 'An Imperfect Jewel: Military Theory and the Military Profession', *Journal of Strategic Studies* 34.6 (2011), pp. 853–77.

Through these questions it is possible to think constructively about the inherent indecisiveness of maritime economic warfare. To theorise the grey zone of a method in sea power and grand strategy is neither a petty distraction nor a war-winning strategy by itself.

The Decisiveness and Feasibility of Maritime Economic Warfare

The works of Alfred Thayer Mahan and Julian Corbett remain the most influential attempts to articulate general theories of sea power. There are four major points of agreements between the two: first, the command of the sea is rarely absolute; second, the command of the sea is not inherently decisive; third, that sea power is a source of support for the primary theatre of warfare on land; and, finally, that seeking battle is often but not always the best course of action.[13] Of most importance to thinking about maritime economic warfare is the second point– that the command of the sea, even if complete, is not a sure route to victory by itself. Such a view is usually associated with winning or deterring battles at sea, yet it is also true for attempts to exploit any degree of command that is secured. Indeed, the fact that 'many students are taught that battleships and colonies make up the heart of Mahan's writing' obscure the more nuanced view Mahan had of the influence of sea power upon the land, akin to Corbett's work.[14] Even with a dominant command of the sea, a decisive victory in war is never assured nor a fait accompli. The realism both scholars demonstrated in reaching these conclusions undercuts the exaggerated claims others, such as Basil Liddell Hart, have made about the potentially 'war winning' effects of maritime economic warfare.[15] Sea-power theorists have constantly grappled with the leverage and pain caused by economic warfare at sea whilst trying to manage expectations over what it can deliver.

Mahan made his claims on the importance of sea power based on the wealth generated by maritime trade, as the 'control of the sea is but one link in the chain of exchange by which wealth accumulates; but it is the central link, which lays under contribution other nations for the benefit of the one holding it, and which, history seems to assert, most surely of all gathers to itself riches'.[16] Mahan claimed that great harm could come from *guerre de course* and would 'not likely be abandoned till war itself shall cease'. Immediately, however, he added that when it is 'regarded as a primary and fundamental measure, sufficient

---

[13] Bowen, 'From the Sea to Outer Space', pp. 14–20.

[14] Benjamin F. Armstrong (ed.), *21st Century Mahan: Sound Military Conclusions for the Modern Era* (Annapolis, MD: Naval Institute Press, 2013), p. 9.

[15] A distinction not drawn often enough; see M. Howard, 'The British Way in Warfare: A Reappraisal', in M. Howard (ed.), *The Causes of War and Other Essays* (Cambridge, MA: Harvard University Press, 1983), pp. 169–87.

[16] Mahan, *Influence of Sea Power*, pp. 225–6.

in itself to crush an enemy, it is probably a delusion'.[17] Indeed, *guerre de course* for Mahan should remain secondary to the main task of contesting the enemy's command of the sea, which would enable bolder actions at sea and in the global economy than opportunistic raiding against merchant shipping would ever allow. After all, it 'was Cromwell's main fleet, not his commerce raiders, that destroyed Dutch trade and made grass grow in the streets of Amsterdam'.[18] In Mahan's view, for any form of maritime economic warfare even to approach exerting a decisive influence it had to consist of the full commercial blockade of the enemy's maritime economy, particularly if that economy was a mature maritime trading system that could weather the blows of opportunistic attacks.[19] Indeed, Mahan's justification for the existence of balanced and large fleets was their ability to defend friendly commerce and attack hostile trade after a good enough command of the sea was secured. Any form of commercial warfare at sea had to be supported by fleets so as to enable large-scale interdictions and blockades if it was to have any substantial influence on a war.[20]

Corbett attempted to square the circle of the apparently useful yet indecisive quality of maritime economic warfare, and in particular commercial blockades. It was a natural matter of course that control of SLOCs should be able to stop the passage of public and private property at sea, and a preponderance at sea usually translated into economic pressure.[21] By closing the enemy's commercial ports, one could 'injure the enemy' with a dominant command of the sea, and a blockade could exhaust them over time – although relying upon such a strategy would inherently limit the extent of the objectives that could be secured through it.[22] Indeed, for Corbett, the control of SLOCs and not destroying enemy naval forces was the priority, and the fleet was subordinate to the objective of commanding the sea – an end which did not always require a battle to secure.[23] Contemporaries of Corbett such Charles Callwell shared his essential vision that the aim of maritime strategy was to secure the benefits of sea command, be they amphibious operations or the imposition of economic costs upon the enemy.[24]

Although Corbett was generally positive regarding the importance of engaging in a commercial blockade of the enemy, he had doubts as to how effective a more

---

[17]   Mahan, *Influence of Sea Power*, p. 539.
[18]   G. Till, *Seapower: A Guide for the Twenty-First Century* (3rd ed.; London: Routledge, 2013), p. 212.
[19]   Till, *Seapower*, p. 58.
[20]   J.T. Sumida, *Inventing Grand Strategy and Teaching Command: The Classic Works of Alfred Thayer Mahan Reconsidered* (Baltimore, MD: Johns Hopkins University Press, 1999), p. 45.
[21]   Corbett, *Principles of Maritime Strategy*, pp. 91, 96–7.
[22]   Corbett, *Principles of Maritime Strategy*, p. 187.
[23]   Corbett, *Principles of Maritime Strategy*, pp. 99–100; Till, *Seapower*, p. 76.
[24]   C.E. Callwell, *Military Operations and Maritime Preponderance: Their Relation and Interdependence* (London: Blackwood and Sons, 1905), p. 59.

modest commerce-raiding strategy could be. It was 'partly because the marine resources of a great maritime power were so huge, partly because a truly effective *guerre de course* would have to be conducted in a barbaric way which could backfire on the perpetrator' (more on which below).[25] In a similar fashion, Mahan noted the difference between raiding Spanish treasure ships in the eighteenth century and the British global trading system of the nineteenth and twentieth centuries. The former was more feasible to dismantle through *guerre de course* due to its concentrated nature, whilst the latter was more difficult to strangle due to its large volume and distributed nature.[26]

The notable departure from this orthodoxy was Raoul Castex, who spent considerable effort attempting to argue that economic warfare at sea *could be* more decisive and relevant to grand strategy. A French admiral and naval strategist in the early to mid-twentieth century, Castex's practical and intellectual life was dominated by the problems posed by France's geostrategic position: caught between the British and German Empires. His thought evolved from debates in French naval circles regarding the role of *guerre de course* throughout the second half of the nineteenth century, during which some more junior naval officers – the so-called *Jeune École* – argued for abandoning France's attachment to large fleets in favour of a strategy of commerce raiding and sea-denial.[27] Castex accepted the damage maritime economic warfare could achieve, but tempered the economic and technological determinism of the *Jeune École* with the Mahanian and Corbettian view that economic warfare had to be subordinated to a wider war plan to make a difference to the general war effort.[28] He viewed such 'ancillary' operations as part of larger strategic manoeuvres intended to make relative weaknesses in some areas enable a strength at a more crucial point. *Guerre de course* therefore fulfilled a classic function of strategic manoeuvre, particularly if the economic losses forced on the enemy triggered a favourable distribution of resources and forces elsewhere.[29] Though thin on specific details, Castex is nonetheless surely correct that inflicting a serious toll on merchant shipping may force the adversary fleet to spend more resources on protecting its own commercial traffic than prosecuting fleet superiority, battle, or amphibious operations. Efforts by one combatant to use the sea will oblige their adversary to incur a large price in order to secure command of that sea. Commerce raiding can thus be seen as a strategic manoeuvre designed to tax the enemy's economy of force.[30]

---

[25]   Till, *Seapower*, p. 214; Corbett, *Principles of Maritime Strategy*, p. 278.

[26]   Mahan, *The Influence of Sea Power*, p. 539.

[27]   On summaries of the Mahanian School and *Jeune École* debates in French naval history, see B. Heuser, *The Evolution of Strategy: Thinking War from Antiquity to the Present* (Cambridge: Cambridge University Press, 2010), pp. 71–5.

[28]   Castex, *Strategic Theories*, pp. 48, 71, 348–9.

[29]   Castex, *Strategic Theories*, pp. 106–19, 136.

[30]   Castex, *Strategic Theories*, pp. 87, 92.

Through placing such methods under the concept of strategic manoeuvre, Castex developed French thinking on maritime economic warfare – either blockade or *guerre de course* – beyond mere opportunism, and without indulging in the excesses of the *Jeune École*. Indeed, he argued explicitly that *guerre de course* could not be useful through opportunism alone. It was most effective if done with or in conjunction with a balanced surface fleet which could more effectively dispute the command of the sea once the enemy's strength had been dissipated meeting *guerre de course* attacks. During a window where a good enough command of the sea was achieved, parts of the fleet could then be used to conduct more ambitious economic warfare operations such as blockade.[31] Callwell also considered the use of such methods from the perspective of the weaker naval power. Because devoting most resources to overthrowing the enemy's fleets was not a realistic option, he advocated a Corbettian fleet-in-being approach by avoiding 'engagements unless some happy chance brings about a local naval superiority, and to injure the antagonist by other means if possible'. Callwell added that commerce raiding can only work if it is a sustained effort supported by a general preponderance of force on the relevant lines of communication, and if the target is particularly vulnerable to such disruptions to maritime supply lines.[32] Therefore, a critical factor in deciding to engage in economic warfare is the vulnerability of the target to it, regardless of the capability of oneself to impose a blockade on them. Here he was in agreement with Mahan, who admitted that the impact of sea power 'must be *supported* ... [it is] unsubstantial and evanescent in itself'. Without an effort to make economic distress relevant to grand strategic objectives, the blows of commerce warfare will be painful but not fatal.[33] These views are more detailed variations on a point famously quipped by Corbett that people live on the land and not the sea, and efforts must therefore be directed towards generating effects on land.[34]

Going further, Raja Menon articulates that the crucial factor in maritime economic warfare, and sanctions more generally, is time. The influence of maritime economic warfare increases if its effects are in sync with the phases of the land war or primary theatre of operations, and the economic effects of blockading takes time to accumulate.[35] This follows precedents from earlier theorists.[36] The theoretical point to draw from this is that the vulnerability of the target to disruptions to seaborne commerce is not only important in itself, but that level of vulnerability fluctuates over time. Sometimes an economy can adapt and continue, if not expand during war, sometimes economic pressure may just

---

[31]   Castex, *Strategic Theories*, pp. 67–75.

[32]   Callwell, *Military Operations*, pp. 55, 61.

[33]   Mahan, *The Influence of Sea Power*, p. 132.

[34]   Corbett, *Principles of Maritime Strategy*, p. 14.

[35]   K. Raja Menon, *Maritime Strategy and Continental Wars* (London: Routledge, 1998), pp. 65, 70.

[36]   Mahan, *The Influence of Sea Power*, p. 191; Callwell, *Military Operations*, p. 37.

require time to 'turn the screw'. Moreover, shifts in global patterns of trade and the efficacy of land-based transportation systems for goods have long influenced perceptions of the utility of maritime economic warfare in particular instances.[37] Indeed, this is what Mahan and Corbett only allude to as the silent, accumulative, and exhausting pressure of sea power if preponderance at sea is achieved.[38]

Yet successfully conducting a long-term blockade contains risks for the navy that implements it. Such a strategy may be 'unglamorous, uncertain, and quite often boring compared to what the army and air force presentations are on the same subject'.[39] Such concern may be particularly true for a continental navy, which may be persistently and disparagingly considered as a 'stepchild' within land-centric strategic cultures.[40] Having to compete for resources in a continental state, a navy may have a difficult time to make its case for budgetary resources if its methods of bringing sea power to bear upon the enemy takes time, rather than through the display of combat power and being able to tally enemy combat vessel losses. Indeed, in contemporary debates on airpower and economic warfare, economic coercion from the air is immediately more visible than naval economic coercion, as the case of Kosovo in 1999 shows. This echoes the drawbacks of a more passive style of warfare often associated with the defence and excessively avoiding battles.[41] It is difficult for a navy to focus its intellectual and budgetary resources on economic warfare as 'no ordnance is expended, no battles are won, and the damage is only visible in statistics', which themselves are often contested.[42] A navy focused on a *guerre de course* strategy may also impose difficulties in inter-service battles as it may be perceived merely as an opportunistic band of raiders rather than a 'proper' battle fleet, as opposed to a potentially coordinated tool of grand strategy that could bring a maritime state to the brink.

These conceptual insights show a tendency amongst theorists to equivocate regarding the influence and decisiveness of maritime economic warfare. Most agree that sea power's economic action is most impactful when enacted through a blockade conducted upon a secure basis of sea command. Yet even in this instance, no theorist will ascribe a decisive quality to it. For weaker navies, most agree that opportunities to engage in commerce warfare or *guerre de course*

---

[37]   P.M. Kennedy, *The Rise and Fall of British Naval Mastery* (London: Allen Lane, 1976), pp. 39–66.

[38]   Mahan, *The Influence of Sea Power*, p. 209; Corbett, *Principles of Maritime Strategy*, pp. 96–7.

[39]   Menon, *Maritime Strategy*, p. 70.

[40]   T. Ropp, 'Continental Doctrines of Seapower', in E.M. Earle, *Makers of Modern Strategy: Military Thought from Machiavelli to Hitler* (Princeton, NJ: Princeton University Press, 1943), p. 446.

[41]   Mahan, *The Influence of Sea Power*, p. 290; Corbett, *Principles of Maritime Strategy*, pp. 103–4.

[42]   Menon, *Maritime Strategy*, p. 83.

to inflict pain on the enemy should still be seized upon, ideally in conjunction with operations elsewhere to maximise their effect. In addition, not all targets of economic warfare will be equally susceptible to its effects, and even when applied efficiently against a vulnerable target it will take time for the full weight of maritime economic warfare to be felt and to begin to undermine the strategic capabilities and will of the adversary. Debates over economic interdependence and the belief that increases in international trade tend to curtail the incidence and duration of future wars have failed to account for these crucial points. The pantheon of theorists explored here show that sea power acts slowly, by debilitating economic pressure – as opposed to a sharp, sudden, but recoverable shock. Contra to suggestions by theorists such as Norman Angel, Ivan Bloch, and even Liddell Hart, maritime economic warfare is thus a less precise and readily used tool of policy than might otherwise be assumed.[43] The degree of contribution that maritime economic warfare can make to victory depends on the degree and persistence of command over the sea possessed, the vulnerability of the target to economic privation from the sea, the duration of the war, and the extent to which it is suited and adapted to the objectives and needs of a specific grand strategy.

Grand Strategic Considerations

Even if a commander conducts maritime economic warfare within the ideals set out by Castex's scheme of strategic manoeuvre, maritime economic warfare may still not be suitable for grand strategic purposes. Economic naval warfare must be in tune with the overall economic situation of the war as well as the international political context – specifically, the court of domestic and international public opinion and the role of neutrals and third parties. Studying and planning for economic warfare cannot remain solely a question of techno-geographic feasibility and perceived strategic utility, and nor is it solely the purview of navies whilst other tools of statecraft and military branches – air forces in particular – can create and exploit economic suffering. Economic warfare at sea is after all only one tool of grand strategy, and one that also exists in the chaotic, passionate, and political universe of war.[44] The success of maritime economic warfare in grand strategy, whether historical or contemporary, cannot be assessed in isolation from the fact that war is an uncertain, emotional, and purposeful activity. Historians and strategists must consider the chaotic nature of war in complicating the tasks of imposing blockade and commerce destroying, but also

---

[43]   N.A. Lambert, *Planning Armageddon: British Economic Warfare in the First World War* (Cambridge, MA: Harvard University Press, 2012), pp. 2, 24, 25, 126–7.
[44]   Carl von Clausewitz, *On War*, trans. O.J. Matthis Jolles (London: Random House, 1943), in Caleb Carr (ed.), *The Book of War* (New York: Modern Library, 2000), p. 282; Corbett, *Principles of Maritime Strategy*, p. 237.

must analyse it with regard to its effects on public opinion at home and abroad and its potential for escalation, as well as whether it actually serves the purpose of war or makes the ultimate objective harder to achieve.

Castex argued that maritime communications serve three general purposes in a conflict: to sustain the economy and by extension harm the enemy's, to move military forces, and to provide internal lines of communications where necessary.[45] Despite this emphasis on the economic importance of commanding the sea for grand strategy – where the actions of battle fleets secure the economic and military transport exploitation of SLOCs and are only the means to that end – Castex was adamant that 'command of the sea is impotent against continental powers if by command of the sea we mean only the ability to interrupt maritime communications'.[46] The ultimate economic and logistical value of commanding the sea is thus subordinate to the ultimate goals of grand strategy.[47] The utility of economic warfare at sea waxes and wanes according to the strategic conditions outlined above, but also whether it suits the war at hand. Corbett argued that whether a war was limited or unlimited would partly determine the applicability of blockade and whether it may actually be all that is required to secure the desired political outcome. He claimed that 'in the naval sphere there may be a life and death struggle for maritime supremacy or hostilities which never rise beyond a blockade'.[48] Corbett implied that maritime blockade was an inherently less escalatory measure than invading and overthrowing an enemy. This raises a key question for maritime economic warfare – whether it helps meet the ultimate political objective.

The reason of war and the political objectives behind a specific war influence whether an economic blockade or commerce raiding is desirable or successful semi-independently of its perceived decisiveness. This is particularly true if the state wishes to or is prepared to engage in a long war. It is of the utmost importance for decision-makers to understand the kind of war they aim to undertake – for example, a plan for a short, sharp war should not rely on economic sanctions and blockade to bring about decisive results in a short time span. Whilst it may be the case that economic warfare alone cannot bring about the enemy's collapse, what if collapse is not the objective? The imposition of sufficient level of pressure may be all that is required to bring the adversary to the negotiating table, in which case economic warfare can be a vital tool.[49] The politics of war can 'move the goalposts' by modifying what is expected of economic warfare, and should always guide the more narrow military-strategic discussion of the decisiveness and feasibility of maritime economic warfare above.

---

[45] Castex, *Strategic Theories*, p. 30.
[46] Castex, *Strategic Theories*, p. 47.
[47] Castex, *Strategic Theories*, pp. 50–3.
[48] Corbett, *Principles of Maritime Strategy*, p. 39.
[49] Callwell, *Military Operations*, p. 183.

In a longer war, economic naval warfare, and blockade in particular, can appear attractive to top-level decision-makers, particularly for civilians keen to avoid casualties, whether sustained in military combat or through the application of air power. Indeed, 'once it [blockade] has been realised that the chances of a quick and decisive victory on land are remote, the emphasis shifts from the solider to the economy'.[50] Corbett argued that when other things are equal it is the longest purse that wins and that financial targets are important.[51] However, 'other things' are never equal in war. Only if a war is sufficiently protracted against a suitable foe can macroeconomic pressure be brought to bear. Menon displays a reluctance to believe that starving a nation of seaborne food supplies can bring an enemy to surrender in any decisive fashion by itself (with the exception of perhaps poorer countries whose primary cereals are imported), yet also claims that 'the effects of a commerce war on the fate of maritime nations is so clear and explicit that it needs no elaboration'.[52] His observations in this regard are by no means clear – both Britain and Japan either saw off or endured for a prolonged period the effects of blockading efforts during the world wars of the twentieth century and provide no direct answer as to how Britain achieved victory or Japan eventually surrendered.[53] Indeed, Mahan noted examples where British trade in the age of sail suffered large losses but the Royal Navy continued to operate effectively against continental military campaigns, and overall British maritime trade continued to grow and operate effectively enough despite major losses.[54] More convincingly, Menon also argued that more-developed societies would be more vulnerable to blockade, as affluent populations become softer targets as their luxuries are whittled away over time, and with the increasing amount of global trade and dependence on seaborne commerce for goods, he sees merits in engaging with the ideas of the *Jeune École*.[55] This is particularly true if the enemy is highly dependent on oil imports via sea lines of communication as oil directly enables war assets and the entire economy, and is harder to substitute or do without, unlike many raw materials, cereals, and finer goods.[56]

A determined actor would also seek to find ways to undermine the economic superiority of their adversary, making their deep pockets less relevant or effective. It must be remembered that even an attempt to avoid or sideline combat cannot escape the centrality of violence and the potential of it in grand strategic

---

[50]   Menon, *Maritime Strategy*, p. 74.

[51]   Corbett, *Principles of Maritime Strategy*, p. 99.

[52]   Menon, *Maritime Strategy*, pp. 69, 74.

[53]   On the question of Japanese surrender in the face of economic deprivation, the Soviet invasion, and the dropping of the atomic bombs, see C. Craig and S. Radchenko, *The Atomic Bomb and the Origins of the Cold War* (New Haven, CT: Yale University Press, 2008), pp. 62–89.

[54]   Mahan, *The Influence of Sea Power*, pp. 317–18, 400.

[55]   Menon, *Maritime Strategy*, pp. 74–7.

[56]   Menon, *Maritime Strategy*, p. 69.

calculations. Indeed, a successful blockade is based on the credible threat that attempting to breach it will result in an engagement whose outcome will be easily predictable. The idea of combat must still underwrite the use of military forces, even if the destruction of enemy forces is not the immediate objective of naval or grand strategy.[57] A timidity in using or threatening violence and bloodshed may indicate a 'weak' political object, or a lack of conviction at the very least, denoting a particularly lacklustre attempt to impose some economic cost on the adversary rather than roll the dice on the battlefield.[58]

Although blockade can be seen as less escalatory or more merciful than major combat operations on enemy territory in some circumstances, the moral dimension of war and its non-linear nature also mean that disproportionate effects are possible.[59] Indeed, the most severe of blockades or commerce destruction can lead to shortages of essential goods and medicines, reaching the effects of the realm of more 'total' warfare upon an entire society, hurting and shortening the lives of the civilian population indiscriminately. This could be intended or not on the part of the blockader. Nevertheless, it warrants caution against the view that causing economic pain is inherently less escalatory than engaging in combat operations. The opposite could just as easily apply. War is a political act, but reason is not a tyrant over the forces of passion and chance; a government and military's control on a war's proceedings is never absolute.[60] Conducting escalatory economic sanctions may trigger an adaptation or escalation on the part of the victim that may make achieving one's grand strategic objectives harder. This is particularly true in the court of public opinion if economic warfare's consequences are seen as a 'most barbarous form of war' and the 'most brutal of all forms of attack'.[61] The passions of war can be aroused just as feasibly from economic warfare as they can from direct combat.

This is illustrated well in the more emotive aspects of maritime economic warfare, which concern the morality of punishing an entire population with increasing degrees of deprivation and risking the ire of third parties and neutral actors. Castex appreciated the 'unceasing' psychological value that 'naval mastery' and blockade could impose on the enemy, and closely echoes the 'pressure' that Mahan mentioned above.[62] This pressure, if applied properly, can be systemic in its consequences – threatening the quality of life of entire populations beyond the target state. Corbett argued that 'by closing his commercial ports, we can injure the enemy the most with our command of the sea. In the long run a rigorous

---

[57]   Clausewitz, *On War*, trans. Jolles, pp. 289–90.
[58]   Clausewitz, *On War*, trans. Jolles, pp. 271–2.
[59]   Clausewitz, *On War*, trans. Jolles, pp. 271–2.
[60]   Clausewitz, *On War*, trans. Jolles, p. 279.
[61]   J.F.C. Fuller, *War and Western Civilization, 1832–1932* (London: Duckworth, 1932), p. 230.
[62]   Castex, *Strategic Theories*, p. 359.

and uninterrupted blockade will exhaust the enemy before it exhausts us'.[63] The resultant deprivations take a toll on the civilian population and raise serious questions over whether such measures are necessary and just. Mahan considered moral power to be a crucial component of any naval commander, and by extent any strategist, meaning a 'courage to assume responsibility for actions whose outcomes were both uncertain and potentially dire'.[64] Depriving an economy and its people of goods and imposing systemic consequences through blockade, or raiding civilian merchant vessels, should not be taken lightly on moral grounds.

Menon, indirectly, raises the potential moral and ethical consequences of such successful blockading actions through his examination of Germany in 1918, where malnutrition and psychological despair were visited upon the German population *in part* through the successful and ongoing four-year campaign of economic warfare and the demands of total war on the economy.[65] However, by and large, the ethical questions of just conduct in maritime economic warfare is not given much attention in the classical sea-power literature. For the contemporary practice of maritime economic warfare, the fact that the more successful a naval blockade is the more it stops the functioning of an entire economy – and as a result increases distress and mortality rates among a target population – raises the ethical concerns most associated with discussion of the ethics of strategic bombing and 'total war' rather than sea-power theory.[66] This is a grand strategic question as to what lengths a naval power will resort in order to win a war – whether it is 'worth it'. Indeed, strategists are reminded to consider Clausewitz's wisdom of whether this kind of tool of maritime grand strategy is apt for that kind of war. Whilst financial sanctions could indeed be more targeted affairs, the blocking of maritime traffic tends to be a blunter instrument and poses difficult ethical questions for self-conceived liberal states and navies as tools in grand strategies.

What is given more attention than ethical concerns is the escalatory potential of maritime economic warfare through bringing in neutral and third parties into the conflict, rather than the humanitarian consequences of economic warfare. To increase the chances of success, an economic blockade or *guerre de course* has to stop all potential sources and transfers of the enemy's wealth and supply from reaching its intended destination, and there can be no exceptions to this even if carried by third parties. Indeed, several of the sea-power theorists agree and go so far as to claim that naval warfare is entirely pointless if contraband, property,

63   Corbett, *Principles of Maritime Strategy*, p. 187.

64   Sumida, *Inventing Grand Strategy*, p. 53.

65   Menon, *Maritime Strategy*.

66   On the ethics of strategic bombing and airpower, see Brett J. Cillessen, 'Embracing the Bomb: Ethics, Morality, and Nuclear Deterrence in the US Air Force, 1945–1955', *Journal of Strategic Studies* 21.1 (1998), pp. 96–134; Arash H. Pashakhanlou, 'Air Power in Humanitarian Intervention: Kosovo and Libya in Comparative Perspective', *Defence Studies* 18.1 (2017), pp. 39–57; Daniel L. Byman and Matthew C. Waxman, 'Kosovo and the Great Air Power Debate', *International Security* 24.4 (2000), pp. 5–38.

and wealth cannot be seized at will by the power that commands the sea. In a more restricted war, perhaps a target can be isolated more easily, and neutral powers kept aside. However, a more unrestricted war or a foe desperate to secure its own survival could escalate the war by bringing in third parties to the fray, with the forces of passion and chance altering the original calculus for war as dictated by the forces of reason. As the seas are the major routes and arteries of the international trading system, maritime economic warfare must account for the question of third parties, especially if their trade routes carry them near a theatre of war or actively seek to supply or trade with the victim of maritime economic warfare. Ensuring the foe does not benefit from the sea – especially from trade and logistics – is for many sea-power theorists the entire rationale for navies. For the sea-power theorists, attempting to ban or curtail the ability to interfere with international trade at sea in a time of war struck at the heart of the purpose of violently commanding the sea. Corbett noted that banning the seizing of maritime commerce was pointless because maritime warfare is designed to take out the obstacle which allows one to cripple enemy trading on the sea.[67] Richmond was adamant on the point that for a maritime power like Britain, disrupting and cutting off an enemy's seaborne trade was the only way it had to exert pressure on its continental adversaries. It was apparent that naval victories were but the means to the end of applying economic pressure to produce peace, as without the ability to make the command of the sea translate into economic pressure, fighting over the control of trading routes was strategically fruitless.[68]

Castex was arguably the most outspoken theorist regarding the necessity of unrestricted economic warfare at sea. At the 1922 Washington Conference he 'caused a sensation by a memorandum approving the German submarine campaign' whilst stressing how it provided the most serious threat to Britain in the First World War.[69] Such views are developed further in Castex's magnum opus. He complained how in 1914:

> some now claimed freedom of the seas to be valid in wartime, that commercial navigation should be free both to belligerents and to neutrals and that adversaries could make war only against armed forces ... In short, all private property being sacrosanct, contraband of war no longer existed and blockade ceased to be legitimate.[70]

At the time, the United States was engaged in an intractable dispute with Britain over this principle – of the right of free trading for neutral parties – which in

---

[67] Corbett, *Principles of Maritime Strategy*, p. 94.
[68] H.W. Richmond, *British Strategy, Military and Economic: A Historical Review and its Contemporary Lessons* (Cambridge: Cambridge University Press, 1941), pp. 117–22.
[69] Ropp, 'Continental Doctrines of Seapower', p. 456.
[70] Castex, *Strategic Theories*, p. 37.

short meant that 'naval war has no point if enemy property can travel without hindrance and if neutrals can supply the enemy or conduct his trade'.[71] Pre-1914 efforts to take private property off Britain's contraband list was a case of some states attempting to 'compensate for their weaker navies with stronger laws'.[72] However, behaviour towards neutrals must reflect their importance –Germany between 1914 and 1918 could not ignore transatlantic trade if it was intent on crippling the British economy, even if targeting this traffic dramatically increased the likelihood of US intervention, which it eventually did.[73]

Mahan noted that dominant naval powers could deny neutrality if they so chose. As a historically dominant sea power in Mahan's time, Britain was reluctant to grant any special immunities to 'neutral' trading at sea during a time of war.[74] Callwell warned that the supposed rights of neutrals were 'liable to be trampled upon' if they were 'unable, or unwilling, to defend them' in the open seas, although in the modern era territorial waters of a neutral party may still be considered 'inviolate'.[75] In that sense neutrals and third parties must be armed to avoid being caught up and abused in another's blockading or *guerre de course* campaigns. Castex reasoned that:

> in peacetime, the sea is free to everyone. In war, it belongs to the strongest, who will chase both his enemy and any unfriendly neutrals from it as far as he is militarily and politically able. Such is the purpose of the navy, and that is the end of the argument ... And we can be sure that, in the next war, the Americans, ready to fight to defend the freedom of their own commerce when they are neutral, will brutally uphold the other point of view when they are belligerents.[76]

A dominating sea power can deem any vessel it desires as a fugitive, and 'by controlling the great common, [close] the highways by which commerce moves to and from the enemy's shores'.[77] However, such acts may not always contain a limited war and could risk an unwanted escalation, incurring the wrath of war's chaos and passion.

---

71   Castex, *Strategic Theories*, p. 39.
72   Menon, *Maritime Strategy*, pp. 78–9.
73   Castex, *Strategic Theories*, pp. 39–40.
74   Mahan, *The Influence of Sea Power*, pp. 138, 540.
75   Callwell, *Military Operations*, p. 44.
76   Castex, *Strategic Theories*, p. 40.
77   Mahan, *The Influence of Sea Power*, p. 138.

Conclusion

A key component of maritime economic warfare are these grand strategic considerations: to what extent can the target be isolated from international trade, to what extent can neutral trade be allowed to continue uninterrupted, and can the blockader and blockaded suffer or exploit the moral, psychological, and ethical consequences of imposing suffering on civilians and an entire society? These questions not only inform considerations on the decisiveness and feasibility of economic warfare as a tool of grand strategy but directly illustrate the centrality of sea power to the international system. Only by considering these grand strategic and recurring factors can economic warfare at sea be assessed and judgements as to whether it helped to achieve the ultimate objectives of grand strategy be reached.

This chapter has shown how the sea-power theorists share many similar insights on the nature of economic warfare and sea power, and together form several points of theoretical wisdom that should be useful for historians, contemporary analysts, and practitioners alike. All have grappled with the problem that the economic pressure wrought from a general command of the sea is neither a decisive course of action nor a folly whose use should be dismissed outright. Its feasibility and decisiveness waxes and wanes according to the vulnerability of the target and the time allowed for it to take effect. Most have agreed that even the most effective blockades or *guerre de course* campaigns – even if supported with a balanced fleet – would exert only indirect effects upon the crucial theatre of a conflict, which lies on the land. Though Mahan and Corbett discuss this at some length, it was down to more continentally minded theorists such as Castex, Callwell, and Menon to highlight the necessities of intelligently incorporating economic warfare at sea into a grand strategy rather than just assuming its contribution and to consider its effects in the realm of international politics where neutral parties and shipping were considered as escalatory risks for the blockading or raiding party. A relative blind spot for sea-power theory, however, is the consideration of the ethical and moral quandary of targeting economies and civilians, which economic warfare of any kind does.

These theoretical constants – or strategic truths – of sea-power theory amount to an 'it depends' answer to the question of how effective maritime economic warfare can be. However, such ambiguity is inherent to the very nature of strategic theory. Theory serves to educate the reader as to rules of thumb for judgement and to think upon what conditions particular tools of strategy could be more useful or appropriate for than others according to the case at hand. Indeed, the practice and study of war should not be left to 'natural talents' alone.[78] A more rigorous and critical application of theory to empirical cases allows warfare to become susceptible to investigation by reason.[79] Like Clausewitz's theory, the

---

[78]   Clausewitz, *On War*, trans. Jolles, p. 337.
[79]   Clausewitz, *On War*, trans. Jolles, p. 346.

strategic truths espoused by the sea-power theorists of economic warfare will not reveal any simple secrets of success, merely structure and inform one's own investigation and planning of it.

CHAPTER THREE

Convoys and Companies: Privatising Economic Warfare at Sea in the Dutch Republic, 1580–1800

Erik Odegard

Introduction

On 25 September 1639 (according to the Gregorian calendar), Lieutenant-Admiral Maerten Harpertsz Tromp, commanding a Dutch naval squadron of 12 ships, intercepted a much larger Spanish force of approximately 60 to 70 vessels sailing westwards through the English Channel. The running fight between the two forces on 25–26 September, known in the historiography as the 'Fight in the Channel', has been heralded as the first time the line-ahead formation was used in European waters.[1] Though Tromp lost one vessel, the *Groot Christoffel*, in the fight, the Spanish fleet was badly mauled and retreated to anchor in the relative safety of the Downs. Tromp resupplied and repaired his vessels in Calais, successfully blocking passage to Mardyck and Dunkirk for the Spanish vessels. Having been resupplied, Tromp moved his fleet to the Downs to blockade the anchorage and keep the Spanish fleet at anchor. Over the coming month, his fleet would be reinforced with vessels from across the Republic, until it would number over 100 ships (including fireships) in late October 1639. The frenzied activity which enabled the Dutch to assemble such a large collection of ships in so short a period of time was described by the well-known poet and classicist Caspar Barlaeus in his inaugural lecture at the Amsterdam *Athenaeum Illustre* later that year: 'The Quays, Harbours and Shipyards of Holland and Zeeland teemed with new equipments on water and at war. Amsterdam … mobilized its three maritime Colleges with astounding speed … It seemed not so much that everywhere ships were being built, but that they grew on their own accord'.[2]

---

[1] J.F. Guilmartin Jr, *Galleons and Galleys* (London: Cassell, 2002), p. 203.
[2] Online bibliography of Caspar Barlaeus, Leiden University, Department of Dutch Language and Literature: www.let.leidenuniv.nl/Dutch/Latijn/BarlaeusOratien1689.html#ZeestrydDuins.

This hints at a crucial aspect of Dutch success at the Downs in 1639, namely the ability rapidly and effectively to mobilise ships for service with the battle fleet from naval organisations other than the state-run admiralties themselves, in this case the East and West India Companies. Indeed, what is most interesting about the Battle of the Downs is not so much the actual battle itself but rather the insight it offers in the organisation of Dutch warfare at sea and the institutional world of economic warfare at sea in the seventeenth-century Dutch Republic.[3] This is a significant point as it runs counter to established narratives of this period being characterised by a steady movement towards central state monopoly over the use of armed force, if not the production and recruitment of military supplies and soldiers.[4] From this perspective, Dutch practice seems chaotic and incoherent, not to say regressive. This is especially the case in the Anglophone literature, which tends to focus on the Anglo-Dutch wars, and to emphasise the weaknesses of the Dutch system when compared to that of more centralised English practice. These Wars, which revolved around a series of battle fleet engagements, exposed weaknesses in the existing Dutch system of naval administration that made it more difficult to concentrate a force of large ships capable of acting as a coherent, integrated fleet. Yet it does not follow that the Dutch systems of organisation and administration were inherently inferior to those of the English, or that they were fatally flawed. Rather, we must appreciate that Dutch naval power in the seventeenth and early eighteenth centuries was organised and adapted to support a different purpose, namely to protect private

---

Oratie over de Zee-strijd, tegen de Spaansche Vloot in DUINS.

... De Kaajen, Havens en Scheepstimmerwerven van Holland en Zeeland woelden en grimmelden van nieuwe toerustingen te water en ten oorlog. Amsterdam, dat onder zijne Landgenoten en nabuuren in scheeprijkheid de kroon spant, en een vruchtbare Moeder en Voedster der Zeevaard is, maakte met een ongeloofelijke snelheyd drie Collegien te water gaande, namelijk die van de Admiraliteit, en het Oost en West-Indisch Huis. ... Het scheen niet dat men van alle kanten schepen timmerde, maar ofze van zelfs groeiden. Gy zoud zeggen dat bomen, balken en masten, in scheeps gedaanten veranderden.

[3]   C.R. Boxer (ed.), *The Journal of Maarten Harpertszoon Tromp Anno 1639* (Cambridge: Cambridge University Press, 1930); N.A.M. Rodger, *The Safeguard of the Sea: A Naval History of Britain*, vol. 1, *660–1649* (London: HarperCollins, 1997), pp. 412–13. Most recently, Richard Blakemore and Elaine Murphy start their narrative of the British Civil Wars at Sea with the humiliation of the Royal Navy at the Downs. R. Blakemore and E. Murphy, *The British Civil Wars at Sea, 1638–1653* (Woodbridge: Boydell Press, 2018), pp. 1–4.

[4]   Recent historiography tends to take Hobbes's *Leviathan* as a starting point to discuss the idea of state control over armed force. See, for example, E. Krahmann, *States, Citizens and the Privatisation of Security* (Cambridge: Cambridge University Press, 2011), pp. 22. For an alternative perspective, see the articles in R. Torres-Sánchez, P. Brandon, and M. 't Hart (eds), 'War and Economy: Rediscovering the Eighteenth-Century Military Entrepreneur', *Business History* 60.1 (2018), pp. 4–22.

enterprise at sea across a range of potentially hostile regions, and in a manner that did not place an exorbitant cost upon the admiralty's (and indirectly the Generality's) purse. Typical activities for Dutch warships until the outbreak of war with England in 1652 were escorting the fishing fleet trawling for herring in the North Sea, escorting merchantmen bound for France, Spain, and the Mediterranean through the Channel, cruising in the North Sea to intercept and escort homeward-bound Indiamen, and the blockade of the port of Dunkirk. For Dunkirk was the base of a fleet of Habsburg warships and privateers whose goal was the destruction of the trade on which the Republic depended. The institutional development of Dutch naval power can only be understood as a response to this campaign of economic warfare.

With these factors in mind, this chapter sets out to do three things. First, it will describe the various organisations that were at one point or another made responsible for Dutch economic warfare at sea besides the admiralty boards. Second, it will consider those proposed organisations which for one reason or another were never created. By including these 'failures' the chapter can more clearly articulate the uses and limits of what may be styled the Dutch strategy of 'outsourcing' naval protection. This strategy was never put down on paper in clear terms: it was a strategy of practice, rather than theory. But its fundamental characteristics can be readily identified, nonetheless. Finally, it will chart the effect of this particular institutional development on Dutch naval organisation after the subsidence of the Dunkirk threat and the First Anglo-Dutch war of 1652–54.

Economic Warfare at Sea and the Dutch Republic, 1600–1800; Naval Administration in the Dutch Republic: The Admiralty Boards

The naval administration of the Dutch Republic in the seventeenth century was a complex organisation.[5] Over the course of the early phases of the Dutch revolt against the Habsburg lord of the Netherlands, five distinct admiralty boards had been created to replace the Habsburg admiralty which had been based in Veere.[6] Within the province of Holland, there were three such organisations: the Meuse, centred in Rotterdam, in Amsterdam, and in the Northern Quarter of Holland – in the ports of Hoorn and Enkhuizen. Additionally, the provinces of Zeeland and Friesland had their own admiralties. These admiralty boards were in charge of procuring their own vessels, armaments, and crews and operated their own shore establishments. They were, nominally at least, institutions of

---

[5]   N.A.M. Rodger, *The Command of the Ocean: A Naval History of Britain*, vol. 2, *1649–1815* (London: Allen Lane with the National Maritime Museum, 2004), pp. 9–10.

[6]   L. Sicking, *Neptune and the Netherlands: State, Economy, and War at Sea in the Renaissance* (Leiden: Brill, 2014), pp. 407–17.

the General Union, rather than purely provincial or urban organisations, though practice often differed between individual boards. Each board was provided with a regular source of income, the 'convoys and licences' – or tolls and tariffs – which were levied on all incoming and outgoing traffic (on water as well as on land) in the entire Republic. Each board had been ascribed a geographic region of the country where it could collect these. In principle, therefore, the admiralty organisation was self-financing and did not require appropriations from the central budget of the Generality. The method of financing meant that the admiralties were directly tied to the general health of the maritime economy: fewer ships paying the levies which supported the navy meant less money available to pay for warships to protect trade. A successful campaign of economic warfare against Dutch shipping would directly undermine the financial health of the admiralties. Indeed, throughout the 1620s, financial difficulties forced the admiralties to take out loans and not all ships actually available could be sent to sea. In the winter of 1621–22, only 8 to ten vessels were actually on their blockading stations at Dunkirk, where 32 were prescribed.[7] From the 1620s onwards, the States-General thus regularly had to vote money to cover the admiralties' debts. Additional funds were sometimes voted to provide for more ships, or to cover expenses that were considered outside the regular activities of the admiralties.[8]

Older historiographies stress the inefficiencies inherent in the multiple admiralties. The naval successes of the Dutch Republic during its 'Golden Age' were explained as having occurred despite rather than because of its federalised admiralty organisation. Jan Glete challenged this perception when he argued that, per capita, the Dutch Republic had the highest military expenses of any European state and the largest standing forces.[9] Rather than being chaotic, the Dutch Republic, according to Glete, was supremely efficient in raising revenues to pay for this concentration of military and naval power. Glete sought to understand Dutch naval success *as a result of* its organisation rather despite it. Dispersed organisation made it easier to aggregate local interests behind central aims. It was easier to raise revenues for the fleet in West-Friesland, for example, if local elites knew that money was going to be spent locally, buying goods from local suppliers and overseen by local admiralty lords. Yet Glete did accept the fundamental premise that the increasing technological complexities of warfare at sea could only be mastered by the centralised state, producing a tendency towards a 'state monopoly of violence at sea' as the seventeenth century wore

---

[7]   J.E. Elias, *Schetsen uit de geschiedenis van ons zeewezen*, vol 1, *1568–1652* (The Hague: Martinus Nijhoff, 1916), p. 98.

[8]   Examples being the building of a fleet of battleships (useless for cruising and convoy work) in the 1650s and 1660s, and equipping a support fleet for Brazil in the 1640s.

[9]   J. Glete, *War and the State in Early Modern Europe: Spain, the Dutch Republic and Sweden as Fiscal-Military States, 1500–1660* (London: Psychology Press 2002), pp. 140–73.

on, and which the Republic was slow to follow.[10] A closer examination of the naval organisation of the Dutch Republic will challenge this point of view. In particular, we will see that Glete's conclusions were the product of the focus he placed upon the admiralties as the core components of the Republic's naval power. In reality, Dutch naval organisation was considerably more complex, as a range of naval organisations were created that did not (nominally at least) fall under the purview of the Generality. Many of these organisations were formed specifically to undertake measures of economic warfare: whether safeguarding Dutch trade or preying upon vulnerable commercial traffic passing through the Channel.

To understand the logic behind these institutions, it is necessary briefly to discuss the main objective of Dutch naval forces during the first half of the seventeenth century: the protection of shipping in the face of a concentrated strategy of economic warfare at sea waged from Dunkirk.

Economic Warfare at Sea and the Non-Admiralty Naval Forces

For much of the seventeenth and eighteenth centuries, maritime warfare for the Dutch Republic in Europe meant a defence of its seaborne trade.[11] A relatively small state with a large – at times, the largest – carrying trade, the Republic's independent statehood was closely entangled with its maritime economy. For much of this period the Republic was the lowest-cost provider of shipping and financial services and could not hope to use force to conquer new markets in Europe. Rather, its economic rivals could use force or the threat of force, to try and reduce the Dutch competitive advantage. The most clearly articulated program of economic warfare directed against the Dutch Republic was the campaign waged by the Habsburg kings of Spain and lords of the Southern Netherlands, who, until 1648, tried to subdue what they saw as the rebellious northern provinces of their *pays de par-deçà*. The Spanish Habsburgs waged a campaign of maritime interdiction from the port of Dunkirk directed against Northern shipping and commerce from the late sixteenth century onwards. This campaign was waged by privateers as well as a royal squadron, but both forces were focused on the destruction of Dutch trade and fishing fleets, rather than naval battles with the Dutch force attempting to blockade Dunkirk.[12] The Count-duke of Olivares, favourite of King Phillip IV until 1643, even proposed creating an Imperial

---

[10]   J. Glete, *Navies and Nations: Warships, Navies and State Building in Europe and America, 1500–1860*, 2 vols (Stockholm: Almqvist & Wiksell, 1993), i. 6–13.

[11]   For a distinction between *naval* and *maritime* warfare, see H. Strachan (ed.), *The Direction of War: Contemporary Strategy in Historical Perspective* (Cambridge: Cambridge University Press, 2013), p. 151.

[12]   R.A. Stradling, *The Armada of Flanders: Spanish Maritime Policy and European War, 1568–1668* (Cambridge: Cambridge University Press 1992), p. 26.

Habsburg naval force in Lübeck and Wismar to destroy the Dutch Baltic trades, referred to in the Netherlands as the mother-trade (*moedernegotie*).[13]

Dealing with these threats required a maritime rather than just a naval strategy: this was a campaign waged as much by armed merchantmen and herring-busses as by the naval squadrons. Navy ships, which in the Dutch case means ships sent to sea by one of the five regional admiralty boards, were employed in one of three tasks: as escorts for incoming and outgoing convoys of merchantmen and the herring fleets; as cruisers patrolling the Channel and North Sea in search of Dunkirkers; and on blockading station at Dunkirk itself. But besides these naval deployments a broader maritime strategy can be discerned. Though it was never articulated in policy papers or formulated from a theoretical perspective, the practice is quite clear: protection at sea was to be provided as much as possible by the economic agents that would most benefit from it. This is distinct from maritime warfare as an entrepreneurial strategy, though the Generality at times did encourage Dutch privateers to go after Dunkirkers, offering attractive prize-money for captured privateers or recaptured Dutch vessels.[14] These operated at times very successfully, but were hampered by slow payment of the promised money.[15]

Besides naval forces supplied by admiralty boards and the private forces of privateers, there was a third solution: allowing lower levels of government and corporate bodies to equip their own naval forces to protect their own shipping. These activities could overlap with equipping ships as a form of military entrepreneurship, as in the case of the West India Company's privateering, but this was not necessarily the case.[16] In many cases, lower government used corporate naval power as a form of defensive warfare. Throughout the first half of the century the Directorate which regulated the herring fisheries, the East and West India Companies, and a number of convoy Directorates organised by

---

[13]   J. Glete, *Swedish Naval Administration, 1521–1721: Resource Flows and Organisational Capabilities* (Leiden: Brill, 2010), pp. 404–5.

[14]   J. Roelevink (ed.), *Resolutiën der Staten-Generaal*, vol. 7, *1 juli 1624–31 december 1625* (The Hague: Instituut voor Nederlandse Geschiedenis, 1994), p. 446 (Resolution of 4 July 1625). Throughout this chapter I will use the word 'Generality' rather than 'central state' or an equivalent term to denote the federal level (though this is in itself a problematic way of phrasing it) of government of the Dutch Republic, since 'central state' in the Dutch context carries with it a level of sovereignty over the provinces that was acutely debated in the Republic.

[15]   A.C. Kersbergen, 'Uit het bedrijf van de Nieuwe Geuzen van Rotterdam', *Rotterdams Jaarboekje* 4.1 (1933), pp. 63–76, 65.

[16]   David Parrott briefly mentions these alternative organisations in *The Business of War: Military Enterprise and Military Revolution in Early Modern Europe* (Cambridge: Cambridge University Press, 2012), pp. 111–12. But the urban convoy Directorates and the herring fisheries convoy should not, perhaps, be considered a form of military entrepreneurship since their object was not to make a profit from their convoy services.

merchants and backed by their municipalities acquired the right to protect their shipping by equipping their own warships. These urban boards were financed by a levy of the merchantmen that profited from the increased protection.

Outside of European waters, the Generality actively tried to make warfare at sea the responsibility of corporate bodies such as the chartered India companies. This divesting of responsibility for warfare outside European waters was systematic: the West India Company was even allowed to give out its own letters of marque and acted as a prize court for prizes captured within its charter area, to the immense chagrin of the admiralties.[17] Nor was this tendency to what we may call 'outsourcing' restricted just to warfare at sea. The Dutch Directors of the Levantine Trade (*Directie van Levantse handel*) were empowered to levy a tax based on the tonnage of ships sailing into the Mediterranean, from which, amongst other things, the salaries of the Dutch ambassador in Istanbul and the consul in Smyrna as well as gifts to the Sultan were financed. The Directors were also responsible for ensuring that all vessels trading into the Mediterranean were properly armed and manned.[18] The importance of the contribution of these organisations to Dutch naval power is best summarised by again looking at the battle of the Downs in October 1639.

Table 3.1 presents the provenance of the Dutch ships at the Downs on 21 October 1639. Of the 96 ships present at the battle, a little over half were ships equipped by the admiralties (51 out of 96).[19] But the remaining vessels were not merely armed merchantmen quickly mobilised and converted for wartime duty. The majority were in fact ships equipped for economic warfare duties by boards and Directorates responsible for maintaining naval forces other than the admiralties themselves. As such, the composition of the fleet at the Downs in October 1639 sheds crucial light on these institutions which were essential in mobilising the 'maritime potential' of the Republic.[20] The figures show that not only were non-admiralty ships crucial for improvising a battle fleet at a moment of major crisis, they also reveal that these ships were most likely equipped each year for rather less glamorous convoy duties.

---

[17] V. Enthoven, *Zeeland en de opkomst van de Republiek: handel en strijd in de Scheldedelta, c.1550–1621* (Leiden: Luctor et Victor, 1996), p. 84.

[18] A.H.H. van den Burgh, *Inventaris van het archief van de Directie van de Levantse Handel en de Navigatie in de Middellandse Zee (1614) 1625–1826 (1828)* (The Hague: Nationaal Archief, 1882), pp. 20–3.

[19] The figure of 51 is a maximum, but could be lower. J.C. de Jonge, *Geschiedenis van het Nederlandse Zeewezen* (1858), vol. 1, mentions 41 *Landsschepen*, but the loss of many admiralty sources makes it impossible to check his figures. It is perfectly possible that ships that are only recorded as coming from Amsterdam were if fact not equipped by the admiralty of that city but by one of the other naval colleges.

[20] For an English-language debate on the use of the term 'maritime potential', proposed by Louis Sicking in Dutch as *scheepsmacht*, see R.J. Blakemore and E. Murphy, *The British Civil Wars at Sea, 1638–1653* (Woodbridge: Boydell Press, 2018), pp. 12–13.

Table 3.1:
Composition of the Dutch Fleet at the Downs, 21 October 1639

Organisation	Number of ships
Admiralty of the Meuse	10
Admiralty of Amsterdam	23
Admiralty of Zeeland	10
Admiralty of Noorderkwartier	6
Admiralty of Friesland	2
East India Company	8
West India Company	3
Buysconvoy/Directorate of Great Fisheries	4
Urban *Directies*	6
Convoy – unspecified	9
Hired merchantmen or privateers	9
Unknown	6
Total	96

Source: The basis of this table is the fleet-list presented in James Bender, *Dutch Warships in the Age of Sail, 1600–1714: Design, Construction, Careers and Fates* (Barnsley: Seaforth, 2014), pp. 52–5. The information from this list has been updated with information from NL-HaNA, 1.01.02 Staten-Generaal, inv. nrs 12561.83, 12561.86.

Battle fleet duty was an anomaly for these ships, their main tasks were uncelebrated convoy duties or, more notoriously, the depredation of enemy shipping in distant waters. As a result, the institutions that provided them have generally been seen in quite a negative light in Dutch historiography. Jaap Bruijn, the doyen of Dutch naval history, for example, called the creation of the urban convoy boards (*directies*) as 'symptomatic of the lack of success of the admiralties and the States General in their struggle with Spain'.[21] Moreover, the failure of these forces in their primary role – negating the Dunkirk privateers – is often seen as conclusive evidence of the unsuitability of the organisation which gave rise to them. Yet this judgement ignores the fact that the Spanish Habsburgs, still one of Europe's premier naval powers, had at their disposal a superb naval base on whose doorstep passed the vast carrying trades and fisheries of the Republic. Moreover, by viewing the Republic's ability to mobilise a battle fleet as the key

---

[21]   J.R. Bruijn, *The Dutch Navy of the Seventeenth and Eighteenth Centuries* (Columbia: University of South Carolina Press, 1990), p. 27.

measure of its maritime strength distorts the reality of the relationship between seaborne trade, the Dutch economy, and the Republic's naval forces.

Continuous, low-level naval activity – primarily the protection of seaborne trade or, to a lesser degree, privateering – was vital to the Dutch maritime economy. As the levies placed upon imports and exports were the primary source of revenue for the admiralties which provided ships for the Republic's navy, the independent organisations which undertook the protection of seaborne trade were thus a vital prop of Dutch naval power, not a source of acute vulnerability to it. Moreover, they also played an important role in reducing the burden placed upon the admiralty fleets by creating a more competitive market for maritime security: merchants could organise their own protection for convoys if the admiralty's offers were too expensive.[22] Some of these organisations therefore form an interesting counterpoint to today's military contractors, which are at times presented as wholly unprecedented in history.[23]

Creation of a Complex Naval Seascape, 1580–1650

The first body to which a part of economic warfare at sea was devolved was the Directorate of the Great Fisheries (*College van de Grote Visserij*). The exact origins of the organisation are unclear, but as early as 1522 the cities of Holland with a stake in the herring fisheries were made responsible for equipping convoys for the fleet.[24] The 'supreme captain' of these convoy vessels was appointed by the Habsburg stadholder on the advice of the six cities involved in the herring fisheries.[25] The responsibility of the cities with a stake in the fisheries to protect the fleet was further codified after the revolt of 1568. The Directorate of the Great Fisheries was a body through which the five cities of Delft, Rotterdam, Schiedam, Brielle, and Enkhuizen regulated the herring fisheries. This organisation privileged these cities over rivals such as Maassluis and Vlaardingen, which operated large fleets of herring-busses as well. The Directorate remained in charge of regulating the Dutch herring fisheries until

---

[22] Interestingly, this counters some criticism levelled at the late Jan Glete's idea of protection as a commodity that could be sold and bought. See J. Nordin, 'The Historian Jan Glete', in A.M. Forssberg, M. Hallenberg, O. Husz, and J. Nordin (eds), *Organizing History: Studies in Honour of Jan Glete* (Lund: Nordic Academic Press, 2011), pp. 21–44, 36–7.

[23] See, for example, Krahman, *States, Citizens and the Privatisation of Security*, pp. 1–2. The comparison with the chartered companies is interesting in this case, since they too did not merely rent out military force but operated sizable shore establishments, and their regular personnel requirements dwarfed those of the admiralties in peacetime years.

[24] Sicking, *Neptune and the Netherlands*, p. 162.

[25] Sicking, *Neptune and the Netherlands*, p. 164.

its abolition in 1857.[26] By enforcing a high standard of produce, the Directorate of the Great Fisheries was able to make Dutch herring a byword for quality. Yet the herring-busses themselves were vulnerable targets for privateers. The large economic interests behind the fisheries dictated defensive measures, which in practice meant convoy. This already took place as early as the 1560s, but by the late 1580s the cities in the Directorate had acquired the privilege of equipping ships to convoy the fleets outside of the regular admiralty organisation, which was at that time still taking shape. In these circumstances, allowing the Directorate of the Great Fisheries to equip and regulate its own escorts was a logical decision, as the admiralty organisation was still nascent and focused on supporting the war on land. The rights granted to the Directorate of the Great Fisheries would be an important precedent when the problem of protecting the merchant and fishing fleets came up again in the 1620s.[27]

The success of the Dunkirk privateers immediately after 1621 meant that from the early 1620s the Directorate would yearly equip six ships to provide convoy for the fishing fleet, independently of the admiralties.[28] This was strengthened in 1625 when the Directorate acquired the right to rule on the conduct of the crews of these escorts, called *buysconvoy* (buss convoys), rather than the admiralty boards. Even warships seconded by the admiralties to the fisheries fleet would come under the authority of the admiral appointed by the Directorate. At the battle of the Downs in 1639, at least four so-called 'buss convoys' were added to the fleet.[29] Though not ideal – there were conflicts between admirals of the fisheries fleet and the 'regular navy', for example – this was an effective method of raising funds for specific naval tasks from the specific interests that benefited from these tasks.[30] As such, the Directorate of the Great Fisheries was to provide a model for a range of other naval organisations that would arise from 1602 onwards, for defensive as well as offensive economic warfare at sea.

---

[26]   B. Poulsen, *Dutch Herring: An Environmental History, c.1600–1860* (Amsterdam: Aksant, 2008), pp. 43–4.

[27]   Elias, *Schetsen uit de geschiedenis van ons zeewezen*, vol. 1, pp. 115–16.

[28]   A.P. van Vliet, *Vissers en kapers: de zeevisserij vanuit het Maasmondgebied en de Duinkerker kapers (ca. 1580–1648)* (The Hague: Hollandse Historische Reeks, 1994), p. 215.

[29]   See Table 3.1.

[30]   Van Vliet mentions a conflict between De With, as admiral of the fisheries fleet, and Van Dorp, then Lieutenant-Admiral of the Meuse in 1632: Van Vliet, *Vissers en kapers*, p. 216.

Three Companies: East India, Guinea, and West India

Warfare at sea was never merely defensive, however. While Dutch ships carrying trades and fisheries required convoys and escorts in European waters, outside Europe the situation was often the reverse. Spanish shipping to and from the Americas and Portuguese shipping in Asia and the Atlantic were attractive targets for Dutch warships, both admiralty and privately owned. Prize-taking served more than merely to enrich the captors. Here, too, a clear strategic imperative is apparent. By intercepting Spanish and Portuguese commerce, especially the homeward-bound ships of the Spanish 'silver fleets' and the Portuguese *Carreira da Índia*, the income of the Habsburg kings of Spain and Portugal was diminished, reducing their ability to finance the war in the Netherlands. Defensive economic warfare in European waters was thus mirrored by an offensive stance in the Americas, Africa, and Asia. This offensive warfare, too, was outsourced from the admiralties to corporations with a profit motive. The experience of the Dutch Republic thus forms an interesting counterpoint to present-day fears over the privatisation of warfare as a uniquely present-day phenomenon.[31]

The Dutch East India Company (*Vereenigde Oost-Indische Compagnie* – VOC), chartered in 1602, took many of the characteristics of the Directorate of the Great Fisheries, but in a different organisational model. Whereas the Directorate was the joint venture of a number of cities through their civic governments, the VOC was, or would become, a joint-stock company with tradeable shares. But the company still shared a federal structure with the Fisheries Directorate, with local branches – chambers – installed in specific cities. This organisational model privileged cities which had firms active in the trade with Asia when the company was founded by a process of merging these 'pre-companies'. In its charter, the VOC was made responsible for its own defence in Asia. This was important as it meant that the company could not call upon the admiralties to protect its trade there, lifting the potentially heavy burden of long-distance warfare from the admiralties' shoulders. In practice, the VOC would be supported by transfers of ships and cannon from the admiralties during the first 20 years of its existence.[32] But the company also built up its own stock of weapons and would build large ships suited to long-distance trade and warfare, as well as a substantial number of ships that were more specifically design with wartime use in mind, called *jachten*.[33]

---

[31]   In March 2018, the lower chamber of the Dutch parliament proposed a law enshrining the right of shipping firms to provide for private contractors to defend their ships: https:// zoek.officielebekendmakingen.nl/kst-34558-2.html (accessed 30 Apr. 2018).

[32]   J.E. Elias, *De vlootbouw in Nederland, 1596–1655* (Amsterdam: Noord-Hollandsche uitgeversmaatschappij, 1933), pp. 32–6.

[33]   R. Parthesius, 'Dutch Ships in Tropical Waters: The Development of the Dutch East India Company (VOC) Shipping Network in Asia 1595–1660', PhD thesis, University of

The success of the VOC in its campaigns in Asia against the Portuguese obscures the fact that in the first instance its use of force was intended for defensive purposes only (or at least so thought many investors and even directors of the company). Victor Enthoven has calculated that in the period 1613–20 the VOC captured prizes worth 2 million guilders, making privateering a crucial source of income. Only by the later 1620s did income from trade clearly outstrip income from privateering.[34] Regardless of initial plans, the company would develop as a potent weapon against the Habsburg foes of the Republic, undermining Portuguese trade in Asia, diminishing the income of the Portuguese crown in Europe, liberating the admiralties to focus on the war in home waters, and building up a fleet second to none in Asian waters. No admiralty warships would venture east of the Cape until the expedition of 1783–86.[35] The success of the VOC as a separate naval institution stimulated attempts by the Generality to make more merchants responsible for their own defence. The first such attempt, the Guinea Company, failed to materialise, but in 1621 the West India Company was made solely responsible for conducting warfare in the Atlantic.

Dutch merchants had been active on the Gold Coast since the late 1500s. A trading lodge at Mouri, east of Elmina, was founded in 1596, but destroyed by the Portuguese in 1610. A fort was built here two years later at the request of the merchants trading in the area. The fort, its garrison, and the attendant patrols by warships were arranged by the Admiralty of Amsterdam. The States-General did attempt to move the merchant interests to align in a single chartered company which could be made responsible for these tasks, but the merchants refused this offer of incorporation. The protection arranged by the admiralty worked efficiently, producing little incentive to alter the arrangements, while incorporation would only trouble the individual merchants with difficult compromises as to division of the trade.[36] Privatisation of the responsibility to protect trade could thus be refused if merchants saw no advantage and if admiralty warships provided adequate protection.

Despite the failure to create a single Guinea Company, responsibility for the maintenance of the fort at Mouri could be offloaded from the responsibility of the Admiralty of Amsterdam after 1621, when the West India Company (WIC) was created.[37] Even more clearly than was the case for the VOC, the WIC was

---

Amsterdam, 2007, pp. 63–74 provides an explanation for the use of the word 'yacht' in the VOC's parlance.

[34]   Enthoven, *Zeeland en de opkomst van de Republiek*, pp. 201–11.

[35]   H. Terpstra, *Het eerste landseskader in de Oostindische wateren* (Leiden: E.J.Brill, 1945).

[36]   S. van Brakel, *De Hollandsche handelscompagnieën der zeventiende eeuw: Hun ontstaan – hunne inrichting* (The Hague: Martinus Nijhoff, 1908), pp. 25–7.

[37]   For the creation and chartering of the WIC, see H. den Heijer, *De geschiedenis van de WIC* (Zutphen: Walburg Pers, 2002), pp. 21–34.

created for warfare. The company charter explicitly stated that one important source of income would be privateering, and in contrast to the VOC the States-General took a direct stake of 750,000 guilders in the new company, in effect becoming a shareholder.[38] In addition, the Generality committed itself to giving the newly created company 16 warships and four war-yachts, a substantial fleet.[39] This transfer of ships came on top of the transfer of the very largest of the states' ships to the VOC in the ill-fated Nassau fleet of 1622–24. In contrast to what Bruijn and before him Elias argued, the decision to offload the substantial number of large warships that had been carefully acquired in the final years of the Truce (1609–21) was taken not by penny-pinching admiralties, but on order of the States-General itself. Though the WIC was confronted with many setbacks from the very beginning, it was nonetheless able to equip impressive fleets for service in American and African waters. The fleet with which Piet Pietersen Heyn captured the New Spain Fleet at Matanzas, Cuba in 1628 numbered 30 ships with 665 cannon and just over 3,500 men (soldiers and sailors).[40] Of this fleet, five ships numbered over 30 guns, with the flagship *Amsterdam* mounting no fewer than 50. Before the construction of *Aemilia* as the new fleet flagship in 1632, there was no ship of this size and armament in service with any of the admiralties. Similarly, the WIC was able to equip substantial fleets to support its operations on the ground after the invasion of Pernambuco in 1630. A ship-list presents no fewer than 31 ships with a total of 2,169 men on board as present 'on the coast of Brazil' as of November 1631.[41] Just three months after the Battle of the Downs (October 1639), in January 1640, the WIC was able to beat back a large Luso-Spanish armada from Brazil in a series of engagements known as the 'battle of Pernambuco'.[42] So not only was the company capable of supporting the admiralties in home waters with a number of large ships, including the *Jupiter* of 40 guns, it was also able to conduct its own naval operations in support of its own economic ends.[43]

---

[38]   A. Bick, 'Governing the Free Sea: The Dutch West India Company and Commercial Politics, 1618–1645', PhD thesis, Princeton University, 2012, p. 121; P.J. van Winter, *De Westindische Compagnie ter kamer Stad en Lande* (The Hague: Martinus Nijhoff, 1978), p. 20.

[39]   A.C. Meijer, '"Liefhebbers des vaderlandts ende beminders van de Commercie": de plannen tot oprichting van een generale Westindische Compagnie gedurende de jaren 1600–1609', *Archief: Mededelingen van het Koninklijk Zeeuwsch Genootschap der Wetenschappen* (1986), pp. 21–70, 54–5.

[40]   L. van Aitzema, *Saken van Staet en Oorlogh in ende omtrent de Vereenigte Nederlanden*, vol. 1, pp. 720–1.

[41]   Nationaal Archief (NL-HaNA), 1.05.01.01, OWIC, inv.nr. 49, 483–4.

[42]   J.C.M. Warnsinck, *Van vlootvoogden en zeeslagen* (Amsterdam: Van Kampen, 1942), pp. 128–59.

[43]   The National Maritime Museum, Greenwich has a rare Van de Velde drawing of the *Jupiter*, one of the few in which a WIC ship is positively identified: PA17251, *Portrait of the* Jupiter.

As well as being made responsible for the conduct of the war in the Atlantic, the WIC was even entrusted with issuing letters of marque and acting as a prize court. From 1632 onwards, the majority of privateering ventures were undertaken not by the WIC itself but by private privateers who received their papers from the company and who had to pay an 20 per cent 'recognition fee' to the company (later reduced to 18 per cent).[44] Ironically, after 1648, some of these privateers operated against Portuguese targets from Spanish ports.[45]

By the summer of 1639, the global division of tasks between admiralties and the chartered companies was working well. Dutch economic interests outside of Europe were aggressively promoted by the chartered companies, leaving the admiralties free to focus on the European theatre. The companies could even reinforce the fleet at the Downs with a total of 11 large ships. But the high costs involved in this aggressive campaign of economic warfare were already taking their toll on the WIC, a problem that would only worsen during the coming decade. While the Republic could wage an aggressive campaign of economic warfare overseas, in European waters the resumption of war in 1621 was putting serious pressure on the economic lifelines of the Republic: the European carrying trades and the herring fisheries. The admiralties were increasingly strained to provide adequate protection for both. The previously mentioned increased role of the Directorate of the Great Fisheries in 1625 was one response to this problem. But the onslaught of economic warfare emanating from Dunkirk would spur more innovative solutions in home waters as well.

Defensive Economic Warfare in Home Waters

With the creation of the chartered trading companies, the admiralties could absolve themselves of any responsibility for conducting the kind of offensive economic warfare that they had conducted in the late sixteenth and early seventeenth centuries, notably the Atlantic voyages of Cornelis van der Does in 1599–1600 and Paules van Caerden's voyage to Brazil in 1605.[46] The admiralties were thus free to focus on one core task: fighting Dunkirk privateers in home waters.

---

[44]   J. Francke, *Utiliteyt voor de Gemeene Saake: De Zeeuwse commissievaart en haar achterban tijdens de Negenjarige Oorlog, 1688–1697* (Middelburg: Koninklijk Zeeuwsch Genootschap der Wetenschappen, 2001), p. 84.

[45]   F. Binder, 'Die Zeeländische Kaperfahrt, 1654–1662', *Archief. Mededelingen van het Koninklijk Zeeuwsch Genootschap der Wetenschappen* (1976), pp. 40–75, 47.

[46]   J.P. Sigmond, *Zeemacht in Holland en Zeeland in de zestiende eeuw* (Hilversum: Verloren, 2013), pp. 282–5; Enthoven, *Zeeland en de opkomst van de Republiek*, pp. 185–8. The Van Caerden mission is perhaps most notable for the painting by Hendrick Cornelisz. Vroom, *De Amsterdamse viermaster 'De Hollandse Tuyn' en andere schepen na terugkeer uit Brazilië onder bevel van Paulus van Caerden*, Rijksmuseum SK-A-1361.

Privateering from Dunkirk had picked up quickly after the resumption of hostilities between the Republic and Spain in 1621. Dunkirk hosted a royal squadron: six ships and 650 men in 1622, rising to 30 ships and 3,700 men at its height in 1635.[47] This combined force of royal warships and privateers was extremely effective: in 1627 alone, royal ships and privateers took 94 vessels and sank another 85.[48] The royal ships were larger, more heavily armed, and were a match for Dutch warships. The royal ships often worked as consorts to smaller and more lightly armed privateers, providing the privateers with crucial support.[49] In one attack on the herring fleet in October–November 1625, more than 84 herring-busses were destroyed, of which 54 were burnt.[50] But not only the herring fleets were targeted, ingoing and outgoing convoys through the English Channel towards France and the Mediterranean were often attacked, as were the vulnerable flutes on the run to Norway and the Baltic.

To combat this effective campaign of economic warfare more measures were required. The historiography is often quite negative of these attempts to stem the losses.[51] This sentiment echoes the frequent criticisms of the convoys and blockading fleets from the period itself. Implicit in this was that the Republic could somehow have done better in protecting its merchant marine and fisheries fleet – both the largest in Europe – from concentrated action by a motivated and well-equipped force operating from a base right on the doorstep of the major fishing grounds and convoy routes. An estimate by the States of Holland in 1636 put the size of the merchant and fishing fleets that needed to be protected in European waters at some 3,700 vessels.[52] Clearly, avoiding losses, even substantial losses, was going to be difficult. But how could the fleets be protected?

The admiralty boards sent ships to sea with three different tasks. In the first place a substantial force of larger ships and lighter cruisers blockaded the port of Dunkirk itself. This was in practice a leaky blockade. Easterly winds which drove the squadron off station also allowed Dunkirkers to slip out to sea. The harbour itself was heavily defended and none of the attempted attacks on the port itself was ever successful.[53] This blockading squadron was often planned at

---

[47]   Stradling, *The Armada of Flanders*, p. 252.

[48]   Stradling, *The Armada of Flanders*, p. 255.

[49]   Stradling, *The Armada of Flanders*, p. 218.

[50]   Van Vliet, *Vissers en kapers*, p. 191.

[51]   J.R. Bruijn, 'The Raison d'Être and the Actual Employment of the Dutch Navy in Early Modern Times', in: N.A.M. Rodger et al. (eds), *Strategy and the Sea: Essays in Honour of John B. Hattendorf* (Woodbridge: Boydell & Brewer, 2016), pp. 76–87, 83.

[52]   J.E. Elias, *Het voorspel van den eersten Engelschen oorlog* (The Hague: Martinus Nijhoff, 1920), pp. 61–2.

[53]   Van Vliet, *Vissers en kapers*, pp. 234–40; J.R. Bruijn, *Varend verleden: de Nederlandse oorlogsvloot in de zeventiende en achttiende eeuw* (Amsterdam: Balans, 1998), p. 33.

32 ships, though frequently smaller in practice.[54] In addition, from 1622 onwards, 19 ships in four squadrons cruised the area between the mouth of the Seine and the Skagerrak. These cruisers could assist threatened convoys and pursue any Dunkirkers they happened across.[55] Providing convoy for merchantmen and fishing vessels was a third task of the admiralties. In order to provide additional convoy escorts, new organisations were implemented or proposed.

Outsourcing Convoys in Home Waters: The Company of Maritime Insurance and the Urban Directorates

The high losses in the convoy battles of the 1620s spurred a number of proposals to improve defence of the merchant fleet. The most radical was the proposal to found a Company of Maritime Insurance in 1628–29. The charter for this proposed company stipulated that it would equip no fewer than 60 convoy escorts, relieving the hard-pressed admiralty boards of this responsibility entirely.[56] In exchange, all incoming and outgoing vessels would be required to take out maritime insurance with the company. Additionally, the company would receive the exclusive privilege to conduct trade on the Northern Coast of Africa and the Levant as far as Smyrna and the Greek islands. The proposal probably originated at least in part with participants in the *Directie van de Levantsche handel*, who were disgruntled with that organisation's inability to protect their trade adequately.[57] Though the Stadholder was enthusiastic, and the proposal could count on a majority of votes in the States-General, the merchants of Holland rallied in opposition against the proposal. The forced insurance would, it was feared, give the company a monopoly of all trade of the Republic. The proposal was revived again in 1634 and 1638, the last time by Stadholder Frederik Hendrik.[58] It was then one of a set of measures intended to stem the losses. The other parts of the plan called for farming out the convoys and licences and centralising the activities of the admiralty boards at Hellevoetsluis, which would serve as the base of the blockading fleet.[59] The Company of Maritime Insurance would be

---

[54]   Van Vliet, *Vissers en kapers*, pp. 224–5.

[55]   Van Vliet, *Vissers en kapers*, pp. 232–4.

[56]   Anon., '1628–29. Concept van eene Compagnie van Assurantie en van haar Octrooi', *Kroniek van het Historisch Genootschap gevestigd te Utrecht*, 23 (Utrecht: Kemink en Zoon, 1867), pp. 138–78.

[57]   Elias, *Schetsen uit de geschiedenis van ons zeewezen*, vol. 1, pp. 122–8. P.W. Klein, *De Trippen in de 17e eeuw: een studie over het ondernemersgedrag op de Hollandse stapelmarkt* (Assen: Van Gorcum, 1965), pp. 317–20.

[58]   P.J. Blok, 'Koopmansadviezen aangaande het plan tot oprichting eener compagnie van assurantie. 1629–1635', *Bijdragen en mededelingen van het Historisch Genootschap gevestigd te Utrecht, een-en-twintigste deel* (Amsterdam, 1900), pp. 1–160, 143–50.

[59]   Bruijn, *Varend verleden*, p. 37.

Table 3.2:
Armament of warships in the Sound under Witte de With in 1645

	Average number of guns	Mode of guns
Admiralty warships (n = 17)	29.4	26
Urban convoy boards (n = 27)	27.3	26

Source: Gerhard Wilhelm Kernkamp, *De sleutels van de Sont: het aandeel van de Republiek in den Deensch-Zweedschen oorloog van 1644–1645* (The Hague: Martinus Nijhoff, 1890), pp. 323–5.

responsible for providing convoy escorts. Finally, trade with the enemy would be prohibited. This last measure sparked the ire of the Amsterdam burgomasters, leading to an open break between the powerful city and the stadholder.[60] The idea would be mooted for a final time in 1653, during the war with England, but was rejected out of hand.[61] The failure of the proposed insurance company shows a limit to the privatisation of warfare at sea. In contrast to the East India and West India Companies, the proposed company would regulate all incoming and outgoing traffic to the Republic, giving it too strong a control over commerce. Though the Generality did not seem intent on establishing a monopoly of violence at sea, merchants' interests became hesitant to hand over naval control of home waters to a for-profit corporation, exposing the limits of the joint-stock chartered company model that had been employed successfully in reducing naval commitments overseas. The conflict over the insurance company exposed a rift between the stadholder, who supported the idea for it would free up funds for a land war, and the city of Amsterdam, which argued wholeheartedly against it. This caused the Amsterdam government strongly to support the latest addition to the Dutch Republic's naval landscape as an alternative to the admiralty boards which were seen as being too much under the control of the central government. These were the urban convoy boards or *Directies*.

The high losses of the 1620s had angered the merchants of the Republic. Criticism of the admiralty organisation and its perceived weaknesses opened the door for merchant interests to defend their own interests. From 1631 onwards, the cities of Amsterdam, Hoorn, Enkhuizen, Edam, and Medemblik, and the Frisian port of Harlingen, acquired the right to equip their own convoy escorts. Four years earlier, the VOC had acquired the right to equip its own cruisers to escort the homeward-bound Indiamen in the North Sea.[62] These urban *Directies* would

---

[60]   J.E. Elias, *De vroedschap van Amsterdam, 1578–1795*, vol. 1 (Amsterdam: Israel, 1963 [1903]), p. lxxxvi.

[61]   Aitzema, *Saken van Staet en Oorlogh*, vol. 1, p. 812.

[62]   Pieter van Dam, *Beschryvinge van de Oostindische Compagnie*, ed. F.W. Stapel (The Hague: Martinus Nijhoff, 1929), vol. 1.2, pp. 561–2.

levy a duty on the merchants of the town and use the income to equip warships.[63] Bruijn styles these as armed merchantmen, but without more specific information on these ships that is an assertion that is hard to test.[64] The minutes of the Rotterdam *Directie*, which styled itself as 'The Directors of the new cruisers of the Meuse', reveal that when the organisation was founded in November 1643 a number of Directors were dispatched to Amsterdam to see if four suitable ships could be procured there. This could refer either to suitable armed merchantmen or privately built warships.[65] There is only scarce information available to compare the vessels of the *Directies* with those of the regular fleet. One point of comparison is provided by the list of ships in the Sound under the command of Witte de With in 1645. Witte de With had sailed to the Sound with a force of 44 ships to escort a large fleet of Dutch merchantmen through the Sound in defiance of new Danish tolls. This was a mixed force, composed of 18 admiralty warships and 28 ships supplied by the various urban convoy boards. If we exclude the flagship *Brederode* from the statistics, since it is clearly an outlier in size and armament, the average armament of the admiralty and urban convoy ships is quite similar (see Table 3.2). Of course, this does not make allowance for the calibre of the guns, but since most Dutch warships mounted quite light guns, the difference would probably not be large. Both admiralties and the urban Directorates primarily acquired small cruisers that could intercept fast-sailing privateers. The need for speed was succinctly put in an evaluation of anti-privateering efforts presented to the Amsterdam city council in 1632: 'for it is seldom seen that a cow catches a hare'.[66]

The convoy boards were seen by some urban governments in Holland (notably Amsterdam) as a better alternative to the Generality-controlled admiralties. It is unsurprising that when the Dutch fleets had to mobilise during the crisis preceding the outbreak of the First Anglo-Dutch war, the urban convoy boards were entrusted with arming and equipping 100 of the supplementary fleet of 150 vessels. The vessels hired by the *Directies* were widely criticised as not being fit for service.[67] But this can be argued to have been due to the immense requirements laid down.[68] It proved impossible to find that many suitable ships, even amongst the Republic's shipping, and even with the habit of private shipbuilders of building

---

[63]   Bruijn, *Varend verleden*, p. 36.

[64]   Bruijn, *Varend verleden*, p. 36.

[65]   Nationaal Archief (NL-HaNA), 1.03.02, Equipering oorlogsschepen, inv.nr. 1, unfoliated, 23 Nov. 1643.

[66]   Elias, *De vlootbouw in Nederland, 1596–1655*, pp. 53–4.

[67]   Aitzema, *Saken van Staet en Oorlogh*, book 32, p. 761.

[68]   The minutes of the Amsterdam *Directie* of 1652 on the Directorate's activities during the outbreak of war with England show that of the 24 ships then in service or fitting out ten were either bought or hired from private merchants and one was hired from the Greenland whaling company. Nationaal Archief (NL-HaNA), 1.03.02, Equipering oorlogsschepen, inv.nr. 3, unfoliated, list of ships fitting out.

warships as speculative investment.[69] The *directies*, which had until the outbreak of war with England, filled an important role in providing protection to Dutch shipping in European waters.

The System Comes Undone

The Dutch system of naval organisation had come under pressure even before the outbreak of war with England in 1652. Throughout the 1640s, the naval forces of the WIC deteriorated rapidly. Offensive economic warfare at sea – privateering – was outsourced to yet other companies after 1632. From the mid-1640s onwards, one of these firms, the *Zeeuwse kaperdirectie*, the Zeeland privateering Directorate, was increasingly important in maintaining an offensive squadron in Brazil. This same firm had previously operated against Dunkirk privateers, but with the fall of Dunkirk there were no longer any prizes to be had in the North Sea.[70] The reliance on other firms in providing local naval power puts the precipitous decline of the WIC in sharp relief. Privateering, previously a mainstay for the Atlantic company, was left to other companies, but the task of defending the conquests of Brazil was increasingly difficult for the WIC to manage. Whereas the WIC had still been able to offer substantial reinforcement to the states' navy in the battle of the Downs in 1639, by 1647 the company was no longer able to protect its much diminished colony in Brazil with sufficient naval forces. Instead, to preserve the Brazilian colony, on which the success of the company depended, the Generality had to intervene. In 1647, a 'secours' was equipped jointly by the admiralties and the company. The Admiralties of Amsterdam (five ships), The Meuse (three ships), and Noorderkwartier (four ships) would equip a fleet of 12 large vessels; the WIC was to equip nine 'yachts' or frigates but could only muster seven.[71] This was the first of several squadrons of admiralty ships that were sent to Brazil in support of the colony until the outbreak of the First Anglo-Dutch war in 1652. Though none of these squadrons was particularly successful, this was the first instance in which the division of tasks between the admiralties and other naval entities broke down. Instead of the WIC supporting the states' fleet, the states' fleet was required to support the company. The outbreak of the First Anglo-Dutch war would put more pressure on the system.

The first Anglo-Dutch war is commonly seen as heralding the moment at which the Dutch republic changed to a more 'modern' mode of naval organisation.[72] The

---

[69] For speculative building, see, for example, Glete, *Swedish Naval Administration*, pp. 398–9.

[70] C.R. Boxer, *The Dutch in Brazil, 1624–1654* (Oxford: Clarendon Press, 1957), pp. 201–2; W.J. van Hoboken, *Witte de With in Brazilie, 1648–1649* (Amsterdam: Noord-Hollandsche Uitgevers Maatschappij, 1955), pp. 67–8; Binder, 'Die Zeeländische Kaperfahrt', p. 41.

[71] Van Hoboken, *Witte de With in Brazilie*, pp. 33–4.

[72] Glete, *Navies and Nations*, i. 180–7.

increased importance of line-of-battle tactics and the increasing technological complexity of sailing warships, it is argued, precluded private organisations from engaging in this activity. The poor performance of the urban convoy boards during the war with England led to their dissolution after the war. In reality, these urban organisations had always been intended to be temporary and their intended goal had never been to equip ships for the battle fleet. By the end of the war, therefore, convoying merchantmen in European waters became the sole responsibility of the admiralty boards, which cooperated with local merchant groups to determine the need for escorts and the routes to be sailed. This left only the East India Company and the Directorate of the Great Fisheries.

The VOC had contributed large ships to the fleet during the First Anglo-Dutch war. These vessels had been much criticised by the admiralty officers, though some did indeed fight well.[73] The VOC had still been able to organise a substantial contribution to the Dutch battle fleet during the Second Anglo-Dutch war, though its role changed during the conflict from organising its own naval forces, in effect operating as a 'sixth admiralty', to providing funds for the admiralty boards to equip twenty ships.[74] During the wars of the 1660s and 1670s, the VOC was still superior to its rivals in Asian waters. But during the wars with France after 1688, the company was hard-pressed in Asia. Though it captured two French Indiamen at the Cape in 1689, it lost three ships to the French in 1689, and 11 the following year.[75] This forced the company to equip more escorts during the late seventeenth century. New regulations on shipbuilding and armament in the 1690s reduced the number of guns on the largest vessels and reduced their calibre. Not only could the Indiamen no longer serve as a reserve for the battle fleet, they were now unable to take on cruisers as well. In European waters, the VOC became more dependent on convoys provided by the admiralty boards. In Asia, the VOC was increasingly outclassed by the naval forces sent to the area by Britain and France during the eighteenth century. But neither did the VOC develop a local cruiser force analogous to the Bombay marine of the East India Company. The problem that the lack of standing naval forces created is well illustrated by the debacles such as the failed attempt to capture the Angria fortress of Vijaydurg in 1739 and the battle of Biderra in 1759.[76] The VOC was in fact unable to defend its own trade

---

[73]   For example, the *Vogel Struijs* under command of Cornelis Adriaensz Cruyck during the first day of the Battle of Portland and the *Mercurius* and *Nassau* during the Bettle of Ter Heijden. See Elias, *De vlootbouw in Nederland, 1596–1655*, p. 90 n. 4.

[74]   For the role of the VOC in the second Anglo-Dutch war, see E. Odegard, 'The Sixth Admiralty: The Dutch East India Company and the Military Revolution at Sea, c.1639–1667', *International Journal of Maritime History*, 26.4 (2014), pp. 269–84.

[75]   Francke, *Utiliteyt voor de Gemeene Saake*, p. 82.

[76]   H.C.M. van de Wetering, 'De VOC en de kapers van Angria: een expeditie naar de westkust van India in 1739', *Tijdschrift voor Zeegeschiedenis* 10.2 (1991), pp. 117–30; H.K. s'Jacob, 'Bedara Revisited: A Reappraisal of the Dutch Expedition of 1759 to

and possessions in Asia, but was saved from more substantial losses by the long period of Dutch neutrality after 1714. When war broke out with Britain in 1780, the best the VOC could do was to propose to modify some of its Indiamen to act as escorts of the outgoing fleets.[77] To prop up the company's position in Asia, it required support by the admiralties. Only the Directorate of the Great Fisheries remained as a separate institution responsible for (defensive) economic warfare at sea until the very end of the Republic in 1795.

Conclusion

During the first half of the seventeenth century, the Dutch republic waged a worldwide maritime war of attrition against its Habsburg foes. In European waters, the Dutch Republic had to defend its carrying trades and fisheries from destruction by privateers and warships operating from Dunkirk. Overseas, in the Americas and in Asia, Dutch ships preyed on Spanish and Portuguese trade. The regular navy of the five admiralty boards was not up to the demands this worldwide conflict posed. Responsibility for warfare at sea was thus consistently outsourced to those organisations which directly stood to benefit from increased protection at sea. The Directorate of the Great Fisheries, the various urban convoy boards, and the East India Company all became responsible for convoying their own vessels in home waters, while the East India and West India Companies were delegated with organising predatory warfare in Asia and the Americas. This is an interesting development in itself, but also underlines that the perceived movement towards a Hobbesian world in which the central state acquired a monopoly of violence was more wishful thinking than practice in this period. Though the idea of a state monopoly on violence was an important conceptual development, it was not yet a reliable description of contemporary practice.

Through this outsourcing and by making the chartered companies responsible for economic warfare in the extra-European world, the admiralties themselves were left free to focus on the war against Dunkirk privateers in home waters. The 150 years from *circa* 1645 onwards witnessed the slow collapse of this system, with first the WIC, then the urban convoy boards, and finally the VOC being either dissolved or no longer able to fulfil their responsibilities. The period from 1650 onwards thus saw the steady increase of the area of operations of admiralty warships, including the Atlantic expeditions of the 1660s and 1670s, which resulted in the conquest of Surinam and the brief (re)occupations of Tobago and New Amsterdam in the 1670s. From the late 1730s onwards, regular convoys

---

Bengal', in J.J.L. Gommans and Om Prakash, *Circumambulations in South Asian History: Essays in Honour of Dirk H.A. Kolff* (Leiden: Brill, 2003), pp. 117–31.

[77]   Nationaal Archief (NL-HaNA), Collectie Meerman van der Goes, 1.10.57, inv.nr. 95 contains a proposal from 1781 to arm the *Schoonderloo* with 40 guns.

to Curaçao were organised to protect shipping from attack by the Spanish *Real Compañía Guipuzcoana*.[78] The VOC was able to hold its own in the seventeenth century, but humiliating defeats were only avoided by the long period of Dutch neutrality in Europe following the War of Spanish Succession. The importance of the fragmented system of naval institutions of the Dutch Republic is that it reminds us that organising economic warfare at sea could be done in several quite distinct ways. The Dutch Republic shows a way in which protection or predation at sea could be made the responsibility of those economic actors whose interests were at stake.

But the slow collapse of this system also holds out a warning. Though the system worked well on several occasions, it was ultimately dependent on the economic viability of the interests concerned. It created a path-dependency that made it difficult to defend the colonial empire in adverse economic circumstances. If, as happened in the case of the WIC and the VOC, the profitability of the particular activity declined, it could no longer afford effective naval forces to protect its interest, thus creating a vicious circle. Though the protection of Dutch commerce in European waters during the second phase of the Eighty Years War can be at least partially attributed to the willingness to experiment with novel institutional solutions, the quick loss of the Atlantic empire *circa* 1650–80, as well as the collapse of the VOC's Asian empire after 1795, can be attributed to the lack of standing naval forces in these areas. The revolution of 1795 in which a unitary 'Batavian' navy was created was thus the culminating of a process of concentration that had been under way for nearly a century and a half.

---

[78]   W. Klooster and G. Oostindie, *Realm between Empires: The Second Dutch Atlantic, 1690–1815* (Ithaca, NY: Cornell University Press, 2018), p. 36.

CHAPTER FOUR

Merchants of Fortune: Negotiating Spanish Neutrality in the American War of Independence

Anna Brinkman

Introduction: 11 Captured Ships

On 17 April 1777, the Spanish Governor of Louisiana, Bernardo de Gálvez, ordered his guards out in the dead of night to proceed quietly up the Mississippi River and seize every British merchant vessel they could find. Eleven ships were taken and escorted down to New Orleans, where they and the goods they contained were confiscated, and the masters and seamen imprisoned under the charge of smuggling. Captain Thomas Lloyd of HMS *Atalanta*, the ranking British naval officer in the area, duly demanded that the British ships be returned, and the sailors released. Lloyd accused the Spanish governor of violating his state's neutrality in Britain's civil war with her North American colonies, and of violating the 1763 peace treaty, which provided for free navigation of the Mississippi river to British ships.[1] Gálvez refused Lloyd's demands. The incident eventually reached the Courts of Madrid and London where British ministers played down the incident and Spanish ministers backed the actions of Gálvez. The interests of the British merchants who sought restitution and compensation for the seizure of their ships were set aside by a British ministry committed to a broader strategy of appeasing the Spanish in order to avoid their involvement in the ongoing conflict in North America. This incident was but one of many attempts by the British to encourage Spain to remain neutral during the American War of Independence, and one of many instances where the Spanish ministry displayed how little effort it put into policing the actions of its colonial officials regardless of their potential consequences for Anglo-Spanish relations. It was also an incident emblematic of the dangers posed by pursuing wartime commerce in the Americas.

Hostilities opened between Britain and her North American colonies in 1775, and were followed by five years of diplomatic ruptures with France, Spain, and Holland respectively. This forced Atlantic merchants to adapt to an increasingly

---

[1] William L. Clements Library, University of Michigan, Shelburne Papers, F1771 Ca: Case of the Seizure of British Vessels by the Spaniards in the River Mississippi, 1778.

hostile and ambiguously regulated maritime environment. After the Seven Years War (1756–63), Anglo-Spanish commerce was, supposedly, regulated in part by the 1763 Treaty of Paris and the Anglo-Spanish Treaty of 1667. Enforcing the treaties in the colonial sphere, however, was an almost impossible task. When 13 of Britain's North American colonies rebelled, declared independence, and began what would come to be known as the War of American Independence, Spain declared itself neutral. However, the war raging in North America opened up lucrative avenues of commerce for Spanish and British merchants operating in the West Indies. The rebels needed military supplies which came from Europe and the West Indies, and the British islands relied upon trade with the mainland American colonies.[2]

Eighteenth-century wars were usually detrimental to merchants and planters. but those adept at seaborne smuggling in Europe or the West Indies were keen to make a profit during North America's rebellion. Smuggling thus quickly became ubiquitous, and a primary concern of British maritime strategy. Ministers in London wanted two things: to isolate the rebels in the North from foreign aid through commerce predation and to prevent a rupture with neutral nations. Smuggling could be detrimental to both aims. A rebellion might be quashed, but a European war breaking out in the Caribbean alongside the rebellion was to be avoided. A key component of Britain's foreign policy and strategy of commerce predation in the Americas, therefore, was to maintain and cultivate Spanish neutrality. Diplomatic incidents between British officials and their Spanish counterparts were to be avoided; and if they could not be prevented, they had to be contained in order to avoid pushing neutral countries into a war. Unfortunately for Britain and Spain, the ability to directly control or oversee the actions of their citizens and representatives at such distance from the metropole was minimal. Britain's policy and strategy were thus at the mercy and whim of officers and subjects who made decisions about legal and illegal activity thousands of miles from London. Contentious incidents involving neutral rights were therefore mostly resolved after the fact by politicians in Europe scrambling to ameliorate the diplomatic damage already done at sea or in the colonies.

By closely examining the incident of the 11 captured British merchant ships in the Mississippi, the essentially reactive nature of Britain's wartime policy becomes clear. Britain's efforts to maintain Spanish neutrality highlight how a neutral nation could wield a great amount of influence and power during a maritime conflict. Neutrality was not a static state of inaction; rather, it was a very dynamic position from which concessions, liberties, and privileges might be won from belligerent powers who were keen not to see a neutral nation become allied with an enemy. By the same token, however, the dynamism of neutrality could plunge neutral actors into uncertainty over the safety of their

---

[2]   A. O'Shaughnessy, *An Empire Divided: The American Revolution and the British Caribbean* (Philadelphia: University of Pennsylvania Press, 2000), p. 143.

trade and the rights of their citizens. When the incident of the 11 ships on the Mississippi is put into the context of a pattern of nefarious mercantile activity by both the Spanish and British, neutrality emerges as a central dynamic concept in maritime economic warfare during the second half of the eighteenth century. Neutral nations were able to wield power during times of war because trade in maritime empires was largely based on mercantilist thinking. France, Spain, and Britain prohibited their colonial trade from being carried in foreign ships, and foreign traders were mostly confined to operating in European ports by measures prohibiting them from trading directly with colonies overseas. Empires whose approach to trade was governed by such mercantilist principles rendered themselves particularly vulnerable to economic warfare, because merchant ships carrying their flag became legal targets for enemy warships in wartime. The most obvious means of avoiding capture by enemy warships was to allow trade to be carried in neutral bottoms, as neutral shipping was not universally acknowledged as a legitimate target in wartime. The legitimacy of targeting neutral shipping would depend on existing bilateral treaties between belligerent and neutral states. Neutral maritime nations could thus profit from wars between maritime empires, and became crucial factors in deciding the viability of a maritime strategy based on economic warfare.

There has been much recent debate over neutrality, and the concept is enjoying a much needed revival as a subject deserving of attention when evaluating the behaviour of empires or states at war. The debate, however, is largely centred on the nineteenth and early twentieth centuries. Scholars such as Maartje Abbenhuis and Stephen Neff show the dynamism of neutrality as it applies to the post-Napoleonic era, but the use of neutrality as a tool of statecraft in this period is often presented as a departure from pre-nineteenth-century concepts, which are depicted as primitive or nascent in comparison.[3] This chapter will illustrate that debates about eighteenth-century neutrality need to be brought out of their current state as a straw man for their nineteenth-century counterparts and analysed in their own right as crucial components of early modern warfare. By illuminating the crucial role neutrality played in both British and Spanish strategy during the American War of Independence, it will argue that maritime economic warfare needs to be viewed as part of a much broader picture of great power relations which were ongoing in both peace and war, and which often occurred below the threshold of inter-state conflict.

---

[3]   See M. Abbenhuis, *An Age of Neutrals: Great Power Politics, 1815–1914* (Cambridge: Cambridge University Press, 2014) and S. Neff, *Justice among Nations: A History of International Law* (Cambridge, MA: Harvard University Press, 2014). For an older but still very relevant work on neutrality in the eighteenth century, see Carl Kulsrud, *Maritime Neutrality to 1780: A History of the Main Principles Governing Neutrality and Belligerency to 1780* (Boston: Little, Brown, and Co., 1936).

Anglo-Spanish Affairs in the Lead-up to War

Britain and Spain pursued very different strategic aims in the 1770s that were largely based on each empire's position in the European balance of power at the end of the Seven Years War. After the peace process in 1763, Britain had acquired additional territory in the Americas, naval hegemony in the Western Atlantic, an inflated national debt, and the unquenched ire of both France and Spain.[4] Spain had suffered the humiliating loss of the vital ports of Havana and Manila (both returned to Spanish control during the peace process in exchange for Florida), had failed to eradicate British settlements on the Moskito and Honduran coasts, failed to take back Gibraltar, and had gained control of New Orleans and Louisiana.[5] In many ways, the losses suffered by Spain in the Seven Years War made it inescapably clear that its empire was not well defended against attacks carried out by a maritime power. The obvious vulnerability of Spain's empire in America, along with a long-held paranoia that Britain's ultimate aim was to invade and take over Spain's American possessions, led King Charles III and his ministers to launch a new era of imperial reforms that would prepare the empire for a new war with Britain.[6] Britain, on the other hand, turned its attention inward after the Seven Years War and focused on increasing revenues through trade and taxation as well as minimising spending. There was no appetite for, and no need to seek, war again with Spain because it did not pose a commercial or maritime threat.[7]

Reforms in both Britain and Spain in the 1760s and 1770s were commercial in nature. However, Spain's reforms also included provisions that slowly prepared Spain for a future war in the Americas. Since the end of the Seven Years War, Spain and its bellicose king had been troubled by the balance of power in the Americas. Active steps were taken to prepare for the next opportune moment in which to engage Britain in another war. These steps included military reforms in the colonies and shipbuilding programmes in Havana, Cartagena de Levante,

---

[4]   Sam Willis, *The Struggle for Sea Power: A Naval History of American Independence* (London: Atlantic Books, 2015), p. 42.

[5]   A. Kuethe and K. Andrien, *The Spanish Atlantic World in the Eighteenth Century: War and the Bourbon Reforms, 1713–1796* (Cambridge: Cambridge University Press, 2014), p. 237 and Yale Law School, Lillian Goldman Law Library, The Avalon Project: Documents in Law, History and Diplomacy, Treaty of Paris 1763: http://avalon.law.yale. edu/18th_century/paris763.asp (accessed 9 May 2017).

[6]   Adrian Pearce, *British Trade with Spanish America 1763–1808* (Liverpool: Liverpool University Press, 2007), p. xxv and Kuethe and Andrien, *The Spanish Atlantic World in the Eighteenth Century*, p. 231.

[7]   Sophus Reinert, 'Rivalry: Greatness in Early Modern Political Economy', in Philip J. Stern and Carl Wennerlind (eds), *Mercantilism Reimagined: Political Economy in Early Modern Britain and its Empire* (Oxford: Oxford University Press, 2014), p. 348.

Cadiz, and El Ferrol. By 1769, Spain had roughly 60 ships of the line.[8] In contrast to Spain's shipbuilding endeavours, reforms in Britain actively weakened British sea power by reducing spending on the navy.[9] Britain instead launched an experiment in free trade in November of 1766, which opened up a number of ports in Jamaica and Dominica to some types of foreign trade and foreign vessels. Spain also experimented with free trade, starting in 1765.[10] Newly opened ports and experiments with breaking monopolies, however, had the side effect of making colonial officials more skittish about illegal trade. These years saw increased tensions between British and Spanish colonial actors involved in maritime trade.[11] By the time the 13 North American colonies rebelled in 1775, merchants in the Caribbean, Spanish America, and North America were ready to take advantage of all the commercial opportunities offered by a colonial civil war. British sea power was not prepared to combat wartime smuggling. Spain, therefore, bided its time for the right moment to wage war against Britain, whilst happily abusing its neutrality to help the rebels and France weaken Britain's war effort.

Spanish 'Neutrality' and Trade with Americans

The rebellion in North America was not self-sustaining from a supply point of view. In order to prosecute the war against Britain, the American rebels had to import supplies from outside of the 13 colonies. Key among these supplies was gunpowder, which came in through the Caribbean islands of Martinique, Cuba, and the Dutch Antilles. The British navy was unable to fight the smuggling prowess of French, Spanish, Dutch, and even British merchants who scrambled to exploit the commercial opportunity of supplying the rebels with powder.[12] They were hindered not only by the lack of British warships in the Caribbean, but also by treaties that greatly limited their ability to stop, search, or capture neutral merchant ships, even if they were carrying supplies to the rebels.[13] One

---

[8]   Kuethe and Andrien, *The Spanish Atlantic World in the Eighteenth Century*, pp. 271 and 282.

[9]   Willis, *The Struggle for Sea Power*, pp. 198–9.

[10]   Pearce, *British Trade with Spanish America*, pp. 51–3; Kuethe and Andrien, *The Spanish Atlantic World*, p. 245. The term 'free trade' is slightly confusing when applied to Spain's 1765 decree because it opened Havana to some inter-island commerce and to direct trade with nine ports in Spain. It did not open Havana to British trade.

[11]   Pearce, *British Trade with Spanish America*, pp. 55 and 61.

[12]   Pearce, *British Trade with Spanish America*, p. 44.

[13]   Kulsrud, *Maritime Neutrality to 1780*, p. 136 and 'Treaty of Peace and Friendship between Great Britain and Spain, Signed at Madrid, 13(23) May 1667', *Oxford Historical Treaties* (Oxford: Oxford University Press): http://opil.ouplaw.com/view/10.1093/law:oht/law-oht-10-CTS-63.regGroup.1/law-oht-10-CTS-63?rskey=pcFr3u&result=3&prd=OHT (accessed 9 May 2017).

of the main conduits of supplies for the American rebels was New Orleans and the Mississippi River. American ships could wait near the mainland coast for supplies to be brought in European ships or, if protection was extended by the Spanish governor of New Orleans, they could wait safely for their smuggled cargo in New Orleans.[14]

British sea power, or the threat of it, was not entirely impotent, however. Though the American rebels, with the help of British, Dutch, French, and Spanish merchants, developed a successful privateering arm that regularly and successfully operated in the Caribbean, the rebellion was hesitant to use American ships for their supply lines out of fear that the British navy would capture them.[15] Asking friendly European governors or merchants for help was a good way to subvert British sea power. One such friendly governor was Bernardo de Gálvez of New Orleans, the same man who ordered the capture of 11 British ships in April of 1777.

In January 1779, the Governor of Virginia, Patrick Henry, wrote to Gálvez to ask that he help Virginia fight the British. Henry's request was not merely a matter of smuggling supplies; his letter called for Gálvez to completely disregard and violate Spanish neutrality. Henry began the letter by thanking the Spanish for all of their help in keeping Virginia supplied and able to fight the war. He followed this with an expression of friendship 'Sensible of the value of that friendship which your nation hath tendered to Virginia, and of the favours received from you, I am anxious to make the best returns in my Power'.[16] He continued the letter by making two suggestions and a request that would help defeat the British but would also require substantial Spanish involvement. Henry first ventured to suggest that it would be beneficial to Spain if British West-Florida were annexed to the Americans because this would prevent the British from continuing their colonial rivalry with Spain and curtail British access to the Mississippi River.[17] The implication in this suggestion was that help from Spain would be useful in order to achieve the annexation. Henry suggested that it would be beneficial to the Americans and Spaniards to build a fort near the mouth of the Ohio River in order to better protect trade between the Americans and New Orleans. He then asked Gálvez for a loan of 150,000 pistoles and whether it would be easiest to advance the money in New Orleans, Havana, or Cadiz.[18]

Henry's letter, along with Gálvez's response, reveals much about Spanish neutrality and about Gálvez as an embodiment of that neutrality: it was full of encouragement and proclamations of friendship but avoided concrete

---

[14]   The National Archives, Kew (TNA), PRO 30/55/8: Patrick Henry to Governor Gálvez, 14 Jan.1779; Willis, *The Struggle for Sea Power*, p. 97.

[15]   Willis, *The Struggle for Sea Power*, p. 99.

[16]   TNA, PRO 30/55/8: Patrick Henry to Governor Gálvez, 14 Jan. 1779.

[17]   British West Florida bordered the Mississippi River and shared a border with Spanish Louisiana.

[18]   TNA, PRO 30/55/8: Patrick Henry to Governor Gálvez, 14 Jan. 1779.

commitment to action that would overtly violate Spanish neutrality. In response to the call for annexing West-Florida, Gálvez wrote that it would be natural for Spain to desire that Britain no longer hold the territory. 'All the advantages that they draw from the Mississippi are those I think will not be mourned by Spain, on the contrary, it is natural that she desires this outcome'.[19] This was a statement that simply affirmed an obvious strategic desire on behalf of Spain, but Gálvez went no further and did not offer any help or advice on how such an annexation might be achieved. Gálvez took a similar stance on the suggestion of building a fort at the mouth of the Ohio River. He affirmed that such a fort would be beneficial to American commerce and would hinder British access to the Mississippi, but he did not offer any meaningful support for the endeavour. The mouth of the Ohio river was in British territory but not in the territory of any of the 13 colonies. Had Spain helped the Americans build a fort there, it would have been a gross violation of their neutrality and an attack on British territory. Henry's request for money was also met with a non-committal statement. Gálvez expressed regret that it was not within his power to grant Henry the money but that he would send Henry's request to Charles III and wait for orders.[20] Whilst Gálvez avoided committing to egregious violations of neutrality, he did not leave Henry devoid of material support. Gálvez had secured munitions and other supplies for the Americans that he had sent through an agent named Olivier Pollock, and he expected that Henry would see them arrive both via the Mississippi and by sea in due course.[21]

The Governor of New Orleans was walking a very thin line in his correspondence with Patrick Henry and one of the most interesting points of their exchange is the fact that neither of them ever alluded to or mentioned Spain's neutrality. Henry's requests for Spanish assistance were mostly framed as endeavours that would benefit Spain in its struggle against Britain and his letter reads like it was written to an ally who was also at war with Britain. Gálvez's letter is carefully worded, but at no point did he apologise or express regret that Spain's neutrality prevented him from helping the Americans. There were no statements that affirmed Spain's commitment to its neutrality. When Gálvez had to dodge Henry's requests he did so by acknowledging that Spain's strategic interests were the same as those of the Americans and that Spain would benefit from the erosion of British territory and power. Gálvez was happy to abuse Spain's neutrality to a point. He sent supplies to the Americans and also maintained a friendly correspondence about combined strategic aims that could be beneficial to Spain once it entered the war.

---

[19]   TNA, PRO 30/55/13: Gálvez to Patrick Henry, received 24 Nov. 1779.
[20]   TNA, PRO 30/55/13: Gálvez to Patrick Henry, received 24 Nov. 1779.
[21]   TNA, PRO 30/55/13: Gálvez to Patrick Henry, received 24 Nov. 1779.

Upholding Neutrality

The situation in Britain's island colonies during the war was unenviable. The island colonies, particularly the Leeward islands, depended on importing provisions from the North American colonies in order to dedicate as much land as possible to the cultivation of cash crops.[22] Without these imports, and with American privateers creating an atmosphere of fear that drove up prices for both goods and insurance, the British West Indies suffered dramatically during the war.[23] However horrible the conditions on the islands, and despite the collapse of many West Indian merchants, the overall impact on Britain's maritime economy before France and Spain became belligerents was minimal.[24] The unstable conditions produced by the war in the Caribbean were, however, good for cultivating a large smuggling trade that included Spanish protection of American smugglers/privateers as well as Spanish trade with British merchants. It was an economy of opportunity in which the rigours of a diplomatic neutrality found little traction. For instance, in 1777, Don Juan de Cabrera, a militia Captain in Cuba, apprehended British smugglers from Jamaica trading with Miguel Pititivey, who lived near Puerto Principe in Cuba.[25] Pititivey complained that he was detained unfairly by Cabrera and the matter was referred to the governor of Cuba, who subsequently referred it to officials in Spain. There is no record of its resolution or of what happened to the British ship, but the incident is emblematic of the illegal trade between British and Spanish subjects that could not be fully controlled by either Spanish or British officials. There were, however, occasions when neutrality was respected, even in the face of opportunities to flaunt it.

Before the incident of the 11 ships in 1777, Gálvez had wanted to attack British shipping in order to defend rebel shipping in the Mississippi, and, in June 1776, he wrote to the Governor of Cuba, the Marques de la Torre, calling for aid. A British frigate had entered the Mississippi river with the intention of capturing rebel shipping. The frigate was entirely within its rights to do this as long as it respected the neutrality of Spanish territory bordering the river by taking any belligerent action outside of cannon shot distance.[26] Governor Gálvez, however, wanted two ships sent from Havana in order to apprehend the British frigate should it threaten Spanish possessions or try to take rebel ships. This was not a direct call to violate Spain's declared neutrality, but by calling for Spanish warships, Gálvez was carelessly increasing the possibility of a diplomatic incident between Britain and Spain. The governor of Havana, less keen perhaps

---

[22]   O'Shaughnessy, *An Empire Divided*, p. 69 and p. 162.

[23]   O'Shaughnessy, *An Empire Divided*, p. 163.

[24]   O'Shaughnessy, *An Empire Divided*, p. 192.

[25]   Archivo General de Indias (AGI), CUBA, 1172: Don Juan Cabrera to the Marqués de la Torre, 1777.

[26]   Yale Law School, Lillian Goldman Law Library, The Avalon Project: Documents in Law, History and Diplomacy, Treaty of Paris 1763.

to violate neutrality, declined to send aid and wrote to the Minister of the Indies in Spain, José de Gálvez (who was Bernardo de Gálvez's uncle), to ask for guidance. He wrote that until he received explicit orders, he would do nothing to violate Spain's neutrality where Britain was concerned.[27] This was not merely a one-off instance of de la Torre respecting Spain's neutrality. A few months later, in August of 1776, Captain Thomas Daseey of HMS *Diligence* arrived at Havana and sent a letter to de la Torre in which he explained that he had been informed that an American privateer that had captured several British ships from Jamaica had left their crews at Havana. He requested that if those crews were still in Havana they be sent aboard his ship so that he might take them back to Jamaica. He also requested that his ship be allowed into Havana's harbour in order to take on water as the Frigate's supply was dangerously low.[28] De la Torre responded to Captain Daseey by saying that the American privateer had indeed left crews at Havana but that they had all made their way off the island in other ships and none were left. He then invited Daseey to anchor in the harbour and take on as much water as was necessary and not to hesitate to ask for any other help that might be needed.[29] De la Torre's treatment of both the American privateer and the British warship was perfectly in keeping with what was ideally to be expected from the Anglo-Spanish treaty of 1667 in that it did not favour either belligerent or profit from the state of war.[30] Governor de la Torre and Governor Gálvez represented two extremes of how Spanish and British colonial officials responded to the outbreak of civil war in Britain's North American colonies. Most officials were somewhere in between strict adherence to Spanish neutrality and a cavalier disregard for fomenting Anglo-Spanish disputes.

From Colonial Officials to the Offices of Ministers

The body of archival evidence on smuggling activity in the Caribbean during the American War of Independence is extensive and full of examples of British and Spanish officials disregarding and upholding Spain's neutrality.[31] There is, unfortunately, a much smaller body of evidence that allows historians to

---

[27]   AGI, CUBA, 1221: Letter to Gálvez from Havana, 4 June 1776.

[28]   AGI, CUBA, 1221: Dassey to Governor of Havana, 14 Aug. 1776.

[29]   AGI, CUBA, 1221: Response from Thomas Dassey to Governor of Havana, 14 Aug. 1776.

[30]   'Treaty of Peace and Friendship between Great Britain and Spain, Signed at Madrid, 13(23) May 1667'.

[31]   See, for instance, Willis, *The Struggle for Sea Power*; D.J. Starkey, *British Privateering Enterprise in the Eighteenth Century* (Exeter: University of Exeter Press, 1990); Pearce, *British Trade with Spanish America 1763–1808*; Juan Hernández Franco, *La gestión política y el pensamiento reformista del Conde de Floridablanca* (Murcia: Universidad de Murcia, 2008).

follow and analyse incidents of breached neutrality from their occurrence in the colonial sphere to the reactions and actions of ministers in Europe. Ministers hoped to limit any political fallout that could jeopardise their strategic goals. One such incident was that of the 11 captured British ships that opened this chapter. The affair is documented from Gálvez's initial action through to its referral to the Court of Spain and the British Secretary of State for the Southern Department. The incident demonstrates very clearly that Britain was sufficiently invested in maintaining Spain's neutrality that ministers were unwilling to pursue justice and redress even under the pressure of the British merchants. It also demonstrates that Spanish ministers were sometimes content to let their officials abuse neutrality and aid the Americans, as long as it did not pull them prematurely into the war.

New Orleans and the Mississippi River were vital to the British, Americans, and Spanish throughout the American war but, in the early months of 1777, maritime activity increased markedly. The Spanish opened New Orleans as a friendly port to American privateers and allowed them to bring in their prizes (a decision that brought into question Spain's commitment to neutrality). In response, the British increased their naval presence on the River.[32] It was during this increased flurry of maritime activity and following Spanish support for American privateers that Governor Gálvez captured 11 British merchant ships in the spring of 1777.

Gálvez had the ships brought to New Orleans and had their crews imprisoned. Captain Thomas Lloyd of HMS *Atalanta,* who was cruising in the area, demanded that they be freed and the ships returned, because Gálvez was in clear violation of Spanish neutrality and the treaties that defined it. Lloyd sent his demand via letter on 27 April 1777, ten days after the British ships had been taken. From the *Atalanta*'s log book it is clear that some of the negotiations over the incident took place in person on board Captain Lloyd's ship. On 11 May 1777, Governor Gálvez himself came on board the *Atalanta* and delivered a letter to Captain Lloyd.[33] His previous letter had been delivered to the ship by an official and it seems unlikely that he would have delivered a letter himself without taking the opportunity to speak to Captain Lloyd face to face. There is seemingly no extant reply from Lloyd to Gálvez's letter from 11 May, so it is possible that he read the letter and then responded verbally to Gálvez. If Gálvez did request a verbal answer from Lloyd, it may have been a cunning attempt to end the affair quickly. In his letter, Gálvez claimed he had ordered the seizure of British vessels because they were involved in contraband trade to Spanish territory and were caught in the process of unloading their cargo into Spanish territory.[34] This was contested by the British account. Gálvez continued by stating that legal proceedings against

32   Willis, *The Struggle for Sea Power*, pp. 277–8.
33   TNA, ADM 52/1586: Log of the *Atalanta*, entry for 11 May 1777.
34   TNA, PRO 30/55/5: Gálvez to Lloyd, 11 May 1777.

the captains and owners of the ships had already begun and that nothing in his actions violated the peace treaty of 1763 or Spanish neutrality. He wrote:

> You think the vessels of his Britanick Majesty aught to be free from arrest and confiscation, not only for suspicion but also when there is clear proofs, I think differently, and my opinion is agreeable to Law as it is certain that the vessels of the one prince as well as those of the other that comply as they ought to do should not be molested or detained in the free navigation of the River Mississippi it is also just that he who separates from the treaty and abuses the Liberty granted to the prejudice of either of the sovereigns or their subjects, are subject to the laws they break.[35]

Gálvez's statement about the laws governing the use of the Mississippi was not particularly clear, and it served only to obfuscate the real point of contention between Lloyd and Gálvez. Both men agreed that the treaty of 1763 granted freedom of navigation in the Mississippi without being molested by the ships of other countries. Both men also agreed that the treaty was considered violated if either country engaged in smuggling or contraband activity that broke the domestic laws of either country.[36] Where the two men differed was in what the British ships were doing at the time of their capture. Lloyd maintained that the British ships had been deliberately captured by Gálvez when they were peacefully navigating between the mouth of the Mississippi and the town of Ibberville.[37] Gálvez maintained that he caught the ships in the act of landing contraband goods. If Lloyd were standing in front of Gálvez reading the letter and responding on the spot, it is not hard to imagine that the British captain would have found it difficult to reply in such a way as to make Gálvez admit that he had violated neutrality and was lying about the British ships. It is of course impossible to determine whether Gálvez was lying and undertook actively to violate Spanish neutrality. However, it is possible to make an educated speculation based on the cargoes of the captured ships and on Gálvez's other conduct throughout the war.

Some of the captured British ships had cargoes of lumber that were supposedly being sent to the British islands in the West Indies.[38] Such a cargo would not have been useful contraband material in Spanish New Orleans because lumber could have been acquired through trade with Spanish colonies or through established routes with North America. The British West Indies, however, and particularly

---

[35] TNA, PRO 30/55/5: Gálvez to Lloyd, 11 May 1777.

[36] Yale Law School, Lillian Goldman Law Library, The Avalon Project: Documents in Law, History and Diplomacy, Treaty of Paris 1763.

[37] CL, Shelburne Papers, F1771 Ca: Case of the Seizure of British Vessels by the Spaniards in the River Mississippi, 1778.

[38] CL, Shelburne Papers, F1771 Ca: Case of the Seizure of British Vessels by the Spaniards in the River Mississippi, 1778.

the naval base in Antigua, would have been in need of lumber in order to support British warships because its usual access to lumber through trade with North America was cut off.[39] It is also unlikely that 11 ships would have been landing contraband at the same time, as it would have greatly increased the likelihood that they be discovered by Spanish authorities or American privateers.

Out of Colonial Hands

Gálvez's conduct during the war, before Spain became a belligerent, marks him as a man who favoured the interests of the American rebels in order to damage Britain's ability to wage war. His correspondence with Governor de la Torre and Patrick Henry, analysed above, demonstrates his purpose very clearly. It is worth pointing to another incident of Gálvez shielding American interests, which also involves Captain Lloyd and the affair of the 11 captured ships. The day after Gálvez went on board HMS *Atalanta*, Captain Lloyd was given information from a passing British merchant that an American privateer was in the river. Lloyd decided to give chase and informed Governor Gálvez by letter that he was leaving New Orleans in order to seek out and capture the American ship. With remarkable celerity, Gálvez sent a boat with a letter for Captain Lloyd that reached the *Atalanta* one hour after she had weighed anchor.[40] In this letter, Gálvez warned Lloyd against breaking the 1763 treaty: 'you will remember the privileges of the territory which all the river enjoys ... any vessel whatever within our cannon shot is consequently (by the treaty of peace and the law of Nations) prohibited to commit hostilities at such a short distance, the same information I will give Captain Barry'.[41] This was an entirely unnecessary reminder given that Lloyd and Gálvez had been discussing Spanish neutrality and the treaty of 1763 the day before. It was also unnecessary to send a boat chasing after the *Atalanta* with the reminder unless it was meant as a warning to Captain Lloyd not to interfere with American shipping in the River because Gálvez would take action. Given the capture of the 11 British merchant ships and his previous request for Spanish warships to protect American trade, it was probably not an idle or veiled threat. Gálvez's behaviour makes it likely that Lloyd's version of events concerning the 11 British merchant ships is the correct version. If so, the Governor of New Orleans was abusing Spanish neutrality and showing partiality to the rebel Americans because he likely made a calculation that the British would not pursue the matter once Captain Lloyd had voiced his objections and been turned away.

Captain Lloyd, however, did not let the matter drop; he conveyed his experience to Governor Chester of British West-Florida, who sent a Colonel

---

39   Willis, *The Struggle for Sea Power*, p. 254.
40   TNA, ADM 52/1586: Log of the *Atalanta,* entry for 12 May 1777.
41   TNA, PRO 30/55/5: Gálvez to Lloyd, 12 May 1777.

Dixon and Mr Stephenson (an influential merchant) to meet the Spanish Governor in order to find out why he had ordered the capture of the British ships in clear violation of the 1763 peace treaty. Gálvez told them:

> it was done in the Heat of Resentment (one of his Britannic Majesty's Sloops of War, the *West-Florida*, having seized a Spanish smuggling-boat, in some of the British Lakes, with twenty or thirty Barrels of Tar;) that had he ... should not have done it; but, having sent an account thereof to the Court of Spain, and cause for representation must be there determined.[42]

By sending an account of the affair to the Court of Spain, Gálvez made it impossible for the dispute to be settled by officials in the West-Indies, and it seems possible that he did this both as a stalling tactic (the court system in Spain that dealt with prize cases was notorious for being slow to adjudicate cases) and as a way to take the decision out of his hands. With pressure coming from British Naval officers and the Governor of West-Florida, Gálvez may have begun to find himself outflanked. By admitting to some level of wrongdoing and removing the decision over the legality of the seizures to Spain, he kept the captured British ships within his jurisdiction and local British pressure at bay.

News of the captures reached London in July of 1777, and one of the merchants whose ships had been captured met with Lord George Germain, Secretary of State for America, to demand of him what was being done to resolve the issue. Germain informed him of Captain Lloyd's actions and of Governor Chester's deputation. In December of 1777, some of the affected merchants delivered a Memorial to Lord Weymouth, the Secretary of State for the Southern Department, known for both drunkenness and laziness, wishing that something might be done by the British government to redress the wrongs done them by the Spanish Governor. The reply, written by Weymouth's secretary, was that nothing could be done until the Spanish courts gave a decision upon the matter.

The decision from the Spanish courts arrived in February of 1778 and it declared that the seizure of the British ships had been justified since they were illegally trading with Spaniards in a Spanish port. The response to the British merchants, given again by Weymouth's secretary, was 'That the state of public affairs was such that, Nothing could be done in this business'.[43] Five months later, after continued pressure from the British merchants, Weymouth wrote to the ministers of Spain to inquire over the affair and a possible resolution. The Court of Spain returned no answer to the ambassador's enquiries. Frustrated,

---

[42]   CL, Shelburne Papers, F1771 Ca: Case of the Seizure of British Vessels by the Spaniards in the River Mississippi, 1778.
[43]   CL, Shelburne Papers, F1771 Ca: Case of the Seizure of British Vessels by the Spaniards in the River Mississippi, 1778.

one of the British merchants made an application for the detention of Spanish property that was taken hostage by one of his ships, *Fortune*, until the Mississippi ships were restored. His application was denied, despite the legality of the action under the Prize Act, because the government thought 'it was not a time to quarrel with the Court of Spain'.[44] From the point of view of the British government, the latter months of 1777 and the start of 1778 were not good times to increase Anglo-Spanish tensions in the political sphere because war with France was looming and colonial actors were doing a fine job of increasing tensions on their own.

In early January 1778, an American warship started to cruise down the Mississippi capturing every British ship it could and plundering British settlements. When the warship reached New Orleans, Gálvez welcomed the Americans and allowed them to sell their prizes in New Orleans. Tensions between the Spanish and the British escalated over this incident until a British naval captain fired into New Orleans. Gálvez's response was brazenly to declare that Spain would continue to observe her neutrality.[45] By the time the Spanish courts in Europe had declared the taking of the 11 merchant ships to have been a legal seizure, and by the time Weymouth's secretary responded to the British merchants clamouring for justice, ministers in both Spain and Britain would have heard about the incident involving the American warship in New Orleans. Spanish ministers were not yet ready to enter the war, and British ministers wanted to keep Spain from entering the war at all. It was not beneficial to either side to blow up the violations of neutrality happening in New Orleans and the Mississippi. Unable to control the actions of their agents in the colonies, British ministers let the matter lie once the Spanish Court passed its decision on the 11 captured ships, and Spanish ministers simply avoided raising the topic at all with British ministers.[46]

When the British owner of the *Fortune* was denied the possibility of keeping Spanish property hostage in July of 1778, Britain was at war with France and, rather foolishly, but hopefully, trying to negotiate a quick end to the Anglo-French conflict with Spain as a mediator. The correspondence between Weymouth and Lord Grantham, Britain's ambassador to Spain, in the summer and autumn of 1778, largely focused on Spain as a possible mediator and not on Spain's abuses of neutrality. Spanish misconduct is rarely mentioned, whereas attempts to show the Spanish that British ministers respected Spain's rights in the Americas are frequently made.[47] One of the few times that Spanish violations

---

[44]   CL, Shelburne Papers, F1771 Ca: Case of the Seizure of British Vessels by the Spaniards in the River Mississippi, 1778.

[45]   Willis, *The Struggle for Sea Power*, p. 279.

[46]   CL, Shelburne Papers, F1771 Ca: Case of the Seizure of British Vessels by the Spaniards in the River Mississippi, 1778.

[47]   New York Public Library, Hardwicke Collection, 129: Letters from Weymouth to Grantham 21 July 1778–11 June 1779.

of neutrality were raised in the Weymouth–Grantham correspondence in 1778 or 1779 was in January 1779, when a group of 15 British merchants who owned two ships captured by the Spanish submitted a petition to Weymouth in the hope that the British government would seek redress for them through the Court of Spain. Weymouth included the petition in his letter, but he also asked Grantham to seek an answer on an inquiry that was submitted to the Spanish Court about the capture of a British ship in the Black River, upon which the Spanish Court had so far been silent. The last request in the letter was about the British ships that were carried into New Orleans and the petition for their release, submitted by the British government, that had also gone unanswered.[48] It is impossible to say if the ships referred to were those captured by Gálvez or those captured by the American privateer, but this appears to be the last time that the captured ships in New Orleans were mentioned in British ministerial correspondence. Weymouth claimed that 'the Spanish Governor at New Orleans cannot justify the violence and injustice of these proceedings',[49] but he ended his letter to the ambassador by writing about Spanish grievances over British captures of Spanish ships:

> Every means are used that strict and speedy justice should be obtained by the subjects of His Catholic Majesty, if they have received any injury; ... it often happens that information cannot be obtained as early as it is wished; and often it is necessary from the constitution of the Country [Britain], that the reparation should be obtained by sentences from the Courts of Law, to which the aggrieved [Spanish] are unwilling to apply in a proper manner.[50]

Weymouth is clearly of the opinion that the incidents in New Orleans were a violation of Spanish neutrality, but he does no more than ask Grantham to enquire after the state of a petition. There are no demands for justice and no demands for an explanation. By contrast, when discussing British abuses of Spanish neutrality, Weymouth takes the time to make very clear that the British government is doing everything it can to make right the actions of its agents abroad, but that Spanish merchants must apply to British law courts for proper redress. Weymouth was not serving the interest of British merchants so much as serving the grander strategic goal of trying to stave off a maritime war with Spain when Britain was already involved in a maritime war with France and the American rebels. The men who lost the most to Britain's strategy were the merchants who dared operate in the uncertain maritime climate of Britain's war with the rebel colonies, where the

---

[48]   British Library, London, Grantham Papers, Add. MS 24166: Weymouth to Grantham. 29 Jan. 1779.

[49]   British Library, London, Grantham Papers, Add. MS 24166: Weymouth to Grantham. 29 Jan. 1779.

[50]   British Library, London, Grantham Papers, Add. MS 24166: Weymouth to Grantham. 29 Jan. 1779.

rewards of commerce could be extremely high, but where the risk of capture, whether legal or not, could lead to devastating results.

Conclusions

The juxtaposition between Lord Weymouth's two statements and the absence of any meaningful action taken by the British government to hold Spain accountable for violations of neutrality illustrate how the British government approached Anglo-Spanish relations during the American war. Knowing that Spanish and British agents in the Americas could not be controlled, British ministers preferred to play down Spanish violations of neutrality and play up British interest in addressing Spanish grievances. This approach was intended to keep Spain out of the war whilst preserving a maritime strategy which contained an element of commerce predation.

It is difficult to discern if Britain's strategy helped to stave off Spanish involvement in the war, or if Spain would always have waited until the summer of 1779, when her navy and alliance with France fully prepared her for a war of revenge. Regardless, British ministers ascribed great importance to neutrality as a vital component of maritime economic warfare and were willing to deny justice to British merchants in order to preserve it. Neutrality during the American War was not stagnant, and all of the officials, ministers, smugglers, and officers who dealt with it were aware of its dynamism and used it to their advantage when possible. Britain faced an overwhelming problem in the American war, in that it had grown powerful enough in the maritime sphere to be seen as a threat by European colonial powers. Spain's concept of neutrality, as an active participant in a sea-power-based war allowed Spain to erode British power until the moment was right for war. Britain, unwilling to create more belligerents, could only hope that encouraging neutrality would keep Spain from becoming an open enemy.

Competition in the maritime sphere at the end of the eighteenth century was fierce, and though mercantilist policies were increasingly relaxed, the breakout of a European war increased opportunities for smugglers and neutral merchants. The increased access of smugglers and neutral merchants to wartime markets was at odds with strategies of economic warfare usually pursued by maritime belligerents. Neutral nations could give enemy trade a safe means of transportation just as belligerent nations attempted to shut down and eradicate the same trade. As such, neutrality and neutral rights were vital components of maritime strategies based on economic warfare at sea. Neutral nations, if not properly negotiated with by belligerents, could too easily turn into enemies.

CHAPTER FIVE

Sea Power and Neutrality: The American Experience in Europe during the French Wars, 1793–1812

Silvia Marzagalli

From the late seventeenth century to the end of Napoleonic Wars in 1815, France and Britain engaged in a long series of conflicts, which culminated in undisputed British naval mastery in the nineteenth century. Sea power was Britain's chief strategic tool during this century-long struggle for colonies, strategic positions, and financial markets.[1] As naval capacity and the volume of seaborne trade grew throughout the period, blockade and belligerents' regulation of international shipping became an increasingly powerful instrument with which to weaken an enemy's economy, and therefore its ability to sustain the war effort. Thus, the capacity to control sea lines of communication and the littoral were crucial in belligerents' war policies. The Revolutionary and Napoleonic Wars, referred to here as the 'French Wars' (1793–1815), represented the zenith of economic warfare at sea opposing France and Great Britain in the long eighteenth century, encompassing ambitious plans such as Britain's attempt to starve France in 1793 or Napoleon's 1806 continental blockade aimed at closing European markets to British trade.[2] France failed in its endeavour to organise a European boycott of Britain. Similarly, the British blockade alone could not defeat France. Targeting

---

[1] I. Hont, *Jealousy of Trade: International Competition and the Nation State in Historical Perspective* (Cambridge, MA: Harvard University Press, 2005); N.A.M. Rodger, *The Command of the Ocean: A Naval History of Britain*, vol. 2, *1649–1815* (London: Allen Lane with the National Maritime Museum, 2004); H.V. Bowen, 'Forum the Contractor State, *c.*1650–1815', *International Journal of Maritime History* 25.1 (2013), pp. 239–74; Roger Knight and Martin Wilcox, *Sustaining the Fleet, 1793–1815: War, the British Navy and the Contractor State* (Woodridge: Boydell, 2010).

[2] P. Pourchasse, 'La guerre de la faim: l'approvisionnement de la République, le blocus britannique, et les bonnes affaires des neutres au cours des guerres révolutionnaires (1793–1795)', HdR thesis, Université de Bretagne Sud, 2013; F. Crouzet, *L'économie britannique et le blocus continental, 1806–1813*, 2 vols (Paris: Presses universitaires de France, 1958; 2nd ed.; Paris: Economica, 1987); K.B. Aaslestad and J. Joor (eds),

enemy trade would not prove decisive, but it nevertheless had significant economic and political consequences.

Whereas the impact of sea power on the economies of belligerents has long attracted considerable interest from historians, the impact on non-belligerent states has only recently begun to receive scholarly attention, and this subject still requires a holistic effort to conceptualise its global ramifications.[3] Our understanding of economic warfare ought to include not only the strategy of belligerents but also an analysis of their systemic effects on all commercial agents and ports. This chapter will demonstrate the importance of such an integrated approach by examining how the United States participated in the reorganisation of trade flows during the French Wars.[4] In particular, it focuses upon how trade and shipping were covered by neutral flags, to the benefit of both neutral and belligerent merchants who sought to elude enemy attempts to conduct economic warfare at sea, as well as the restrictions imposed by their own government on international trade. The recourse to neutral shipping induced belligerent powers to try repeatedly to reduce neutrals' capacity to connect markets. Britain and France recognised in the course of the French Wars that neutral shipping was seriously reducing the impact of their economic war against the enemy. Restricting neutral shipping, however, affected also their own economy, and international relations. Belligerents' policies, thus, evolved pragmatically over time and space. In August 1794, for instance, Britain gave up the 1793 plan to starve France through an extension of the notion of 'contraband of war' to

---

Revisiting Napoleon's Continental System: Local, Regional and European Experiences (Basingstoke: Palgrave Macmillan, 2015).

[3]   Most existing literature focuses on the theoretical and juridical notion of neutrality. See K. Stapelbroek (ed.), *Trade and War: The Neutrality of Commerce in the Inter-State System*, COLLeGIUM: Studies across Disciplines in the Humanities and Social Sciences, 10 (2011): www.helsinki.fi/collegium/journal/volumes/volume_10/index.htm (accessed 5 Nov. 2019); A. Alimento (ed.), *War Trade and Neutrality: Europe and the Mediterranean in the Seventeenth and Eighteenth Centuries* (Milan: Franco Angeli, 2011). By contrast, the Forum tried to show how agents concretely negotiated and took advantage of neutrality. See S. Marzagalli and L. Müller (eds), '"In Apparent Disagreement with All Law of Nations in the World": Negotiating Neutrality for Shipping and Trade during the French Revolutionary Wars', *International Journal of Maritime History* 28.1 (2016), pp. 108–92. See also J. Eloranta, E. Golson, P. Hedberg, and M.C. Moreira (eds), *Small and Medium Powers in Global History: Trade, Conflicts, and Neutrality from the 18th to the 20th Centuries* (London: Routledge, 2019).

[4]   Historians considering American trade statistics and the increase in American shipping during the French Wars have rightly stressed the importance of the European wars in fostering American growth. See A. Clauder, *American Commerce as Affected by the Wars of the French Revolution and Napoleon, 1793–1812* (Philadelphia: University of Pennsylvania, 1972 [1932]); Douglass C. North, 'The United States Balance of Payments, 1790–1860', *Trends in American Economy in the Nineteenth Century*, Studies in Income and Wealth 24 (Princeton, NJ: Princeton University Press, 1960), pp. 573–627.

foodstuffs, as this plan was causing increasing tensions with the Scandinavians, was too costly, and ultimately ineffective given the different stratagems neutral captains adopted to protect their cargo.[5] But in 1798 Britain did not hesitate to capture and condemn Swedish convoyed merchant ships, provoking the formation of the Second League of Armed Neutrality in 1800 and a vivid international debate over neutral rights.[6] Similarly, France alternated severe attacks on neutral trade, as in 1798, and great tolerance in the following years up to 1807. A glimpse at the US experience demonstrates the fragility of neutrality, but also its powerful disruptive effects on belligerents' policies of economic warfare.

The United States managed to preserve its neutrality with regard to European belligerents between 1793 and 1812, considerably longer than any European power. By analysing how belligerent merchants took advantage of business opportunities under the cover of neutrality, this chapter contends that the impact of economic warfare depended on belligerents' attitudes and the evolving position of a neutral state relative to the hostile powers. Although in the late years of the Napoleonic Wars neutral shipping was hardly possible, the evolution was not linear over time. Geography mattered too, with the interests and strengths of a belligerent in a particular maritime region impacting on their ability or will to restrain trade. For instance, while US shipping to the French Empire and the Italian Peninsula was low in 1810–11, it boomed in the Baltic, under British protection, with Russia absorbing 10 per cent of total US exports.[7] The chapter begins by briefly presenting the role of neutral merchant fleets in warfare. It then examines the case of the United States, explaining how American shipping and trade developed in two distinct areas in Europe during this era.

The Reorganisation of International Trade Routes in Times of War under Neutral Flags

Two elements are helpful in comprehending the importance of the economic opportunities that neutrality represented to maritime countries and appreciating the considerable changes induced by European wars on international shipping and trade. The first indicator is the evolution of the merchant fleets of neutral powers. All neutral merchant fleets underwent considerable expansion during the French Wars. For example, the Hamburg fleet increased from 159 units in 1788 to 280 in 1799, while the number of Greek vessels under the Ottoman flag had doubled by the end of the Napoleonic Wars. Between 1793 and 1807,

---

[5]   Pourchasse, 'La guerre de la faim'.

[6]   L. Müller, 'Swedish Merchant Shipping in Troubled Times: The French Revolutionary Wars and Sweden's Neutrality, 1793–1801', *International Journal of Maritime History* 28.1 (2016), pp. 147–64.

[7]   A.W. Crosby, 'America, Russia, Hemp and Napoleon: A Study of Trade between the United States and Russia, 1783–1814', PhD thesis, Boston University, 1961, pp. 1–2.

thousand tons

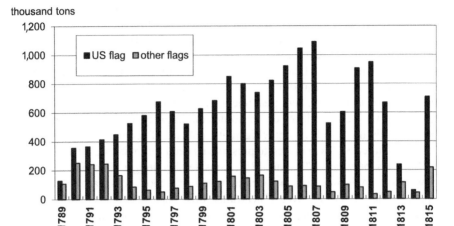

Figure 5.1: Tonnage of ships entering the United States from foreign countries, 1789–1815
Source: Adam Seybert, *Statistical Annals of the United States of America: Founded on Official Documents* (Philadelphia: Dobson, 1818; reprinted New York: Kelley, 1967); *Historical Statistics of the United States: Colonial Times to 1970* (Washington, DC: US Bureau of Census, 1975), Table Q507-08.

both the Danes and the Swedes sent twice as many ships to southern Europe as in the period 1784–92. The total tonnage of US ships entering American ports tripled between 1790 and 1807 (Figure 5.1). This growth was the result of both the construction of new ships and the purchase of second-hand vessels. The second-hand market was boosted chiefly by the sale of prizes taken by belligerent navies and privateers, as well as the sales of vessels – fictitious or real – belonging to belligerents. Although these figures cannot be directly compared across states, taken individually they all express a very clear trend concerning the impact of conflicts on neutral fleets.[8]

---

[8]   This and the following sections expand Silvia Marzagalli, 'La navigation américaine pendant le French Wars (1793–1815): une simple reconfiguration des circuits commerciaux par neutres interposés?' in Eric Schnakenbourg (ed.), *Neutres et neutralité dans l'espace atlantique durant le long XVIIIᵉ siècle (1700–1820): une approche globale/ Neutrals and Neutrality in the Atlantic World during the Long Eighteenth Century (1700–1820): A Global Approach* (Bécherel: Les Perseides, 2015), pp. 131–56. For data on the fleets, see Otto Mathies, *Hamburgs Reederei, 1814–1914* (Hamburg: Friederichsen & Co., 1924); G. Harlaftis, 'The "Eastern Invasion": Greeks in Mediterranean Trade and Shipping in the Eighteenth and Early Nineteenth Centuries', in Maria Fusaro, Colin Heywood, and Mohamed-Salah Omri (eds), *Trade and Cultural Exchange in the Early Modern Mediterranean: Braudel's Maritime Legacy* (London: I.B. Tauris, 2010), p. 232.

While the number of neutral vessels increased in wartime, they did not monopolise international transport. The British merchant fleet underwent a 70 per cent increase between 1793 and 1815, driven in part by demand for transport services for the Royal Navy and the British army, which employed 10 per cent of merchant ships.[9] More important, however, was British merchants' unrivalled penetration of markets across the globe. For trade with the European continent, however, British merchants relied primarily on neutral bottoms. France faced a tougher dilemma. Its ships were increasingly vulnerable to capture by the Royal Navy and enemy privateers.[10] French shipowners soon confined their vessels' activity to regional costal trade, leaving most long-haul shipping to foreign bottoms. In 1795, a mere two years after war against Britain broke out, only 40 per cent of ships entering Marseilles sailed under the French flag compared with 64 per cent in 1787. The total number of entries had decreased by 31 per cent.[11] French ports on the Atlantic coast experienced a similar phenomenon: while roughly 70 larger ships owned in Nantes sailed within a large area encompassing Amsterdam and Trieste between 1790 and 1791, their range shrank to Brest in the north and Bordeaux in the south between 1798 and 1800, and their average tonnage decreased by half.[12]

The second indicator of the opportunities that the French Wars offered to American merchants and shipowners is provided by the dramatic increase of US foreign trade (Figure 5.2). This trend was largely – although not exclusively – driven by the rapid and sustained growth of the re-export trade, mainly comprising West Indian colonial goods, but also European and Asian products which now passed through American ports before reaching their intended destination.

Economic warfare at sea forced both belligerent and neutral merchants to reorganise trade routes. This was necessary to protect ships and cargoes from the risk of capture. Belligerents traditionally seized all enemy ships they could find

---

D.H. Andersen and P. Pourchasse, 'La navigation des flottes de l'Europe du Nord vers la Méditerranée (XVIIᵉ–XVIIIᵉ siècles)', in A. Bartolomei and S. Marzagalli (eds), 'La Méditerranée dans les circulations atlantiques au XVIIIᵉ siècle', *Revue d'histoire maritime*, 13 (2011), pp. 21–44.

[9]   R. Knight, *Britain against Napoleon: The Organization of Victory, 1793–1815* (London: Allen Lane, 2013), p. 180.

[10]   On British privateering, see D.J. Starkey, *British Privateering Enterprise in the Eighteenth Century* (Exeter: University of Exeter Press, 1990).

[11]   Data provided by the Health Office in Marseilles (Archives départementales des Bouches-du-Rhône, 200 E 543, 550 and 551) and available online: http://navigocorpus.org.

[12]   Karine Audran, 'Les négoces portuaires bretons sous la révolution et l'Empire; bilan et stratégies: Saint-Malo, Morlaix, Brest, Lorient et Nantes, 1789–1815', PhD thesis, Université de Bretagne-Sud, 2007, vol. 1, pp. 274–83 and vol. 2, p. 249. In 1790, coastal traders from Nantes averaged 117 *tonneaux*; in 1799–1800 (year VIII), 59 *tonneaux* (the French ordinance of 1681 made the *tonneau de mer* a unit equivalent to 1.44 cubic metres).

$US 000s

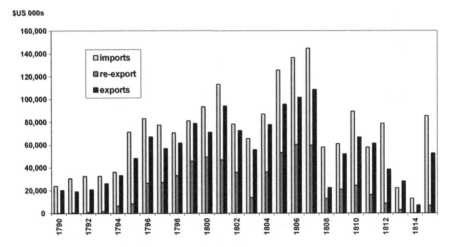

Figure 5.2: US foreign trade, 1790–1815
Source: Douglass C. North, 'The United States Balance of Payments, 1790–1860', *Trends in American Economy in the Nineteenth Century* (Princeton, NJ: Princeton University Press, 1960), pp. 573–627.

at sea. From 1756 onwards British courts also condemned cargoes belonging to enemy subjects carried by neutral bottoms. They even forbade neutral ships from sailing between ports where trade was not permitted under their flag in time of peace. France, for instance, forbade (or, after 1767, severely restricted) foreign access to its colonies in peacetimes. It also prevented foreign ships from carrying goods between two French ports. In turn, Britain considered neutral ships on these trade routes to be valid prizes.

Belligerents were aware that, despite the restrictions they imposed, neutral shipowners were taking an increasing share of global shipping. They feared that this might benefit their opponent and thus affect their war effort at sea. As the French Wars progressed, belligerents increasingly sought to reduce neutral participation to their enemy's trade.[13] Prior to 1805 it had been sufficient for a ship to make a stopover in a neutral port without unloading cargo or paying duties in order to then progress unimpeded to an enemy port. Now the British Admiralty and Vice-Admiralty courts began to condemn neutral ships regardless of a stopover if they could prove that the cargo had always been intended for an enemy port.[14] It therefore became necessary to muddy the waters by changing

---

[13]   The most comprehensive study of the evolution of British and French commercial legislation against the enemy in Crouzet, *L'économie britannique*.
[14]   B. Perkins, 'Sir William Scott and the *Essex*', *William and Mary Quarterly* 13 (1956), pp. 169–83.

ship and crew, or by selling the cargo to other merchants on site. This factor increased the number of intermediaries, but also served to concentrate trade in those ports such as New York, Philadelphia, Baltimore, Boston, or Salem.[15] On the east coast of the United States it was possible to compose a cargo adapted to different markets across the world at almost any time of year, and to find a freight for ships which came in.

Belligerents' policies, therefore, forced merchants and shipowners to consistently reorganise trade routes and commercial networks, and to adapt to changing legislation. Such reorganisation was possible because demand remained high. The French Wars occurred after a century of sharp trade growth, which was sustained by changes in consumption patterns, increasing market integration, and population expansion.[16] Also, the beginnings of industrialisation made certain products more accessible because they were now less expensive, widening the base of potential consumers. The French Wars did not reduce demand in British manufactured goods or colonial goods, but they consistently changed the trade patterns that linked the three regions comprising the Atlantic world – Europe, Africa, and the Americas – so to connect producers and consumers despite the fetters imposed by belligerents. Although the intensity of market integration in the eighteenth century and the role of the Atlantic world in the economic advances experienced in the nineteenth century remain hotly contested among historians,[17] there is no doubt that the sharp increase in colonial trade in the eighteenth century benefited ports in western Europe, as well as agriculture and industry in this region. On the eve of the French Revolution, around 1,400 ships a year crossed the Atlantic to trade between the Americas and Europe, while approximately 220 slave traders delivered labour to the plantation system in the

---

[15]   The same phenomenon of booming staple ports in neutral countries can be found in Leghorn until 1808, when this Mediterranean port was annexed to the French Empire by Napoleon.

[16]   J. de Vries, *The Industrious Revolution: Consumer Behavior and the Household Economy, 1650 to the Present* (New York: Cambridge University Press, 2008).

[17]   P. Emmer, 'In Search of a System: The Atlantic Economy, 1500–1800', in Horst Pietschmann (ed.), *Atlantic History: History of the Atlantic System* (Göttingen: Vandenhoeck & Ruprecht, 2002), pp. 169–78 strongly denies the existence of anything like an 'Atlantic system' and asserts that the existence of the vast majority of early modern inhabitants of Europe, Africa, and the Americas did not depend on the Atlantic economy. Patrick O'Brian, 'European Economic Development: The Contribution of the Periphery', *Economic History Review* 35.1 (1982), pp. 1–18 stressed the importance of national markets for European economic growth. See also P. Verley, *L'échelle du Monde: essai sur l'industrialisation de l'Occident* (Paris: Gallimard, 1997); K. Pomeranz, *The Great Divergence: China, Europe, and the Making of the Modern World Economy* (Princeton, NJ: Princeton University Press, 2000); D. Acemoglu, S. Johnson, and J. Robinson, 'The Rise of Europe: Atlantic Trade, Institutional Change, and Economic Growth', *American Economic Review* 95.33 (2005), pp. 546–79; J. de Vries, 'The Limits of Globalization in the Early Modern World', *Economic History Review* 63.3 (2010), pp. 710–33.

West Indies and North America.[18] Additionally, 1,400 Newfoundland fishing ships fed both Europeans and West Indian slaves with cod. Beside this transatlantic trade, shipments of tobacco, sugar, coffee, and indigo to Europe sustained a significant inter-European and Mediterranean re-export trade, which represented up to 80 per cent of the sugar imported to France, while slave traders re-exported fabrics previously imported from Asia. This system was irrigated by Spanish silver and Portuguese gold.

The maritime activities induced by Atlantic transatlantic trade, and to a large extent framed by imperial rules imposed upon colonies, were therefore far from negligible. They were also tremendously vulnerable during wars between the major European colonial powers. The regular flow of ships and goods in peacetime was disrupted, with merchant ships seized by enemy forces. Merchants, however, had developed means for successfully circumventing restrictions imposed by belligerents. Apart from convoying, which in the French case became problematic after 1794, recourse to neutral shipping was the simplest way to maintain transatlantic and European trade. More broadly, contemporary entrepreneurs proved quick to react to changes in colonial production and trade. By 1807, the quantity of sugar imported from the Americas to Europe was greater than in 1789 despite the collapse of sugar production in Saint-Domingue, the most significant production hub for sugar following the 1791 slave revolt, which resulted in the independence of Haiti in 1804. Responses to European demand for sugar, therefore, required both a reorganisation of production in other areas and the capacity to tackle the difficulties posed by trading with the colonies of a belligerent.[19]

In peacetime, this integrated productive and commercial system was placed under the theoretical control of the European mother countries, which strove to channel the flows of goods within their empire. Imperial legislation aimed at reserving to the crown's subjects – and indirectly to the sovereign – the main benefits of colonial trade. In practice, however, restrictive legislation could not be carried out effectively. Smuggling was endemic[20] – not least because European powers encouraged penetration into other empires – and European colonial trade policies were frequently suspended in the eighteenth century whenever

---

[18]   On average, 219 slave traders sailed between North America and the West Indies during the period 1787–89; see www.slavevoyages.org.

[19]   E.B. Schumpeter, *English Overseas Trade Statistics, 1697–1808* (Oxford: Clarendon Press, 1960), table XVIII; T. Pitkin, *A Statistical View of the Commerce of the United States* (Hartfort, CT: Charles Hosmer, 1816; repr. 1967; new ed.; New Haven, CT: Durrie & Peck, 1835), chap. 3, table 4; J. Meyer, *Histoire du sucre* (Paris: Desjonquères, 1989).

[20]   See J. Tarrade, *Le commerce colonial de la France à la fin de l'Ancien Régime*, 2 vols (Paris: Presses universitaires de France, 1972). On the illegal sugar trade, see C. Schnakenbourg, 'Les sucreries de la Guadeloupe dans la seconde moitié du XVIIIᵉ siècle (1760–1790): contribution à l'étude de la crise de l'économie coloniale à la fin de l'ancien régime', Thèse d'état, Université de Paris II, 1972; W. Klooster, *Illicit Riches: Dutch Trade in the Caribbean, 1648–1795* (Leiden: KITLV Press, 1998).

colonial supply became difficult. Under these circumstances the mother country often opened its colonial ports to neutral carriers. Such was the case in the Revolutionary Wars. War broke out between France and Britain on 1 February 1793. Less than three weeks later, on 19 February, France authorised neutral carriers to sail to and trade with its colonies.

Past experience showed that it was possible to maintain trade flows during wartime and to revert to peacetime patterns as soon as the conflict ceased. On average, eighteenth-century French merchants enjoyed peace and protection from foreign competition in the West Indies only half of the time. Otherwise, France and Britain were at war and their property at sea was under the threat of British privateering. It is therefore understandable as to why French merchants and shipowners who had built their fortunes on colonial trade in the eighteenth century strove to maintain it during the French Wars. In the case of Bordeaux, the recourse to US ships was the most successful strategy.

US Shipping: An Effective Means to Sustain Trade in Bordeaux

To understand more precisely how economic warfare at sea, as conducted by belligerent states, could be circumvented, I will focus next on how American merchants and captains contributed to the reorganisation of trade in Bordeaux, a traditional exporter of wine which became, starting in the 1730s, the main French port for West Indian trade. On the eve of the French Revolution, Bordeaux merchants constituted a large community. They had heavily invested in colonial trade, most notably in Saint-Domingue, as well as in the import–export trade to northern Europe. The latter was connected to transatlantic shipping as it enabled a consistent re-export trade of colonial goods.[21] When profits in colonial trade declined in the 1780s due to increasing competition, several large Bordeaux merchant houses bolstered their investments in the slave trade and markets in the Indian Ocean, while others shifted their operations into finance and contracting.[22]

---

[21]   On Bordeaux's eighteenth-century trade, see P. Butel, *Les négociants bordelais, l'Europe et les Îles au XVIIIe siècle* (Paris: Aubier-Montaigne, 1996 [1974]). On this part of the chapter, see also S. Marzagalli, 'Was Warfare Necessary for the Functioning of Eighteenth-Century Colonial Systems? Some Reflections on the Necessity of Cross-Imperial and Foreign Trade in the French Case', in Cátia A.P. Antunes and Amelia Polónia (eds), *Beyond Empires: Global, Self-Organizing, Cross-Imperial Networks, 1500–1800* (Leiden: Brill, 2016), pp. 253–77.

[22]   Éric Saugera, *Bordeaux, port négrier: XVIIe–XIXe siècles. Chronologie, économie, idéologie* (Paris: Karthala, 1995); P. Butel, 'Réorientations du négoce français à la fin du XVIIIe siècle, les Monneron et l'océan Indien', in Paul Butel and Louis M. Cullen (eds), *Négoce et industrie en France et en Irlande aux XVIIIe et XIXe siècles* (Paris: Editions du CNRS, 1980), pp. 65–73. For a case study of a major merchant family moving toward banking, plantation, and contracting, see S. Marzagalli, 'Limites et opportunités

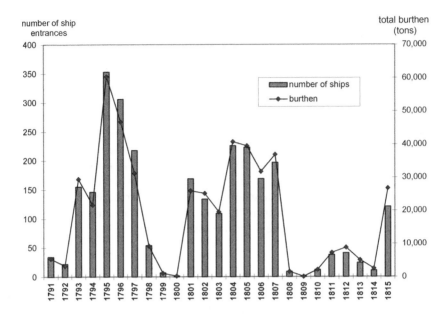

Figure 5.3: Entrances of US ships in Bordeaux, 1791–1815
Source: Compiled from various sources. See S. Marzagalli, *Bordeaux et les États-Unis, 1776–1815: politique et stratégies négociantes dans la genèse d'un réseau commercial* (Geneva: Droz, 2015). Data for the second half of 1792 is incomplete. All ship data are available at http://navigocorpus.org/.

None, however, took advantage of the opening of the North American markets once the United States became independent. Bordeaux merchants considered Americans more like competitors in the West Indies than as potential commercial partners. Some Bordeaux shipowners eventually tried to make their West Indian ventures more profitable by including a stopover in the United States, but the monopoly of the General Farm in France prevented the development of a lively trade in tobacco, which was the only significant North American product the French market would be interested in. In the early 1790s there were few indicators of the coming importance of American shipping for Bordeaux; its evolution demonstrates that the impetus for connecting the two markets came only with the outbreak of war between France and Great Britain in 1793 (Figure 5.3).[23]

The declaration of war made it extremely difficult to maintain trade between Bordeaux and the Caribbean. This proffered a huge opportunity to North

---

dans l'Atlantique français au XVIIIe siècle: le cas de la maison Gradis de Bordeaux', *Outre-Mers* 362–3 (2009), pp. 87–110.
[23]   S. Marzagalli, *Bordeaux et les États-Unis, 1776–1815: politique et stratégies négociantes dans la genèse d'un réseau commercial* (Geneva: Droz, 2015).

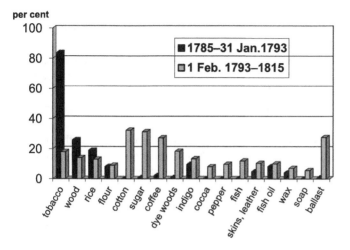

per cent

■1785–31 Jan.1793
□1 Feb. 1793–1815

Figure 5.4: Frequency of specific goods in the cargoes imported at Bordeaux on US ships before and during the French wars
Source: S. Marzagalli, *Bordeaux et les États-Unis, 1776–1815: politique et stratégies négociantes dans la genèse d'un réseau commercial* (Geneva: Droz, 2015), p. 220. Data refer to the presence of the given commodities on board the total of incoming ships and not on their volume or value.

American entrepreneurs, who took over a considerable part of the Bordeaux import and export trade, including its colonial trade. The first indicator of this change is quantitative. The arrival of American ships in Bordeaux increased from fewer than 40 in the early 1790s to 150 per year in 1793 and 1794 – despite an embargo that paralysed the port of Bordeaux for six months following the Federalist revolt – and reached 350 by 1795. Although this proved the zenith of American shipping at Bordeaux, fuelled by the extremely poor crop of 1794, the average of the years 1803–7 was still over 200 arrivals per year, which remained five times greater than in peacetime. However, by 1808, the combined effect of both belligerents' harsher forms of economic warfare, coupled with President Thomas Jefferson's economic embargo, considerably reduced American shipping in Bordeaux.

The most important indicator of the ongoing transformation in trading patterns, however, emerges through an analysis of the nature of the cargoes imported and exported by American vessels. Prior to the outbreak of war, American ships came to Bordeaux chiefly to deliver tobacco, rice, cereals, and wood. From 1793, however, colonial goods, namely sugar, coffee, indigo, and dye wood, were carried by half of inbound ships. These ships also carried Asian spices, notably pepper, as well as fish and fish oil (Figure 5.4). The changes were just as significant on the export side: whereas before 1793 American ships often cleared Bordeaux on ballast to look elsewhere for a return cargo, once war between France and Britain was under way they systematically departed with a cargo, mostly consisting of wine, brandy, and miscellaneous manufactured goods. Essentially, American ships had taken over Bordeaux's traditional export trade. Moreover, the demand for carrying services in Bordeaux was large enough to attract a substantial number of neutral vessels arriving on ballast. This entirely new phenomenon concerned a quarter of the American ships entering Bordeaux between 1793 and 1815.

An additional indicator of the transformations provoked by the war is provided by the changes in the commercial areas covered by American ships calling at Bordeaux. Whereas before the outbreak of hostilities they confined themselves to round-trip itineraries between the east coast of the United States and Bordeaux, including no more than a stopover in Portugal on the return trip to load salt, after 1793 they connected Bordeaux to all of the ports with which it had been accustomed to trade in the past, ranging from the Indian Ocean to the Americas, and from the Iberian Peninsula to the Baltic.[24] During the French Wars, therefore, American ships provided transport services that went well beyond the framework of the bilateral trade which had characterised US relations with Bordeaux between the end of the War of American Independence and the beginning of the French Revolution. The ports of the United States became the warehouses not only of French (and other belligerent) West Indian trade but also for products coming from the Indian Ocean, and even of Bordeaux cargoes bound for other European ports.[25]

Within the framework of this major reorganisation of trade flows in times of war, it is necessary to understand the underlying merchant networks which sustained it by examining the ways in which neutral Americans merchants and captains took advantage of opportunities created by the war. Belligerents faced considerable obstacles to the efficient implementation of their economic policies because neutral shipping helped their enemies to carry on those trades which they aimed to disrupt; most notably trade on the enemy's account. The trade to Bordeaux under the flag of the United States was carried by a very large number

---

[24]   The majority of American ships arriving at (or departing from) Bordeaux between 1793 and 1815 came from (or were sent to) a port in the United States. But, as mentioned above, this was required because belligerents' legislation enforced an intermediate stopover in a neutral country. For maps of the spatial changes over time of American shipping in Bordeaux, see Marzagalli, *Bordeaux et les* États-Unis, pp. 183–5 and Marzagalli, 'Was Warfare Necessary?' pp. 267–8.

[25]   The final destination of cargoes carried from Bordeaux to the United States can be tracked in the registers of drawback. See, for instance, Peabody Essex Museum (Salem), Newburyport Custom House # 282, # 497: 'Abstract of drawbacks of duties payable in the district of Newburyport on goods, wares, and merchandise exported from the United States, 1804–1805'. The brig *Three Brothers* imported a cargo of brandy which was re-exported to Copenhagen (20 Apr. 1805).

of firms and ships. Virtually every American shipowner possessed a ship which called at Bordeaux during the French Wars at one time or another. By contrast, there was a strong concentration of the consignment business in Bordeaux. Whereas 62 per cent of incoming American ships in 1796 were consigned to the captain, a supercargo, or one of her shipowners on board (a clear sign of the non-existence of previously established American trade networks with the French port), this was the case for only 5 per cent of ships arriving in 1807.[26] By that point, seven firms in Bordeaux were the consignees of half of the incoming American ships, with the rest scattered among 40 different houses dealing with fewer than five cargoes each. This situation is representative of the whole period: between 1795 and 1815, half of the consignments were handled by 13 different firms in Bordeaux,[27] nine of which were a partnership of American merchants who had moved to Bordeaux. At first sight, then, American merchants were controlling and disproportionately benefiting from all business related to shipping and trade carried in Bordeaux under the neutral flag of the United States.

Yet matters were more complex. A part of this business, in fact, covered French interests. There were at least three different and complementary methods Bordeaux merchants adopted to maintain their business during the war, making use of neutral merchants and captains. The first consisted of procuring a neutral flag to their ships and continuing to operate them under neutral cover. In 1789, Bordeaux merchants owned a fleet of 235 different ships (average burthen: 286 *tonneaux*), which engaged in colonial trade and the slave trade.[28] Once war had made it impossible for these vessels to sail under French colours, many Bordeaux shipowners sold them to foreigners. A consistent part of these sales, however, was fake. French legislation authorised French ships to sail with two sets of papers in wartime, provided that the shipowner deposited a sum equivalent to the value of the ship with the authorities. The new set of neutral papers was then delivered by foreign consuls to a subject of their country, declaring he had purchased the vessels. Consuls were aware that such sales were suspicious, but some took advantage of it by accepting bribes in return, while others felt that they had no means to prove the abuse of their flag. In 1804, the American Consul in Bordeaux wrote to the state secretary to inform him of the difficulties he had run into when trying to protect the American flag from the suspicion of covering French property:

---

[26]   Marzagalli, *Bordeaux et les États-Unis*, p. 348.

[27]   As the composition of merchant firms evolved over time, I considered here as 'a firm' the successive denominations of a house, provided that at least one of its partners stayed in business.

[28]   Data provided by J.-C. Bats to Navigocorpus, an online database I created. This data, which comes from the Admiralty office of Bordeaux (Archives départementales de la Gironde, 6B) will be incorporated into Navigocorpus at a later stage. See http://navigocorpus.org/ and http://navigocorpus.hypotheses.org/.

I have already mentioned to you the delicate situation I am frequently placed in by refusing to grant Consular Certificates to vessels purchased here by Americans on French account: the merchants of this city really believe that they render a great service to our commerce when they find means of putting their ships under the flag of the United States. They even tell me it is my duty, and the wish of my Government, that I assist them in this particular, and when they find persuasion will not answer, they generally finish by offering me from one thousand to five thousand francs, according to the magnitude of the object.[29]

Over 100 Bordeaux ships obtained the ability to sail under a foreign flag after the war started anew in 1803 and at least half of them took the US flag. Yet these vessels continued to sail on French account.

A second option which Bordeaux merchants resorted to so as to maintain their seaborne trade was to freight neutral ships and put their own goods under the cover of a neutral merchant. It was relatively easy to obtain papers from a neutral merchant stating that he owned the cargo, but this required complete trust.[30] Although it is impossible to quantify the number of cargoes belonging to French merchants on American ships sailing to and from Bordeaux, the archival record shows a wide variety of instances in which this occurred, suggesting that this was a frequent practice. The commission rate varied between 2 per cent and 5 per cent, which was reasonable for a French merchant given that insurance rates were considerably lower for neutral cargoes and ships. It was also beneficial to the neutral partner, who undertook virtually no risk.[31]

Thirdly, part of the 'American' shipping and trade to Bordeaux was carried by Bordeaux merchants who had settled in the United States during the French Wars and had been naturalised so as to benefit from neutral status. Many of them belonged to merchant families which had been heavily involved in colonial trade before the war and some of these emigrants had fled from Saint-Domingue, where approximately 40 per cent of colonists originated from Bordeaux and the

---

[29]   National Archives and Record Administration (NARA), RG 59, Bordeaux consulate, T 164, reel 2, United States consul W. Lee to the State Secretary, Bordeaux, 29 Nov. 1804.

[30]   In 1798, for instance, the *Live Oak* sailed from Bordeaux to Boston with a cargo of wine and brandy belonging jointly to Strobel & Martini of Bordeaux, to Joseph Russell of Boston, and to Joseph Woodward, the last declaring on a private agreement 'that notwithstanding I shall appear on the documents concerning that Shipment as sole Proprietor thereof, I am in reality only one third concerned'. NARA, RG84, Bordeaux consulate, 234, no. 3, transcription of the agreement at the Consulate of Bordeaux, 12 Mar. 1800. Private agreement of 7 Aug. 1798.

[31]   In May 1793, the insurance premium on an American ship for a transatlantic journey was 5 per cent, whereas French, British, Dutch, and Spanish ships could not purchase insurance below a rate of 25 per cent. Historical Society of Pennsylvania, Reed and Forde papers, letterbook, letter to Coxe & Clarke of New Orleans, 26 May 1793.

surrounding region. Isaac Roget, for instance, a Sephardim Bordeaux merchant, settled in New York in the summer of 1793 with his brother Abraham Junior, who had moved from Port-au-Prince in the early 1790s. Isaac obtained American citizenship and regularly sent ships to Bordeaux under the American flag. He also acted as an intermediary in colonial trade to the French West Indies, moving goods from Bordeaux or Nantes.[32] Other Bordeaux merchants with colonial relations settled across the major ports of the US east coast, including François/Francis Coppinger in Philadelphia and Jean/John Carrère in Baltimore.

In fact, the main 'American' shipowners involved in the Bordeaux trade during the war were naturalised Frenchmen. Between 1804 and 1813, Pierre-Auguste Guestier, a Bordeaux merchant living in Baltimore, fitted out 10 ventures bound from Baltimore to Bordeaux, where his brother Daniel lived.[33] The two brothers owned plantations in Saint-Domingue.[34] As mentioned above, however, the Bordeaux market was largely open and the degree of concentration in the shipping business low: with 40 ventures, Stephen Jumel, one of many Bordeaux merchants who had fled Saint-Domingue in 1793, was the most important American shipowner trading to Bordeaux at this time. Yet his ventures represented a mere 7.5 per cent of all American ships fitted out in New York and sent to Bordeaux during the French Wars, at a time when the French port was the second most important European destination for New York, below Liverpool but ahead of London and Amsterdam.[35]

These examples confirm that early modern merchants used citizenship according to their interest. It is therefore highly problematic to impose upon them an exclusive 'nationality' that would determine their status of neutral or belligerent. Rather, neutrality should be conceived as a potentially advantageous status which all merchants might wish to take advantage of in times of war. It was precisely this blurred character of many business agents that made belligerent courts so suspicious toward trade carried under neutral flags.[36] In judging the validity of a prize carried in a neutral port, plurality of rights and jurisdictions complicated matters even further. The multinational character of many well-connected merchant families, the means through which merchants agreed to cover an enemy's cargoes and ships, and the ability to take on crews of

---

[32] For details of this venture, see S. Marzagalli, 'Establishing Transatlantic Trade Networks in Time of War: Bordeaux and the United States, 1793–1815', *Business History Review* 79.4 (2005), p. 824.

[33] Daniel Guestier's son founded the firm Barton & Guestier.

[34] Centre des Archives Diplomatiques, Nantes, Actes notariés, Philadelphie, 1, p. 6, power of attorney, 22 messidor IV (10 Aug. 1796) Auguste to Daniel Guestier.

[35] M.H. Luke, *The Port of New York, 1800–1810: The Foreign Trade and Business Community* (New York: New York University, 1953), Appendix (data for 1800–10).

[36] On the evolution of British jurisprudence on this issue, see J.H. Bourguignon, *Sir William Scott, Lord Stowell, Judge of the High Court of Admiralty, 1798–1828* (Cambridge: Cambridge University Press, 1987).

different nationalities blurred the dividing line between neutral and belligerent maritime trade. This made economic warfare at sea difficult to implement. In response, between 1805 and 1807, belligerents restricted neutral trade altogether so as to damage their enemy's economy, and accepted the cost of such a decision for their own economy.

The French Wars led to an unprecedented assault on neutral shipping and trade, culminating with the Milan decrees and the British Orders in Council in late 1807. Meanwhile, Napoleon had annexed an increasing number of European ports and extensive portions of European coastlines and imposed strict controls over navigation. By 1808, using neutral shipping was virtually impossible in European waters. Although the economic consequences for belligerents induced them to soften restrictions in subsequent years, neutral trade only partially recovered. For a consistent part of the French Wars, however, the war on enemy trade proved largely ineffective because of the activities of neutral ships and merchants.

The enormous capacity of neutrality in covering enemy trade which emerges from the case study of Bordeaux must not be generalised. The use merchants and captains could make of their neutrality was in fact the result of specific conditions at a given time, in a given context. Neutrality did not determine a clear-cut status defined by legal categories, implying identical consequence for any given neutral flag across the world's oceans. The use of neutrality to protect individuals, ships, and goods depended on the capacity of agents, settled both in belligerent and neutral countries, to reorganise efficiently trade networks. In doing so, they had to cope with belligerents' restrictions and their capacity to implement them, but also to face competition on specific markets and regional hostility. The success of belligerents' economic warfare depended on this complex web of relations and factors.

Sailing in Difficult Waters: US Shipping in the Mediterranean

In the Mediterranean, the United States struggled to assert their neutrality *vis-à-vis* the Barbary States in the 1790s and to obtain indemnity against French and Spanish privateers during the 'Quasi-War' in the late 1790s. American shipping bound for this region could take advantage of the French Wars only once these issues had been resolved, albeit temporarily, through peace treaties with the Barbary regencies (1795–97) and France (1800).[37] Furthermore, American merchants and captains encountered stiff competition from other neutral fleets

---

[37]   On these issues, see R.J. Allison, *The Crescent Obscured: The United States and the Muslim World, 1776–1815* (2nd ed.; Chicago: University of Chicago Press, 2000); F. Lambert, *The Barbary Wars: American Independence in the Atlantic World* (New York: Hill & Wang, 2005); A. DeConde, *The Quasi-War: The Politics and Diplomacy of the Undeclared War with France, 1797–1801* (New York: Scribner, 1966).

(Swedish, Danish, Ragusan, Ottoman) in the Mediterranean. These fleets relied on previously established merchant and credit networks and did not face war with Tripoli between 1801 and 1805.

Whereas the number of American ships entering Bordeaux was consistently higher than those of any other neutral flag, this was not the case in the Mediterranean, where Scandinavian ships dominated until 1808. Nevertheless, American vessels played an increasing part in Mediterranean shipping.[38] In 1807, American ships had paid 144 visits to Leghorn, the most important Mediterranean port for American interests. This was achieved by 124 different vessels. This was higher than the total number of North American ships sailing to the entire Mediterranean Sea prior to the American Revolution or during the early 1790s. American captains were now introducing West Indian goods – produced in belligerents' colonies – to the Mediterranean, and re-exporting Mediterranean products worldwide. They also carried on an increasingly inter-Mediterranean trade, mostly on behalf of local (belligerent) merchants: 40 per cent of American ships entering Marseilles in the first half of 1807, for instance, came from another Mediterranean port. Whereas in 1804 only one vessel in four left Leghorn for another Mediterranean port, one ship in two did so in 1807.[39] The relevance of the freight business helps to explain why, unlike at Bordeaux, no large American merchant firm settled in a major Mediterranean port.[40] American captains relied either on pre-existing British merchant networks or on American consuls, half of whom were not American subjects, to find a freight.[41]

This preliminary study points to a limited direct American involvement in the Mediterranean import–export trade. In this area, Americans might have

---

[38]   For more details, see S. Marzagalli, 'The United States and the Mediterranean during the French Wars (1793–1815), in Eloranta et al., *Small and Medium Powers in Global History*, pp. 52–72.

[39]   NARA, RG 84, Leghorn, Entrances and clearances of American ships, and RG 84, Bordeaux (*sic* for Marseilles), 225.

[40]   The main merchant in Leghorn dealing with American ships and cargoes was the former member of the British factory Webb, of the firm Webb, Holmes & Co., who managed to stay in the city during the Napoleonic era. See Michela D'Angelo, *Mercanti inglesi a Livorno, 1573–1737: alle origini di una British factory* (Messina: Istituto di Studi Storici Gaetano Salvemini, 2004), p. 229. James Holmes was the US consul in Belfast from 1796 to 1815.

[41]   S. Marzagalli, 'Le réseau consulaire des États-Unis en Méditerranée (1790–1815): logiques étatiques, logiques marchandes?' in Arnaud Bartolomei, Guillaume Calafat, Mathieu Grenet, and Jörg Ulbert (eds), *De l'utilité commerciale des consuls: l'institution consulaire et les marchands dans le monde méditerranéen (XVIIᵉ–XIXᵉ siècles)* (Rome and Madrid: Casa de Velázquez-EFR, 2017), pp. 295–307. The development of the US consular service in the Mediterranean in the 1790s and 1800s – with a quarter of American consular posts being in this region – attests the importance of the Mediterranean to the United States.

primarily provided shipping services to cover the trade of merchants belonging to belligerent countries, and connected areas under enemy influence, such as Sicily with Marseilles. This explains the vulnerability of American shipping to privateering. Ulane Bonnel has listed 134 American ships taken in the Mediterranean and in the Strait of Gibraltar and subsequently sentenced by the Conseil de Prises in Paris, a court entrusted with the judgment of neutral prizes after 1800. Bonnel's figure underestimates the total, both because the sources of this court are incomplete and because many neutral ships were condemned during the Quasi-War. The British meanwhile condemned approximately 100 American ships at the Vice-Admiralty courts of Malta and Gibraltar during the French Wars.[42] Although the number of US ships sailing in the Mediterranean at that time is unknown – ranging from a few dozen in the mid-1790s to perhaps as many as 500 in 1807[43] – they were proportionally more likely to be captured than in Bordeaux, where the British seized 64 American ships on their way to or from the French port between 1793 and 1812, while neither Spaniards nor French harassed such trade, except in a few instances (during the 'Quasi-War' and between 1808 and 1809). With more than 2,570 entries made by more than 1,660 different American ships in this period, Bordeaux rose to unprecedented significance for US trade; while the share of the Mediterranean for American shipping was just as relevant as in colonial times: in absolute figures it increased fivefold.[44] Economic wars reshaped connections across markets, but they also altered peacetime hierarchies among ports and areas. The impact of economic warfare at sea on neutrals varied depending on the nature of the trade they carried and on the degree of hostility of belligerent navies and privateers in the area.

Conclusion

These two case studies demonstrate the necessity to consider more closely the area and time frame under scrutiny when assessing the effects of economic warfare upon maritime trade. The comparison between American shipping

---

[42]   See S. Marzagalli, '"However Illegal, Extraordinary or Almost Incredible Such Conduct Might Be": Americans and Neutrality Issues in the Mediterranean during the French Wars', *International Journal of Maritime History* 28.1 (2016), pp. 118–32.

[43]   For existing data and the basis on which estimates for 1807 were calculated, see Marzagalli, 'The United States and the Mediterranean'.

[44]   See Marzagalli, *Bordeaux et les États-Unis*, pp. 472–6 and navigocorpus.org. On the relevance of the Mediterranean for the Thirteen Colonies – one-sixth of their total balance of trade in 1768–72 – see John J. McCusker, 'Worth a War? The Importance of Trade between British America and The Mediterranean', in Silvia Marzagalli, John J. McCusker, and Jim Sofka (eds), *Rough Waters: American Involvement with the Mediterranean in the Eighteenth and Nineteenth Centuries* (St John's, Newfoundland: International Maritime Economic History Association, 2010), pp. 7–24.

in Bordeaux and in the Mediterranean also demonstrates that neutral status *vis-à-vis* European belligerents encompassed different kinds of business and adjusted to different purposes for belligerent merchants. For most of the conflict American ships helped Bordeaux to sustain its trade with both the West Indies and northern Europe. In the Mediterranean, neutral ships took over regional trade routes between enemy ports. The fortunes of American shipping were specific to each area, although the late 1807–8 downturn was clear everywhere. Each neutral flag presents its own chronology and geographical specificities.

The various uses and abuses of the protection neutrality could grant to maritime activities (faked flags, faked property, broken voyages, and change of subjecthood) had traditionally enabled belligerents to carry on trade in wartime. Eighteenth-century French merchants were able to maintain some speculative business during conflicts, and to revert to peacetime colonial trade rapidly once hostilities had concluded. During the French Wars, they maintained some trade with the French colonies, although they lost the profitable re-export trade to northern Europe and the eastern Mediterranean. Neutral carriers were the chief beneficiaries. The French Wars, however, represented a turning point in the ordinary reversal from warfare to peace trade conditions. In 1815, French merchants could not counter the loss of Saint-Domingue and Mauritius Island, nor the rise of colonial production in other areas, notably in the British West Indies, Brazil, and Cuba. The success of the 1791 slave revolt in Saint-Domingue was made possible because of the revolutionary context and not because of war, although the latter might have affected the results of French military response in 1802–3. The 1794 abolition of slavery (which was restored by Napoleon in 1802) severely affected French colonial production in the remaining colonies for a decade, while Haitian independence in 1804 made the collapse irreversible. Less diversified than other empires –with Saint-Domingue providing three-quarters of total colonial imports in 1789 – the French empire was also more vulnerable. In contrast with past experience, the French did not recover their unchallenged position on the world markets for sugar and coffee after 1815. These long-lasting changes were more damaging to France than Britain's naval blockade during the French Wars.

The French Wars also prove that blockades, convoys, and battles were no longer sufficient to secure control of sea routes. The activity of neutral ships during the French Wars represented a major challenge to economic warfare policies, to which the two major belligerent powers responded differently. Both took major steps in 1807 to reduce the protection neutral carriers offered to enemy interests. While British sea power made it possible to conceive of a total subjection of maritime trade to British control, obliging all ships to take a licence or sail to a British port, France lacked the naval forces to control global sea routes. Napoleon chose to shut continental markets to British trade and products. In order to do so, he severely restricted access to neutrals in the ports under his influence, a decision which had fatal consequences in terms of political support among both his allies and his subjects. In affecting neutral rights more radically than before, the French Wars also served to shift contemporary perceptions of the law of nations and deeply affected international relations. As Thomas Paine put it in 1801, Britain was denying 'the right of commerce and the liberty of the seas'.[45] Economic warfare depended on sea power, but the use of sea power with regard to neutral shipping raised once again the old question of the liberty of the sea.

---

[45]   Müller, 'Swedish Merchant', pp. 162–3.

CHAPTER SIX

The Achievement and Cost of the British Convoy System, 1803–1815

Roger Knight

After the fleet actions of Trafalgar in 1805 and San Domingo in 1806, warfare at sea against France became an economic struggle, reinforced when Napoleon Bonaparte imposed the Continental Blockade in 1807. British strategy was framed by the blockade of the enemy ports, preventing French naval and merchant ships from putting to sea. In time, the same restrictions had to be imposed on France's allies: the Dutch, the Danes, and finally the Americans.[1] However, no blockade could prevent the escape of some enemy ships and the convoy system ensured the safety of merchant ships and transports. Progress in this sort of war is difficult to measure, marked only by the landfalls of thousands of unhindered and unremarked convoyed merchantmen, and the unloading of their cargoes.

Protecting merchant ships by convoy had been part of British naval strategy since the seventeenth century and had its origins in the medieval period.[2] A successful convoy – and most of them were – is and was a historical non-event. The convoy arrived at its destination, the escorts reprovisioned, assembled the next convoy, and returned with it to its original port. It has never been the stuff of exciting history. Convoys have thus been included in the national narrative only when a disaster occurred, but they were far from being tedious to those concerned at the time, for they were complex to organise in the age of sail when keeping merchant ships together was difficult. It was a dangerous and exacting business, carried out by skilful crews of both naval and merchant ships, operating for long periods under severe strain.

---

[1] Blockade has a large historiography. For this period, see S. Marzagalli, 'Napoleon's Continental Blockade: An Effective Substitute to Naval Weakness?' in B.A. Elleman and S.C.M. Paine (eds), *Naval Blockades and Seapower: Strategies and Counter-Strategies, 1805–2005* (London: Routledge, 2006), pp. 23–33; Brian Arthur, *How Britain Won the War of 1812: The Royal Navy's Blockades of the United States, 1812–1815* (Woodbridge: Boydell Press, 2011).

[2] D.W. Waters, 'Notes on the Convoy System of Naval Warfare Thirteen to Twentieth Centuries', Part 1: 'Convoy in the Sail Era, 1204–1874'. Typescript, Historical Admiralty, December 1957, pp. 1–4.

Blockade and convoy was a powerful combination. The British blockade of continental ports caused extensive and lasting damage to the economies of south-west France, Holland, and Hamburg, while the Danish state was bankrupted by January 1813. In 1814, the ports of the United States were similarly closed down once enough British warships could be stationed off the North American coast.[3] Throughout these years, convoys kept trade flowing upon which the British economy and credit depended. The protection of trade was thus at the heart of British strategy. This was particularly so between 1807 and 1814, when the war reached an unprecedented size and intensity – at sea as well as on land.[4] Convoys ensured the safe arrival of very-high-value cargoes to Heligoland and Malta, and elsewhere, from which the goods could be smuggled onto the continental mainland, whittling away at the effectiveness of Napoleon's Continental System. They supplied and supported naval operations in the Baltic for seven ice-free months a year. They maintained military operations in the Peninsula for 12 months of the year, and between 1812 and 1814 supplied the naval blockade and military operations against the United States.

The success of the convoy system was vital for two reasons. Virtually all strategic war materials came from outside Britain. Shipbuilding oak in Britain was almost exhausted; the royal forests contributed only an average of just under 4,000 loads a year to the dockyards in these years, not quite enough to build one second-rate ship of the line.[5] Home-produced iron could not satisfy demand and the most malleable iron (and therefore highly prized) came from Sweden. Hardwoods, mast timber, hemp and tar had to come from the Baltic, sulphur from Sicily, and saltpetre from India, all cargoes which had to be protected. At intervals, specie and munitions had to be conveyed to allies in Northern Europe and the Peninsula. Convoys also had to ensure the safety of the coastal trade, which moved bulk commodities such as coal and grain from the north, salt from Cheshire, naval stores from dockyard to dockyard or to and from the London markets. Although the Grand Union and Grand Junction canals were opened

---

[3]   James M. Witt, 'Smuggling and Blockade-Running during the Anglo-Danish War from 1807 to 1814', in K.B. Aaslestad and J. Joor (eds), *Revisiting Napoleon's Continental System: Local, Regional and European Experiences* (Basingstoke: Palgrave Macmillan, 2015), p. 164; Arthur, *How Britain Won the War of 1812*, pp. 161–203.

[4]   P. Crowhurst, *The Defence of British Trade, 1689–1815* (Folkestone: Dawson, 1977); P. Crowhurst, *The French War on Trade: Privateering, 1793–1815* (Aldershot: Scolar Press, 1989); J. Henderson, *Frigates, Sloops & Brigs* (Barnsley: Pen and Sword, 2005 [1970]); R.K. Sutcliffe, *British Expeditionary Warfare and the Defeat of Napoleon, 1793–1815* (Woodbridge: Boydell & Brewer, 2016); N.A.M. Rodger, *The Command of the Ocean: A Naval History of Britain*, vol. 2, *1649–1815* (London: Allen Lane with the National Maritime Museum, 2004), chap. 35; A.N. Ryan, 'The Defence of British Trade with the Baltic, 1808–1813', *English Historical Review* 74 (1959), pp. 443–66.

[5]   Roger Morriss, *The Royal Dockyards during the Revolutionary and Napoleonic Wars* (Leicester: Leicester University Press, 1983), pp. 79–81.

by 1806, they were not fully operational, but slowly they began to take a small proportion of commodities away from coastal routes, and danger.[6]

The second vital role for convoys was to ensure the worldwide safety and smooth-running of trade. The increasing imports of raw materials, of which cotton led the way, and very large amounts of sugar, coffee and other tropical products came from the Caribbean, a significant proportion of which was re-exported to Continental Europe through smuggling hubs such as Heligoland or Malta. The growth of luxury imports from India and China increased the size and number of East India Company convoys, which gathered at St Helena from where the East Indiamen were escorted home by a ship of the line. A burgeoning export of textiles, such as cotton goods, linen, silks, as well as ironware and other metalware, steadily grew. Coal, lead, and other raw materials continued to be distributed worldwide.[7]

Industrial development was also dependent on the trade which convoys made possible. For instance, sawn and square-hewn timber from the Baltic was required for all kinds of construction work necessary for domestic and industrial development: pine baulks for housebuilding and spruce for piling canals and docks and for shaft-timbering and pit props in quarries and mines. Later in the war the British government raised anti-French import duties on timber from Continental Europe. By 1811, the imports of squared timber from British North America totalled 154,000 loads, surpassing imports from the Baltic.[8] Timber ships in North Atlantic convoys increased commensurately. The safe movement of raw materials and manufactured goods kept the British economy healthy and financial confidence sufficiently high for the Treasury to reap enough taxes and loans to pay for the war. Some of these profits were made a long way away from Britain. In the Mediterranean, where British shipping flourished in the cross trades between ports, merchant ships rarely came back to home waters. According to a recent analysis, the Napoleonic War acted as a 'lever of growth' to British trade.[9] This activity enabled Britain's economy to survive and to prosper.

Finally, convoys served to support the British, Portuguese, and Spanish armies in the south of Europe, while they ensured that specie and munitions

---

[6]   R. Knight, *Britain against Napoleon: The Organization of Victory, 1793–1815* (London: Allen Lane, 2013), pp. 164–75. The exceptions were cattle which came on the hoof to Smithfield. Pigs could not travel overland without losing weight and were reared intensively on waste barley and malt from breweries and distilleries in London.

[7]   R. Davis, *The Industrial Revolution and British Overseas Trade* (Leicester: Leicester University Press, 1979), chaps 1–3, pp. 96–7.

[8]   C.E. Fayle, 'The Employment of British Shipping', in C. Northcote Parkinson (ed.), *The Trade Winds: A Study of British Overseas Trade during the French Wars, 1793–1815* (London, George Allen & Unwin, 1948), p. 84.

[9]   Katerina Galani, *British Shipping in the Mediterranean during the Napoleonic Wars: The Untold Story of a Successful Adaptation* (Leiden: Brill, 2017), p. 224.

reached the allied armies of Russia, Prussia, and Austria, which fought their way towards Paris in 1814, with a final reprise at Waterloo in 1815.[10]

Understanding Convoys

In recent years, British convoys in the Napoleonic War have received relatively little attention. By definition they were ephemeral, and few records are available. One exception is the correspondence generated by the cooperation between the admiralty board and the Committee of Lloyd's. The Convoy Acts of 1798 and 1803 made it compulsory for ships in most trades to sail with a convoy. While government indemnity covered enemy capture, insurance for other marine causes was the responsibility of the owner through Lloyd's. Their insurance terms were dependent upon a ship being under the protection of a warship. These differential premiums did much to enforce admiralty policy and, in David Waters' words, Lloyd's was 'in effect, the Operational Research Department of the State'.[11] Both organisations therefore needed to know the departure date of a convoy, its destination, the name of the chief escort ship and its captain, and the number of ships in the convoy. Much of this outline information survives.[12] There are few accounts from the merchant ships themselves, though the business papers of Michael Henley and Son in the National Maritime Museum are a striking exception. The paucity of merchants' records contrasts with the great quantity of their ships. The number recorded in Lloyd's Registers was 15,425 in 1805 and 15,174 in 1810. By the end of the war in 1815 this had increased, remarkably, to 17,062 ships, an indication of the health of the British economy.[13]

The traditional view of naval officers is that they did not much like convoy duty and that they treated the masters of the ships which they escorted with disdain. Sailing at close quarters in confined waters, often in heavy weather, fog, and darkness, was stressful and dangerous. Tempers on both sides could easily

---

[10]   J. Davey, *In Nelson's Wake: The Navy and the Napoleonic Wars* (New Haven, CT: Yale University Press, 2015); C.D. Hall, *British Strategy in the Napoleonic War, 1803–1815* (Manchester: Manchester University Press, 1992); C.D. Hall, *Wellington's Navy: Sea Power and the Peninsular War, 1807–1814* (London: Chatham Publishing, 2004); J. Davey, *The Transformation of British Naval Strategy: Seapower and Supply in Northern Europe, 1808–1812* (Woodbridge: Boydell Press, 2012).

[11]   Waters, 'Notes on the Convoy System', p. iv.

[12]   The National Archives, Kew (TNA), ADM 7/64 is the most complete of several volumes, made out daily by Admiralty clerks and in some haste. Statistics require careful calculation to avoid double counting. Sadly, only numbers of ships per convoy were recorded, with little evidence of total tonnage.

[13]   See www.duanaire.ie/maritime/lloydsregister. See also Sutcliffe, *British Expeditionary Warfare*, pp. 88–94 and Appendix 1 for a higher figure for the numbers and tonnage of merchant ships, taken from parliamentary and customs records.

fray. The navy was hierarchical, with a tradition of instant obedience, while the masters of merchantmen were independently minded and driven by commercial profit. While there is plenty of evidence for social condescension from naval officers towards merchant shipmasters, relationships varied widely.[14] At one end of the spectrum could be found instances of aristocratic naval arrogance, illustrated by numerous examples of complaints to the admiralty from the Committee of Lloyd's. In June 1808, for instance, Captain Sir Charles Brisbane came home with a convoy returning from the West Indies, flushed with his success in capturing the Dutch colony of Curaçao. His frigate, HMS *Arethusa*, crowded on all sail, so that 'the heavy sailing ships found it impossible to keep up ... several vessels in attempting to do so carried away their topmasts and sails and several were left behind'. The masters of 53 of the merchantmen complained to Lloyd's, stating that 'on no former occasion have they ever seen a convoying ship uniformly carrying so great a press of sail'. By the time the *Arethusa* reached the Soundings she had only 78 of 129 ships in sight.[15] Brisbane avoided public censure and was soon after appointed as Governor of the West Indies island of St Vincent.

In October 1814, the Committee forwarded another complaint to the admiralty, this time from a William Driscoll, the owner of the *Carlbury*, a West Indiaman returning with a Jamaican convoy, accusing its commodore, Captain Lord Torrington, of neglect. The merchant ship was sailing eight knots at the time when she was approached by an American privateer. Torrington was accused of ignoring the warning shot fired by the *Carlbury*, making no attempt to come to her rescue, and she was duly captured. Torrington's counter claim was that the master desired 'to be taken, and as a proof asserts the owner to be an American'. 'I assure the Committee', Driscoll added, 'that no American ever did or does hold any part of the *Carlbury* with me, and from being well known at Lloyds, I need not enter into any further explanation, the remark sufficiently falsifies itself'.[16] The war was nearing its end and Torrington did not serve again.

This sort of performance hardly helped the complex convoy system to work. Yet complaints and friction leave more evidence than when things go smoothly and for much of the time convoys were uneventful. A casual reference in a letter home by the conscientious Cuthbert Collingwood at the beginning of the French Revolutionary War provides a contrasting indicator, this time of a convoy from England to Leghorn. 'I seldom slept more than two hours at a time all the way

---

[14]   For the wide variety of types of masters in coasting and foreign trades, and friction between them, see Simon P. Ville, *English Shipowning during the Industrial Revolution: Michael Henley and Son, London Shipowners 1770–1830* (Manchester: Manchester University Press, 1987), chap. 4.

[15]   TNA, ADM 1/1544, 29 June; ADM 1/4580, 13 Aug. 1808, quoted in C. Dowling, 'The Convoy System and the West Indian Trade 1803–1815', DPhil. thesis, University of Oxford, 1965, pp. 306–7.

[16]   TNA, ADM 1/3994, 1 Oct. 1814, William Driscoll to the Committee of Lloyds.

out, and took such care of my charge that no one was missing. All the Masters came on board my ship to thank me for my care and attention to their safety'.[17]

Similarly, the logs of many convoy escorts reveal close cooperation: the surgeon on board the naval vessel would attend cases of sickness aboard merchant vessels, or the carpenter and crew might be sent to help escorted vessels with gear failure. Small warships would tow those ships with mast or sail difficulties, a particular problem with the prevalence of two-masted merchant brigs which were helpless if they lost the use of their foremast. Convoys were kept together by warships which towed vessels when they fell away to leeward, sometimes for days on end.

In spite of some tensions, the number of British convoys and their size continued to grow sharply because the government saw that it was in the country's interests to extend British naval protection to neutral ships. The size and frequency of convoys increased threefold from the last six years of the French Revolutionary War (1796–1801) to the last six years of the Napoleonic War (1809–14). After 1808, the annual average increased to 279 convoys escorting just under 10,000 ships. The total of convoyed ship voyages for this six-year period was 57,448.[18] These figures do not include the many coastal convoys that sailed around Britain throughout the year.

Ultimately, Britain's worldwide convoy system was governed by the weather. Traditional trade patterns were scheduled by harvests, such as that of the currant crop and other Mediterranean produce from the Levant to be transported westwards through the Straits of Gibraltar and then to England or on to northern Europe. The sugar crop in the West Indies was of the first financial importance. Convoys from the islands set off with the processed harvest in the early summer, giving the hurricane season between July and September as wide a berth as possible. The monsoons ordered the timing of passages from India and China across the Indian Ocean and winter ice had to be avoided in both the Baltic and Canada. In the late years of the Napoleonic Wars convoys set off from Britain to the West Indies, the Mediterranean, South America, and the East Indies every month, though sailings from December to February were few. North Sea convoys were much more frequent and those transporting military stores to the Peninsula were ordered off 'as necessary'.[19]

---

[17]   Edward Hughes (ed.), *The Private Correspondence of Admiral Lord Collingwood* (London: Navy Records Society, 1957), p. 71, Collingwood to Sir Edward Blackett, 31 Aug. 1795.

[18]   Between 1796 and 1801, 508 convoys escorted 21,500 convoyed ships. See R.W. Avery, 'The Naval Protection of Britain's Maritime Trade, 1793–1802', DPhil. thesis, University of Oxford, 1983, Appendices). Between 1808 and 1814, 1,674 convoys escorted 57,448 merchant ships (TNA, ADM 7/64, 'Schedule of Convoy Lists Sent to Lloyds', 1808–14). This was 3.4 times the number of convoys and 2.6 the number of ships for the same length of time in the French Revolutionary War.

[19]   G. Marcus, *A Naval History of England: The Age of Nelson* (Sheffield: Applebaum, 1971), p. 391.

The Baltic route saw by far the most merchant ships convoyed, accounting for between two and three times more ships than the next most numerous destination, the West Indies. Convoys across the North Sea and thence to the Baltic were satisfying the ever-increasing demand for naval stores. The large size and frequency of convoys was increased by the need for transports and victuallers to service Admiral Saumarez's fleet which operated in The Baltic from 1808 to 1812.[20] After the West Indies, in descending order, were the Mediterranean and the near Continent. A small number of large ships were convoyed to South America, including those which continued their voyage to India via the Cape after taking on provisions and water. The two busiest years were 1809 and 1813. During the former, troops and stores were required in the Peninsula to reinforce Wellington's army at the same time as the amphibious expeditions to Walcheren. This demand was surpassed in 1813, when Wellington's army moved north towards the French border while the conflict against America expanded. Operations in the new world required enormous quantities of naval stores at Halifax as well as troop reinforcements and army supplies.

The steady growth from 1807 in the demand for convoy protection coincided with the growing shortage of skilled seamen. The navy always had to compete with the crews of more highly paid merchant ships; when trade was good and seamen were scarce, a merchant seaman could receive as much as four times the pay of a naval rating.[21] By the last few years of the war, the country's overall manpower resources became a major problem and both the army and navy found recruiting increasingly difficult. Fierce fighting in the campaigns in the Peninsula led in 1809 to the deaths of over 16,000 soldiers, an annual figure which hardly decreased until 1814. With the additional demands of the American War in 1813 and 1814, the situation worsened.[22]

Part of the navy's solution to this problem was to employ smaller warships. They could be built faster and more cheaply and could sail with a smaller crew, so were less labour intensive. In addition, 32 ships were purchased between 1803 and 1815 and converted into warships. In 1803, 92 hired armed vessels were chartered, though their numbers declined slowly as they were replaced by prizes. The smaller hired ships were used for convoying in European waters as their contracts specified that they should not be used for foreign-going convoys.[23] The traditional size of warship to escort a convoy had been a frigate, of perhaps 32

---

[20]   See Davey, *Transformation of British Naval Strategy*.

[21]   Ville, *English Shipowning*, Appendices 1–4.

[22]   Together with discharges and desertions, annual army casualties reached over 24,000, an average which was maintained for the rest of the war. See K. Linch, *Britain and Wellington's Army: Recruitment, Society and Tradition, 1807–15* (Basingstoke: Palgrave Macmillan, 2011), p. 34; Knight, *Britain against Napoleon*, pp. 436–8.

[23]   Between 1803 and 1815, 518 new warships of all classes and sizes were built, measuring 323,000 tons, trebling the tonnage which had been built in the French Revolutionary War (Knight, *Britain against Napoleon*, pp. 357–9). See R. Winfield,

or 36 guns, and this size of ship was generally used in the French Revolutionary War. But, as we have seen, they could prove too fast for heavily loaded merchant ships, and, besides, frigates were needed for other important tasks. As a result, brig sloops, ship sloops, gun brigs, and gun boats were often called upon to perform well above their design specification. This weakness was exposed by winter storms, for the continuous supply of land operations demanded a year-round service. In the last years of the war there were usually about 400 small warships in commission, totalling 100,000 tons burthen.[24] This represented 60 per cent of all British warships in commission and approximately 25 per cent by tonnage.[25]

The 18-gun brigs, the larger of these small vessels, were seaworthy and effective, but smaller ships divided opinion in the Navy for many years, and the smallest, the four-gun schooners, were palpably too small. William James in the 1820s referred to their 'flimsy and diminutive frames' and called them 'tom-tit cruisers'.[26] Senior naval officers did not like them. Captain Graham Moore, a distinguished captain of a much bigger frigate, complained bitterly in 1810 about reconstituting his frigate's crew, which had been temporarily dispersed into gun boats. The men had been at sea 'until the end of December, very much exposed to the weather, not a little to the enemy's shot, and cooped up in a vessel not much bigger or better than a sentry box. It has ruined my ship's crew'.[27] The 10-gun brigs were notoriously wet and bad sea keepers and nicknamed 'bathing machines'.[28] It was amongst such vessels that the weather casualties were worst.

British Warships in the Age of Sail, 1793–1817 (London: Chatham Publishing, 2005), pp. 369–71, 391–3.

[24]   W. James, *A Naval History of Great Britain*, 6 vols (London: Richard Bentley & Son, 1878), v. 455 (Annual Abstract 1813). In 1809, there were 419 (113,246 tons) in commission, with 9 (3,067 tons) building or ordered. In 1813, 344 (94,955 tons), but in this year there were as many as 33 (10,394 tons) building or ordered, a sign that building and maintenance were not keeping pace with the number of escorts required to meet the extra demand of the American war (James, *Naval History*, iv. 480).

[25]   In 1809, there were 684 warships of all classes rated for sea service in commission, totalling 431,651 tons. Of these, 419 (61%) were sloops and smaller warships, totalling 113,246 tons (26.2%). In 1813, there were 570 of all warships in commission, measuring 404,303 tons: 344 sloops etc. (60%) measuring 94,955 tons (23.4%) were in commission. (James, *Naval History*, iv. 480).

[26]   James, *Naval History*, iv. 335.

[27]   J. Gore (ed.), *Creevey's Life and Times: A Further Selection from the Correspondence of Thomas Creevey* (London: John Murray, 1934), pp. 45–6, Moore to Creevey, 9 Jan. 1810.

[28]   Frederick Marryat, *Peter Simple* 'Author's edition' (London: George Routledge and Sons, 1896), p. 76.

The Effectiveness of Convoy

Any assessment of the effectiveness of the convoy system needs to consider not only losses by capture but also those due to weather and navigational error. The number of merchant ships lost is very difficult to assess accurately owing to the lack of available documents, and we are left with estimates. For instance, C.B. Norman, in a book of 1887, reprinted several times, calculates a total of 5,092 British merchant losses between 1804 and 1813, a rate of 471 a year between 1804 and 1806, and averaging 525 a year after 1807.[29] Mahan's 1893 analysis of merchant losses are still broadly accepted. Tonnage losses amounted to approximately 1.5 per cent to 2 per cent, which allowed Britain to trade with a comfortable margin. Subsequent historians have agreed, even if they might distance themselves from his exact figures.[30] Waters concludes that 645 ships were lost annually, or 54 ships a month.[31] Yet even his figures are probably an underestimate, concerned as they are, in the main, only with losses to French privateers, and take no account of captures after 1807 by Danish, Norwegian, and Dutch privateers. In the Anglo-Danish war, 556 Danish ships were commissioned as privateers; hostilities in the approaches to the Baltic were bitter and protracted.[32] In the two and a half years of the American war, 228 American privateers, armed with 906 guns and crewed by 8,974 men and boys, threatened British trade. The power and speed of American privateers and the skill of their crews were much feared by the British navy and the masters of merchant ships in their charge.[33]

Privateers were not the only danger to British shipping. In 1810, Napoleon seized a large number of British and neutral ships in Swedish, Prussian, and other ports, causing Lloyd's great financial loss.[34] Another hazard was posed by hostile warships, particularly when they operated as a squadron. Encountering windless conditions off an enemy coastline could leave a convoy vulnerable. In June 1808,

---

[29]   C.B. Norman, *The Corsairs of France* (London: Sampson Low, 1887), Appendices XXII, XIX. For the same years, Norman estimates that 282 French privateers were captured by British ships. Lack of references weaken the authority of his figures. See Crowhurst, *French War on Trade*, chap. 3.

[30]   A.T. Mahan, *The Influence of Sea Power upon the French Revolution and Empire, 1793–1812*, 2 vols (London: Sampson Low, Marston and Co., 1893), ii. 223–9; Crowhurst, *French War on Trade*, p. 31; Rodger, *Command of the Ocean*, pp. 559–60.

[31]   Waters, 'Notes on the Convoy System', p. 32.

[32]   Witt, 'Smuggling and Blockade-Running', p. 157.

[33]   A.T. Mahan, *Sea Power in its Relation to the War of 1812*, 2 vols (London: Sampson Low, Marston, 1905), vol. 1, chap. 14; Arthur, *How Britain Won the War of 1812*, Appendix A, table 1 (pp. 198–9) gives a comprehensive list of American merchant ships captured by British warships and addresses the complication of recaptured privateer prizes.

[34]   C. Wright and C.E. Fayle, *A History of Lloyd's* (London: Macmillan, 1928), p. 262; Davey, *In Nelson's Wake*, pp. 249–50.

a convoy of 70 merchant ships off Norway was attacked by 25 oared gunboats, which carried off 25 of the merchantmen. Baltic convoys navigating the Sound or the Great and Little Belt were frequently attacked by these small vessels, although Admirals Saumarez and Keats were largely successful in countering them by carefully positioning ships of the line and smaller vessels in the Great Belt to cover the convoys. By far the worst loss in the Baltic war occurred off the Skaw, the northerly point of Jutland, on 19 July 1810. A convoy of 47 merchant ships, protected only by a small brig, encountered five Danish brigs. They chased off the escort and captured the entire convoy.[35] In 1813 and 1814, small, powerfully manned ships of the American navy cruised against merchant ships sailing without convoy in home waters. They took their prizes to French ports, damaging British home trades and causing a furore amongst British merchants, who petitioned Parliament to stop these casualties. To provide an estimate of total losses over the last seven years of the war, it is necessary to calculate the combined depredations of hostile warships and privateers of four nations over thousands of square miles, together with the casualties of ships and seamen from weather and navigational error. This figure should then finally be offset against the financial gains from prizes and prize goods captured by British warships.[36]

None of these casualty estimates takes account of losses caused by the weather, though in the case of warships courts martial records give very accurate statistics.[37] They were proportionally much higher than merchant ships. It is difficult to be sure of what sixth rates, sloops, and brigs were doing at the point of capture or destruction, for their duties frequently changed between convoy, blockade, and cruising against privateers. Blockading an enemy port in all weathers was the most dangerous task, but convoying was never easy, especially in the North Atlantic and American waters between 1812 and 1814. For this station alone, captures and hostile weather combined to produce 28 losses, nearly ten warships a year. Though most were lost to weather, there were fierce contests with American small warships. In June 1814, for instance, HMS *Reindeer* – a brig sloop of 18 guns – was cruising in the Western Approaches, and fell in with the 22-gun American warship *Wasp*. Out of the *Reindeer*'s complement of 118 men, 25 were killed and 42 wounded before she surrendered. She was in such a bad state that the Americans sank their prize.[38]

---

[35]   Tim Voelcker, *Admiral Saumarez versus Napoleon: The Baltic, 1807–12* (Woodbridge: Boydell Press, 2008), pp. 78, 87; Davey, *In Nelson's Wake*, pp. 243, 247; Ryan, 'The Defence of British Trade', p. 453.

[36]   Mahan, *Influence of Sea Power*, ii. 226; R. Hill, *The Prizes of War: The Naval Prize System in the Napoleonic Wars, 1793–1815* (Stroud: Alan Sutton, 1998), p. 246 estimates that ship captures totalled £30 million (1793–1815).

[37]   Naval loss statistics are taken from a comprehensive compilation of naval courts martial by D.J. Hepper, *British Warship Losses in the Age of Sail, 1650–1859* (Rotherfield: Jean Boudriot Publications, 1994).

[38]   W. James, *Naval Occurrences of the War of 1812* (London: Conway Maritime Press,

Very often problems were encountered by excessive delay in assembling convoys before sailing, leading them into deadly weather, particularly from the Baltic. Usually this was due to difficulties in loading cargoes; merchants often considered that the commercial gain was worth the risk of late sailing and, above all, to avoid the considerable expense of retaining a ship and crew in the Baltic ice for the winter. In late 1808, a convoy of 13 merchantmen, which did not leave Karlskrona until 22 December, was lost in the ice with its escorting brigs. Admiral Saumarez set his face against such late sailing in future years.[39] However, there was little the admiral could do in November 1811 when a homeward Baltic convoy of 72 ships sheltered from a gale at the mouth of the Great Belt. Thirty were driven on shore, and the escorting 98-gun *St George*, returning to Britain after a summer in the Baltic, was damaged by a merchant ship. She was driven aground and damaged her masts, though she reached Wingo Sound near Gothenburg. Jury rigged, she set out again but hit another storm, and three ships of the line – *St George*, *Defence*, and *Cressy* – were wrecked on the Jutland shore. The 74-gun *Hero* was also lost at the same time off the coast of Holland. Two thousand naval officers and seamen perished. Lack of navigational expertise and local knowledge of the waters were the reasons for this disaster. The merchant ships fared much better: 'the masters being well acquainted with the strong currents that set in on the coasts of Jutland and Holland, and by steering well off, they saved themselves from the destruction that befell the unfortunate *Hero*', as the *Hull Packet* newspaper reported.[40]

Edward Pelham Brenton was a captain at this time, and his experience of North Sea and Channel pilots led him to have scathing opinions on them, which he included in his *Life of St Vincent* some 25 years later. He also listed 15 ships of the line shipwrecked during the French wars and estimated in all that 4,000 officers and seamen had drowned because of them during the war: 'The fault is with our own Government, which never yet gave encouragement to young officers to become pilots. We should have a corps of that class in the navy ... They should be good surveyors, and eligible to the highest offices in the service'.[41] The same charge could be levelled at the general inability to calculate longitude, thirty years after Cook's voyages, due to the failure to provide expensive chronometers

---

2004 [1817]), pp. 177–80. Privateers generally did not attack warship escorts, for they were too heavily armed.

[39]   Ryan, 'The Defence of British Trade', pp. 450, 453. The warships were the *Fama*, 18 guns, 315 tons, which went onshore on Bornholm, and the *Salorman* cutter, 10 guns, 121 tons, wrecked near Ystad: loss of life was minimal in both shipwrecks, the crews being helped ashore by local people (Hepper, *Warship Losses*, p. 127).

[40]   *Hull Packet*, 28 Jan. 1812.

[41]   Edward Pelham Brenton, *Life and Correspondence of John, Earl of St Vincent*, 2 vols (London: Henry Colburn, 1838), ii. 335–7. His list of ships, 'from memory', is not accurate, though the logic of his argument was very sound.

and the training to use them.[42] These instruments were little known on smaller warships. An exception was William Henry Dillon, who escorted a convoy in 1814 to Newfoundland as captain of the frigate *Horatio* and who had a privately purchased chronometer and a barometer. He correctly predicted the appearance of the coast and astonished his officers. 'Consequently all confidence was placed on it', he wrote, 'and those who had made their remarks upon the timepiece and barometer were silenced. The officers then told me of the uneasiness many of them felt at being subjected to the "forebodings of the Chronometer and Barometer".[43]

Severe winter conditions could not be avoided. The importance and urgency of supplying Wellington's campaign in the Peninsula did not allow convoys, escorted by small warships, to escape the predictably wild winter weather in the Bay of Biscay. South and westerly gales meant that not all of them got to the Peninsula. Risky winter convoys resulted in much higher insurance rates in the autumn and particularly in January, February, and March of any year. At Lloyd's the rates were a guinea and a half per cent in summer, but in winter they increased to 12 guineas. Many underwriters were so wary of insuring ships in winter that they withdrew from Lloyd's in winter months and, as a result, merchants and shipowners found it difficult to insure their vessels and cargoes.[44]

The loss of small warships on West Indies convoys was also consistently high. On one occasion in 1809 two convoys ran into each other. Commander Benjamin Clement, the captain of HMS *Favourite*, an 18-gun ship sloop of 427 tons, left Port Royal on 17 July with 32 ships, accompanied by the *Pike* schooner. A month later, in mid-Atlantic, they were 'in fine weather and well collected'. But then:

> It came on to blow hard and increased with such violence that with everything furled by two o'clock in the morning our Topmasts were blown away and it increased to a perfect Hurricane ... At Daylight a

---

[42]   The radical improvements from 1808 overseen by Thomas Hurd as Hydrographer of the Navy in the updating and distribution of charts, sailing directions, and Notices to Mariners were a major contributory factor in the successes of British trade protection. Manufacturing chronometers and training officers in their use in a rapidly expanding navy was beyond the Hydrographic Office's resources. Valuable merchantmen, however, carried them. See A. Webb, 'The Expansion of British Naval Hydrographic Administration, 1808–1829', PhD thesis, University of Exeter, 2010, pp. 275–84.

[43]   Sir William Henry Dillon, *A Narrative of my Professional Adventures (1790–1839)*, ed. Michael A. Lewis, 2 vols (London: Navy Records Society, 1956), ii. 295.

[44]   This situation was examined by a Parliamentary Select Committee on Marine Insurance in 1810, but it brought no change to the system. See Wright and Fayle, *History of Lloyd's*, chap. 11. It should be remembered that those merchants smuggling very high value goods into Europe through Heligoland and Malta did so without any insurance at all. See J. Ruger, *Heligoland: Britain, Germany and the Struggle for the North Sea* (Oxford: Oxford University Press, 2017), pp. 25–30.

most shocking s[cen]e presented itself, nothing but dismasted Ships in every direction. On counting the Convoy found from the numbers that we had got a Strange Fleet among us. On its moderating towards noon we exchanged numbers with *H.M.S. Captain* and found by Telegraph that we had at so unfortunate a time met the Tortola Fleet. I much fear as the Fleets were on different tacks at daylight that some of the Ships have been run down and many foundered.[45]

Clement could only see 16 of his 32 ships and there was no sign of the *Pike* schooner. He put himself under the command of the captain of the *Captain* (74), conforming to naval instructions, and towed a merchant ship in continual bad weather for several days until a jury rig was completed for her. He then picked up more stray ships and made his way to the Downs. The *Captain* took the rest of the convoy through the St George's Channel. In his postscript to his letter to the admiralty, Clement almost casually mentioned: 'I have on board between six and seven hundred thousand Dollars Government and above one hundred Thousand belonging to individuals'.[46]

The total losses of small British warships between 1803 and 1815 stood at 298.[47] Some 189 were wrecked or foundered, mostly in winter; 48 foundered in unknown circumstances. Only in the case of two such sinkings were any of the crew rescued, so little is known of the causes of these disasters. The remaining 90 ships were captured or sunk, while six were expended as fire ships. Winter weather and hurricanes were responsible for the loss of far more ships, cargoes, and lives than those captured by enemy warships or privateers. In all, 5,000 officers and seamen were killed or drowned, of which 3,000 disappeared in ships which foundered. From the evidence of courts martial, we can add another 5,000 taken prisoner, either by privateers or through their ships being wrecked on a hostile coast. Total small warship manpower losses were likely to have exceeded 10,000 skilled seamen, averaging about 880 per year over the 11 full years between 1804 and 1814.[48] This figure does not include merchant seamen

---

45   TNA, ADM 1/1654, Commander Benjamin Clement to the Secretary of the Admiralty, 18 Sept. 1809.

46   TNA, ADM 1/1654, Commander Benjamin Clement to the Secretary of the Admiralty, 18 Sept. 1809.

47   This represented an annual average of 25.8, though there was a high level of annual variation. The lightest casualties occurred in 1810 with only ten, the worst was 1807 when 36 vessels were lost.

48   Hepper, *Warship Losses*, pp. 101–52. The figures are calculated from the complements which were fairly standard, set by the Navy Board on tonnage. To compensate for those ships under complement, 10 per cent has been subtracted, but there was far less room for undermanning than in a larger warship. At the start of the war, at least, most of the seamen in small warships were able seamen rather than ordinary seamen or landsmen. See N.A.M. Rodger, *The Wooden World: An Anatomy of the Georgian Navy* (London:

casualties, which remain unknown. By contrast, those aboard ships of the line killed in all the great fleet battles of the French Revolutionary and Napoleonic Wars total 1,364: less than a third of those who lost their lives in small warships.[49]

Little of the Napoleonic war experience was known or remembered in the nineteenth century, when British control of the seas was never challenged, and convoys were hardly employed. When Sir John Knox Laughton wrote an article, 'On Convoy', in 1894, he admitted his ignorance: 'our naval histories tell us little, except when from time to time they have to chronicle some great disaster'. And, of the present day, he concluded that 'a system of commanding certain appointed stations in force ... will so greatly reduce the opportunities of an enemy's cruisers, that the need for escort or convoy will but seldom occur, and then in a modified degree'. This statement reflected the majority opinion of late Victorian naval strategists.[50]

Conclusions

The margin of victory in this long struggle with Napoleonic France and her allies was much closer than historians have hitherto portrayed. The long view of the percentages of losses over many years glosses over critical periods. The two most dangerous were the convoy war against Denmark between 1809 and 1810 and convoying in the North Atlantic in 1812 and 1813. During the former period, Britain found it very difficult to get its ships into the Baltic. Yet the episode ended in the bankruptcy of Denmark because of the blockade. Pressure in the North Atlantic, meanwhile, was relieved only by the first abdication of Napoleon in April 1814. In that year, the admiralty was able to move 15 per cent of its warships westwards from the Mediterranean, Baltic, and Home Waters to the American theatre. The US government was very near to bankruptcy by the time of the Treaty of Ghent – again the result of blockade.[51]

Collins, 1986), Appendix 1 (p. 351) and Winfield, *Warships in the Ages of Sail*. Estimated figures, less 10 per cent, are 5,115 killed, 5,582 taken prisoner. Of these losses from foundering (again less 10 per cent), are 2,979, 65 per cent of all officers and seamen killed or drowned.

[49]   The battle casualties for the *Glorious First of June* (290), *Cape St Vincent* (73), *Nile* (218), *Copenhagen* (254), *Trafalgar* (449), *San Domingo* (74), and *Aix Roads* (6): total 1,364 (James, *Naval History*, i. 435; iv. 101, 414; Roger Knight, *The Pursuit of Victory: The Life and Achievement of Horatio Nelson* (London: Allen Lane, 2005), pp. 250, 297, 381).

[50]   Sir John Knox Laughton, 'On Convoy', in T.A. Brassey (ed.), *The Naval Annual 1894* (Portsmouth: Brasseys, 1894), pp. 225, 241; Waters, 'Notes on Convoy', p. 14.

[51]   Kevin D. McCranie, 'The War of 1812 in the Ongoing Napoleonic Wars: The Response of Britain's Royal Navy', *Journal of Military History* 76 (2012), p. 1085; Arthur, *How Britain Won the War of 1812*, pp. 199–203.

Britain won the convoy battle for two main reasons. A strong alliance in London between the admiralty and Lloyd's enabled senior officers and administrators to reconcile strategic and commercial pressures. Disasters were few but not wholly avoidable. In 1804, the *Apollo* frigate went ashore on the coast of Portugal, wrecking 27 West Indiamen in the process. In the same year, the East India trade and Lloyd's were fortunate to have escaped disastrous losses when Nathaniel Dance hoodwinked Admiral Linois off the coast of Malaya into believing that his convoy of East Indiamen were ships of the line, and the French squadron withdrew. In 1810, as we have seen, 47 merchant ships were lost to the Danes through inadequate protection. But these disasters were not sufficient to shake confidence and the admiralty and Lloyd's maintained a close relationship. In 1814, the Chairman of Lloyds wrote in a letter to the admiralty: 'Effectual protection can only be given to British commerce by a rigid adherence to the convoy system'.[52] The other major factor was the skill and seamanship of both the convoy commodores and the merchant masters. Convoy commodores were specialised officers familiar with the headlands, sandbanks, currents, tides, and safe anchorages of a particular convoy route, and did this for year after year. Their pilotage experience largely overcame the problems of navigation, so reducing the number of disasters at sea, reinforcing the bond between the admiralty and Lloyd's.

By taking only a partial account of the evidence, Mahan and other historians have mistakenly portrayed Britain as enjoying a comfortable margin in its trade protection role. With the acute shortage of seamen, and the consequent lack of ships, this is by no means true. Warships had been built and were ready but could not be manned. The admiralty was well aware of the scale of the losses of both ships and men in merchant and naval ships. The loss of skilled naval and merchant manpower had a significant impact on the conduct of the war. Warship commanders became desperate for men, taking them from their own privateers, which, according to Mahan, 'feared a British ship of war more than it did an enemy of equal force'.[53] This caused real anxiety in the admiralty, which feared that worldwide operations might have to be limited. In early 1813, Lord Melville, the First Lord, admitted in a letter to Wellington that none of his commanders-in-chief had enough frigates and smaller vessels: 'It may be a question at the close of the year for the government, and not merely the admiralty, to consider whether we must not endeavour to limit the scale and extent of naval operations'.[54]

Yet warships were not withdrawn, and the convoys were not broken. British merchants achieved ascendancy while the war was being fought, and not just in

---

[52]   Rodger, *Command of the Ocean*, p. 559, quoting Dowling, 'Convoy System', p. 1, from TNA: ADM 1/3994, 19 Sept. 1814.

[53]   Mahan, *War of 1812*, p. 232.

[54]   McCranie, 'The War of 1812', p. 1094, quoting A.W. Wellington, *Supplementary Despatches, Correspondence and Memoranda of Field Marshall Arthur Duke of Wellington, 1750–1850*, 15 vols (London: Murray, 1858–72), viii. 144–7.

trade to and from Britain. They dominated, for instance, the cross trades in the Mediterranean. While the armies of Europe were settling the political balance of the continent, Royal Navy escorts and the merchant ships which they protected laid the foundations of British global trade dominance. This war was won by a category of naval officers and seamen unknown today. Their careers at sea were truncated by the end of the war. Some took prizes, but these ships were not rich merchantmen, rather privateers and gunboats, so fortunes were not made. No sea battles are celebrated because there were none. What conflict there was, if recorded, came to be described, as noted by the historian A.N. Ryan, as 'minor operations'. As he concluded, 'a more inappropriate term for a struggle upon which depended the control of vital trade routes would be difficult to find'.[55]

---

[55]   Ryan, 'Defence of British Trade', p. 466.

CHAPTER SEVEN

The Royal Navy and Economic Warfare against the United States during the War of 1812

John B. Hattendorf

Stimulated by the beginning of the bicentenary events to mark the War of 1812 eight years ago in 2012, historians in Britain, Canada, and the United States began to re-examine that conflict.[1] Over the period of just a few years, academic historians stripped away many, if not all, of the different nationalistic and patriotic interpretations, along with their accompanying myths and legends. At the same time, the complementary scholarship that accompanied the bicentenaries of the wars of the French Revolution and Empire added depth and context to our understanding of the simultaneously fought War of 1812. This wealth of new original research created, for the first time, the basis for an international scholarly consensus about the War of 1812.[2] A significant aspect of this re-evaluation centred on a reconsideration of the British naval blockade of the

---

[1] For this author's contribution to and new historical reflections on the very beginning of the bicentenary events, see J.B. Hattendorf, 'The Third Allen Villiers Memorial Lecture: The War of 1812 in International Perspective', *Mariner's Mirror* 99.1 (2013), pp. 5–22.

[2] The principal works that created the new international consensus on the naval-economic aspects were: Brian Arthur, *How Britain Won the War of 1812: The Royal Navy's Blockades of the United States, 1812–1815* (Woodbridge: Boydell Press, 2011); Brian Arthur, 'Sir John Borlase Warren and the Royal Navy's Blockade of the United States in the War of 1812', in Brian Vale (ed.), *The Naval Miscellany*, vol. 8 (London: Routledge for the Navy Records Society, 2017), pp. 205–46; T. Bickham, *The Weight of Vengeance: The United States, the British Empire, and the War of 1812* (New York: Oxford University Press, 2012); Faye M. Kert, *Prize and Prejudice: Privateering and Naval Prize in Atlantic Canada in the War of 1812*, Research in Maritime History 11 (St John's, Newfoundland: International Maritime Economic History Association, 1997; Faye M. Kert, *Trimming Yankee Sails: Pirates and Privateers of New Brunswick* (Fredericton, NB: Goose Lane Editions, 2015); Faye M. Kert, *Privateering: Patriots and Profits in the War of 1812* (Baltimore, MD: Johns Hopkins Universty Press, 2015); Roger Knight, *Britain against Napoleon: The Organization of Victory, 1793–1815* (London: Allen Lane, 2013); Andrew Lambert, *The Challenge: Britain against America in the Naval War of 1812* (London: Faber and Faber, 2012); and N.A.M. Rodger, *The Command of the Ocean:*

United States, and of British privateers as weapons of economic warfare in what had previously been dismissed as a 'strategical sideshow'.[3]

This new body of historiography has potentially wide-ranging implications for our broader understanding of maritime strategic theory, important elements of which were based on the wars fought between 1793 and 1815. In particular, the late nineteenth-century American naval historian Alfred Thayer Mahan used the experience of this period to argue that commercial blockade was a very effective weapon for stronger maritime belligerents. When faced with an effective blockade of commercial ports, Mahan argued, the only practicable response available to a weaker nation was commerce-raiding on the high seas – an intrinsically less-effective measure.[4] Furthermore, Mahan claimed that as a measure for the destruction of an enemy's commerce, privateering activity was designed more for the personal gain of the privateers than for injuring the enemy.[5] This chapter will explore the relationship between the new historiography on the War of 1812 and its relationship to maritime strategic theory, questioning Mahanian orthodoxy regarding the efficacy of maritime economic warfare.[6]

The Background to War

The causes of the War of 1812 stretched back to the Peace of Paris in 1783 and the shift it created from the American colonies being an integral part of British commerce and maritime trade to becoming an independent and potentially rival trading power. Both sides shared responsibility for the troubled relations which followed. In British politics, a vocal group resented the terms upon which the Shelburne and Rockingham governments had granted American independence. They believed that the boundaries and terms of American independence had been far too generous. This group would have been happy to have seen the new American republic fail. On other the side, Americans showed little remorse. Their best hope for improving relations between the two countries lay in expanding trade relations,[7] but the fundamental underlying challenge was to create mutual

---

A Naval History of Britain, vol. 2, *1649–1815* (London: Allen Lane with the National Maritime Museum, 2004).

[3]   Reflecting widespread views, the quote is from P.M. Kennedy, *The Rise and Fall of the Great Powers: Economic Change and Military Conflict from 1500 to 2000* (New York: Vintage, 1986), p. 137.

[4]   A.T. Mahan, *Sea Power in its Relation to the War of 1812*, 2 vols (Boston: Little Brown, 1893), i. 288.

[5]   Mahan, *Sea Power in its Relation to the War of 1812*, ii. 241.

[6]   The two sections that follow are slightly revised versions of similar parts of Hattendorf, 'The Third Allen Villiers Memorial Lecture', pp. 9–13.

[7]   See Charles R. Ritcheson, *Aftermath of Revolution: British Policy toward the United States* (New York: Norton, 1971).

respect between the two nations and to appreciate the need for stability in their bilateral relations.[8]

For nearly a decade from the time of the collapse of Peace of Amiens in 1803 and the resumption of war between Britain and France, the neutral United States found itself in an increasingly challenging position between two warring powers. The immediate objective of Americans was to continue their profitable neutral maritime trade with both belligerents. Both the French and the British soon realised that this trade, while neutral, was nevertheless to the advantage of the other side. Each suspected the Americans of duplicity and favouring their enemy, despite the protests of American merchants that they were merely looking for commercial gain. The British suspected that Americans were supporters of France, recalling French military and naval support for Americans between 1778 and 1781 as well as their republican ideals. For the British, the continued American support for Bonaparte, the arch-dictator and destroyer of liberty, was going too far.[9] For some in Britain, the new American republic had proved to be both erratic and irritating. Incessant American protests about impressment and neutral rights were annoying and seemed entirely out of place in the critical situation at hand as Britain struggled to survive, often alone, against overwhelming French military power. Not only that, Americans seemed overly ambitious, greedy, and avaricious, wanting to expand their territory at the expense of Britain and Spain as well as increase American trade and profit.

The Economic Conflict Preceding Naval Warfare

In the years before the declaration of war that brought state navies into the application of economic warfare in 1812, a form of economic warfare had already arisen between Britain and the United States. Between 1790 and 1807, merchant shipping was a growing factor in the US economy. The American merchant fleet had grown from a total of 478,000 tons in 1790 to 1,269,000 in 1807. Over the same period, the percentage of American shipping engaged in foreign trade fluctuated from 72 per cent and declined slightly to 66 per cent as tonnage increased substantially. The value of American exports also rose from $US20.2 million to $US108.3 million. Of this trade, Britain and the British Empire accounted for 43 per cent of the total. During these years the expansion of the American economy was primarily in the growth of its maritime commerce. Domestic production did not match maritime growth, rising only from $US19.9 million to $US35.8 million.[10]

---

[8]   See Bradford Perkins, *The First Rapprochement: England and the United States, 1793–1805* (Berkeley: University of California Press, 1967).

[9]   Bickham, *Weight of Vengeance*, pp. 49–75.

[10]   Lance E. Davis and Stanley L. Engerman, *Naval Blockades in Peace and War: An*

Napoleon's Berlin Decree in November 1806 established the Continental System that banned all trade with Britain and all British goods from entering French-held territories in Europe. Napoleon intended this action to institute economic warfare, but the Continental System was initially noted for the evasion of its provisions. On 11 November 1807, Britain issued a new Order in Council regarding the blockade of France in its series of pronouncements on the subject that stretched back to 1793. Under this decree, Britain banned all French trade with Britain, with British allies, and all French trade with neutrals, and the Royal Navy began a commercial blockade against France. It also began to inspect merchant ships in port and at sea for possible contraband shipments that could aid France. In this process, it seized vessels that failed to submit to inspection. In response to this Order in Council, the emperor issued his Milan Decree in December 1807 to enforce and to extend his earlier measures. Under the new regulation all European nations were forbidden from trading with Britain. He decreed additionally that all neutrals were forbidden to trade with Britain and that any vessel sailing from a British occupied area was liable to seizure, as was any neutral ship that submitted to British search. The neutral United States found itself caught in the vice of economic warfare between France and Britain. The opposing decrees and orders of France and Britain were not intended to starve their enemies and had numerous loopholes for evasion. They were, as one historian has described them, 'exaggerated applications of traditional mercantilist principles designed to wreck each belligerent's commerce and to drain each other's specie'.[11] Nevertheless, both France and Britain had violated the rights of neutrals in their war against each other and Americans thought one was no better than the other. The United States wanted to have no part in their conflict. However, the British Order in Council of 1807 came to symbolise for Americans all that had been wrong with Britain's relationship with the new republic since its independence.[12] In response to this situation, the United States did much more economic damage to itself than either the French or British had done.

At this point, neither Britain nor the United States wanted to go to war with one another. Initially, the United States had reacted with Congress passing in 1806 (but then suspending for nine months) the Non-Importation Act. This act prohibited only the importation from Britain of goods readily made in the United States, but it was designed to signify a potentially stronger American response and to back the American diplomatic effort to stop both impressment and violations of neutral rights.[13] These efforts on the part of the Americans

---

Economic History since 1750 (Cambridge: Cambridge University Press, 2006), tables 3.4–3.8 (pp. 78–83).

[11]   Gordon S. Wood, *Empire of Liberty: A History of the Early Republic, 1789–1815* (Oxford: Oxford University Press, 2009), p. 646.

[12]   Bickham, *Weight of Vengeance*, pp. 26–7.

[13]   Wood, *Empire of Liberty*, pp. 644–6.

had little, if any, effect on British policy. For American leaders this was a frustrating situation. In 1806, some Americans began to worry that the other nations in the world were misinterpreting the American desire to avoid the horrors of war and to maintain peace with others. President Thomas Jefferson noted that the impression that 'our government is entirely in Quaker principles, and will turn the left cheek when the right has been smitten must be corrected when just occasion arises, or we shall become the plunder of all nations'.[14]

With such thoughts in mind, Americans became more and more suspicious of British intentions. British encouragement for Native Americans in the Northwest Territory seemed to pose threats to the United States as did the increasing British impressment campaign. On top of these impressions, Americans took Admiral Lord Gambier's highly successful attack on Denmark in September 1807 and the subsequent seizure of Denmark's neutral navy as a threat to all neutrals. Hesitating to enter directly into a war at this point, President Jefferson and others wanted to make some 'New World' republican rebuke to the monarchs of the 'Old World'. At this point, they turned again to economic warfare as an answer short of outright military and naval action. By late 1807, the purposefully delayed provisions of the 1806 Non-Importation Act were in place. On 18 December 1807, President Jefferson announced a new policy of expanded economic retaliation. In the most significant example in American history of ideology driving public policy, Congress passed, on 22 December 1807, on the Jefferson administration's recommendation, the Embargo Act. This act prohibited all American flag merchant ships from departing on international trading voyages. While it did not prevent British vessels from bringing goods to the United States, it prohibited them from loading American product for export. It is unclear as to what degree the Jefferson administration had carefully thought through the economic implications of this act for the country, but their dominating purpose behind it was ideological.[15] Jefferson declared:

> the ocean presents a field only where no harvest is to be reaped, but that of danger, spoliation and of disgrace.
>
> Under such circumstances the best to be done is what has been done; a dignified retirement within ourselves; a watchful preservation of our resources; and a demonstration to the world that we possess a virtue and a patriotism that can take any shape that will best suit the occasion.[16]

---

[14]   Quoted in Wood, *Empire of Liberty*, p. 649.

[15]   Wood, *Empire of Liberty*, pp. 650–4.

[16]   Quoted in Dumas Malone, *Jefferson the President, Second Term, 1805–1809* (Boston: Little Brown, 1974), p. 488.

Going further, Jefferson explained:

> It is singularly fortunate that an embargo, whilst it guards our essential resources, will have the collateral effect of making it to the interest of all nations to change the system which has driven our commerce from the ocean.[17]

For American leaders in power at the time, the Embargo seems to be a most useful tool to bring forward the ideals of the Enlightenment to reform the character of international relations.[18] While the Embargo had little effect in gaining that goal, it proved disastrous for America's maritime economy.

Jefferson's experiment in using the Embargo as a means of peaceful coercion to remove British and French restrictions on neutral trade paralleled his exaggerated estimate of America's influence in the world. He even imagined that he could use it to acquire East and West Florida from Spain. On the home front, the Embargo raised support for the opposition Federalist Party, particularly among New Englanders, whose maritime enterprises the Embargo had ruined. The merchant fleet in Massachusetts alone accounted for 40 per cent of the tonnage sailing under the American flag. The first year of the Embargo saw merchants there lose $US15 million in freight revenue, a sum equal to the Federal Government's entire income in 1808. By the end of 1808, it was clear that the Embargo was failing to produce the results that the Jefferson administration desired. Nevertheless, in January 1809, Congress passed an act to try to enforce its provisions further, but political forces ensured the end of the Embargo on 4 March 1809, the day that James Madison succeeded his fellow Republican, Thomas Jefferson, as President. Nevertheless, before leaving office, Jefferson replaced the Embargo with the Non-Intercourse Act of 1 March 1809. This act lifted the Embargo on all countries except for Britain and France and places under their control. After a brief and relatively small lift in exports and imports, the new measure with the new Republican administration and the Republican-controlled Congress continued policies that resulted in stultifying the American economy.[19]

The Outbreak of War

From the American point of view, there was neither a single stated cause nor catalyst that moved President James Madison to suggest war to Congress in June 1812. Madison's official war message involved 12 pages of meandering complaints that even he admitted was not a complete list. Britain's interference

---

[17]   Malone, *Jefferson the President, Second Term*, p. 488.

[18]   Wood, *Empire of Liberty*, p. 652.

[19]   Wood, *Empire of Liberty*, pp. 649–58.

with American neutral trade through its Orders in Council and its continuing policy of stopping and searching for British subjects to impress into British naval service from American vessels topped the list of Madison's complaints, followed by the charge that British agents were stirring up Native American tribes against the United States. But these and the others were all effects of a more profound cause that Americans felt. The real concern was to obtain respect for America's independent sovereignty and honour as a nation. It was an ideological and practical protest against an attitude that Americans did not like.[20]

On the domestic side, 1812 was also a presidential election year and Madison was manoeuvring to retain his Republican party's position in power. The declaration of war was made on 18 June 1812, with the presidential election in November. Madison was not a strong leader, and he certainly preferred peaceful methods to obtain his ends. During the first three years of his administration he had attempted to use peaceful methods, and he had been the guiding light behind the restrictive trade system that had been in place between 1806 and 1811.[21] By 1812, it was clear that American trade restrictions needed additional force behind them if they were to bring the results that Madison and the Republicans wanted.

While Madison's formal declaration of war stated that the central issues were maritime, this does not seem to have been the case. In Nicole Eustace's research on the cultural history of the war among the general American population, she shows that the Republican's pro-war rhetoric came from an entirely different context than the ones that military, naval, and maritime professionals – and historians of these topics – typically think. This interpretation helps to explain why this war has been so baffling and difficult to interpret. The three main issues – impressment, trade, and Indians – were cast regarding internal and domestic development. Impressment became a prevalent issue in American ideology: Britain attempted to rupture American families by taking away American men, put them in a kind of slavery on board British warships, so destroying the rights of American citizens.[22] A major issue of the day in America was a critique of Thomas Malthus's work on population. Supporters of the Madison administration supported an American edition of the work by printers who were the major figures in the pro-war effort. A significant part of the pro-war rhetoric linked to a rebuttal of the conclusion that Malthus drew, while at the same time celebrating and encouraging further American domestic population growth.[23] Pro-war polemicists linked all this to Republican ideas about personal freedom that were implicit in American citizenship, trying to show that British policies

---

[20]   Bickham, *Weight of Vengeance*, pp. 20–48.

[21]   D.R. Hickey, *The War of 1812: A Forgotten Conflict*, bicentennial edition (Urbana: University of Illinois Press, 2012), pp. 18–22, 34.

[22]   Nicole Eustace, *1812: War and the Passions of Patriotism* (Philadelphia: University of Pennsylvania Press, 2012). On comparing impressment to slavery, see pp. 81, 84, 173–7; on family separations, see pp. 78, 81–5, 114; on violations of citizenship rights, see pp. 88–92.

[23]   Eustace, *1812*, pp. 14–15.

on impressment, neutral trade, and dealings with Native Americans contravened rights to happy families with numerous children, population growth, and the expansion of American settlement into Native American lands.[24]

The opposing political party, the Federalists, made it clear that they did not think that any of these issues was worth fighting a war about, every one of whom voted against the war in Congress.[25] But, despite political opposition, Republicans passed the declaration of war that Madison requested. When the news eventually arrived in Washington that, in June 1812, almost simultaneously with the American declaration of war, Britain had scrapped the offending system established by the Orders in Council, it made little difference.

The Conduct of War to Achieve American National Objectives

By comparison with other wars in history, American political objectives for fighting the War of 1812 were very limited. It was not intended to be a war for survival, and the more substantial proportion of the population was not involved. With a total population of about 7.2 million in 1810, more than 500,000 served in some fighting capacity during the War of 1812, but of these only 57,000 were regular soldiers, and casualties were only 2,260 (or half of 1 per cent).[26] While Madison's objectives in opening the conflict were not intended to involve any issue of American national survival, he and his administration seemed to be unaware that war raised problems of chance that can have unintended consequences concerning a war's results.

In comparison with other nations that have started wars, the United States was curiously unprepared to undertake such a venture. The US Army since 1808 had an authorised strength of 10,000 officers and men, but in 1811 had just above half that number in service. In Canada, the British Army numbered 5,600 regulars, with more than 250,000 in other parts of the world. There was no senior leader in the small American military establishment, uniformed or civilian, who anyone would describe as having a 'genius for war'[27] – the intuitive ability to size up a comprehensive view of a military situation, maintain courage and determination in the face of mortal danger, and keep a presence of mind in dealing with the unexpected.[28] Few had extensive military experience, and those that did had

[24] Eustace, *1812*, pp. 14–15. On American Indians as British allies, see pp. 3, 23, 46, 128, 143, 149, 157, 207, 226–7; on taking Indian lands, see pp. 20–1, 23, 31, 70, 78, 113, 137, 139, 154, 161, 193, 209; as a war aim, see pp. 146, 212, 225, 234.

[25] J.C.A. Stagg, *The War of 1812: Conflict for a Continent* (Cambridge: Cambridge University Press, 2012), p. 46.

[26] Eustace, *1812*, p. x.

[27] Carl von Clausewitz, *On War*, trans. Michael E. Howard and Peter Paret (Princeton, NJ: Princeton University Press, 1976), Book I, chap. 3, pp. 103–4.

[28] J.T. Sumida, *Decoding Clausewitz: A New Approach to* On War (Lawrence:

little, if any, professional military education. The US Military Academy at West Point had been established only ten years before, in 1802, and only 89 officers had graduated, so all of them were still quite junior. Senior officers who had served 30 years before, in the American Revolution – generals such as Henry Dearborn, Thomas Pinckney, and William Hull – soon demonstrated that they were no longer capable combat commanders.[29]

At the end of 1811, the US Navy had only 15 vessels in active service, with five others laid up in reserve requiring six months for mobilisation.[30] Although a tiny force with which to challenge the world's largest navy, its officers and men had relatively more recent combat experience than their American army counterparts, having fought in the Quasi-War with the French Republic, 1798–1800, and the First Barbary War against Tripoli and Algiers, 1800–5. Since then it had kept a small force on active service at sea. This force was an essential factor for American naval readiness, but the roles of the navy had been a continuing political debate over the previous two decades. There were two opposing views, one that has been called the 'navalists', whose vision was to use American naval force as an arbiter in world politics. Their concept for a navy would allow it to serve as a continuous deterrent to aggression as well as show America's power abroad while her ships protected American commerce and overseas interests. Another group, now called the 'anti-navalists', was not against having a navy but saw different uses for it. They argued that the navalists' vision was impractical and far too costly. Their navy would be a sea-going militia force, smaller in size, with vessels whose capabilities were limited to a very few vessels operating singly on distant stations with the emphasis in home waters on coastal protection and the suppression of piracy.[31] For much of the first century of the United States' existence, the anti-navalists held sway over American naval policy, but there remained a constant tension between the two viewpoints. By and large over the country's first 100 years – with the notable exception of the War of 1812 – American leaders were satisfied to accept the benefits that came indirectly to the United States from the Royal Navy's exercise of global naval power, while Americans focused on westward expansion across the North American continent. As Andrew Jackson told the American people in his inaugural address

---

University Press of Kansas, 2008), p. 131.

[29]   Stagg, *War of 1812*, p. 54.

[30]   William S. Dudley, Michael J. Crawford, and Christine F. Hughes (eds), *The Naval War of 1812: A Documentary History*, vol. 1, *1812* (Washington, DC: Naval Historical Center, 1985), pp. 56–7: Secretary of the Navy to the Chairman of the Naval Committee, 3 Dec. 1811.

[31]   C.L. Symonds, *Navalists and Antinavalists: The Naval Policy Debate in the United States, 1785–1827* (Newark: University of Delaware Press, 1980), pp. 11–25. Joseph Payne Slaughter II, 'A Navy in the New Republic: Strategic Visions of the US Navy, 1783–1812', MA thesis, University of Maryland, 2006, argues that there were five competing visions, not two.

as President in 1829, the United States had 'need of no more ships of war than are requisite to the protection of commerce'.[32]

It was this type of thinking that had led to building the large frigates in the early 1790s that eventually made their mark in the opening six months of the Anglo-American War of 1812. Starting from the concept of the typical French and British frigates of the 1790s, American shipbuilders sought to design six frigates for a small navy that would be an overmatch for possible opponents. Thus, in the case of the 44-gun American frigates, they applied the scantlings of a 74-gun ship to a frigate that, in the words of USS *Constitution*'s builder, 'in blowing weather would be an overmatch for double-deck ships, and in light winds to evade coming to action'.[33]

Professional military and naval officers commonly remark that when a war occurs you must come as you are. So it was with the United States in 1812. Yet this was a peculiar position to adopt for a country whose leaders had precipitated a confrontation with the United Kingdom. President Madison had tried to make some military preparations in early November 1811, only seven months before he declared war. He asked Congress to double the authorised size of the Army to 20,000 men. When Congress responded by increasing it to 35,000, Madison thought it impossible to reach that number in a short time. It also created a permanent establishment that he did not want, given his ideological opposition to having a standing army. Some months later, in April 1812, and more in line with Madison's political biases, Congress authorised the President to call up 100,000 militia men for six months of Federal service. By June, however, the Army had succeeded only in recruiting 5,000 more men, and the War Department was unable to report to Congress exact figures on the country's military strength.[34] In November 1811, Madison had also asked Congress to increase the strength of the Navy, taking action that he hoped British leaders would notice in building 12 74-gun ships of the line and 20 more frigates. In response, Congress sent the opposite message to Britain than Madison intended. Congress defeated the bill, agreeing only to gather shipbuilding timber for ten frigates over the next three years.[35]

On the British side there were 5,600 troops in Upper and Lower Canada – although the Americans had overestimated British strength at 12,000 – backed by a nation already fully mobilised for war albeit otherwise occupied in Europe.[36] For Britain, it was a question of diverting already active forces and military equipment, while for the United States it was a matter of creating and training armed troops and obtaining military equipment. Choosing to be an aggressor, the

---

[32]   Quoted in Payne Slaughter II, 'A Navy in the New Republic', p. 235.

[33]   Joshua Humphries, *c.*1794, quoted in T.G. Martin, *A Most Fortunate Ship: A Narrative History of Old Ironsides* (Annapolis, MD: Naval Institute Press, 1997), p. 4.

[34]   G.C. Daughan, *1812: The Navy's War* (New York: Basic Books, 2011), pp. 27–8.

[35]   Daughan, *1812*, p. 29.

[36]   Stagg, *War of 1812*, pp. 49, 118.

United States had few geostrategic options. It could not directly strike the British Isles, so any attack on Britain was limited to a land attack on Canada, the nearest British territory. Even nearby Bermuda, the Bahamas, and the West Indian colonies were too difficult for Americans to attack without an amphibious force. At sea, America's small navy could hardly put forward a compelling challenge to Britain's command of the world's oceans. Britain could supply more than £20 million for its navy, muster 138,204 seamen in actual service,[37] in a 1,000-ship navy with some 120 to 150 ships of the line.[38]

As a minor and insignificant naval and military power in the global perspective of that time, the United States could not hope to win a power struggle with Britain through armed force. It could only do what minor powers can always attempt to do with their small military forces in a disagreement with a major power: irritate and embarrass the major power by the occasional local victory, use unconventional weapons, challenge local control in distant areas, engage in a propaganda campaign, attack enemy trade and logistics as a means to increase enemy costs, and hope for political opposition to the expense of a protracted war as a means to pressure the enemy government to come to acceptable terms.

As the instigator of the war, the United States put itself on the strategic offensive to achieve its goals in trying to force Britain to act against its policy interests and thereby was in the weaker position not only regarding its armed forces but also regarding its strategic position. Carl von Clausewitz reminded his readers that there is an inherent tendency of an attacker to falter due to the logistical need to launch an attack in stages, diminishing the attacker's strength as he advances and requiring an attacker to assume weaker defensive positions in the course of an attack.[39] An offence can be successful if its first effect completely shatters the will of the defender, but it can equally 'steel the enemy's resolve and stiffen his resistance'.[40] The psychological issues involved in efficiently carrying out successful attacks are so complex and varied that commanders often fail in undertaking them, either stopping short of the objective or overshooting it.[41] On the strategic defensive, Britain held the stronger position. In this strategic position, the defender allows the attacker to wear itself down through its attacks to bring about general exhaustion, at which point a defensive counter-attack can be efficiently used.[42]

---

[37]   Rodger, *Command of the Ocean*, pp. 639, 645.

[38]   Rodger, *Command of the Ocean*, pp. 639, 645, 308; R. Winfield, *British Warships in the Age of Sail, 1793–1817* (London: Chatham Publishing, 2005), p. xiv.

[39]   Clausewitz, *On War*, trans. Howard and Paret, Book VII, chap. 2, p. 524; chap. 3, p. 526; chap. 4, p. 527; chap. 22, 572.

[40]   Clausewitz, *On War*, trans. Howard and Paret, Book VII, chap. 22, p. 573.

[41]   Clausewitz, *On War*, trans. Howard and Paret, Book VII, chap. 22, p. 573.

[42]   Clausewitz, *On War*, trans. Howard and Paret, Book I, chap. 2, pp. 90–1, 93.

The Royal Navy's Economic Blockade

In the period leading up to the outbreak of war, the United States had demonstrated to the world that her policies were of more considerable damage to herself than they could be to France or Britain. In light of this fact, America's economic threats against others carried increasingly less weight. By spring 1812, the ministry in London wanted to avoid war, yet began to make prudent preparations. One of the most critical naval considerations made at this time was to create a naval command in the western Atlantic that could effectively deal with this contingency.

The news of the American declaration of war reached London on 30 June, nearly six weeks after it appeared. In response to the declaration of war, the British government took the first step in economic warfare by banning British ships from trading with the United States and revoking licences to sail to North America without convoy.[43] The admiralty instructed Admiral Sir John Boralse Warren to proceed to North American waters to take under his command not only the ships of the North American Station based at Halifax but also those in the West Indies previously under separate flag officers in the Leeward Islands and at Jamaica. This organisation allowed Warren to manage the whole scope of the naval war in the western Atlantic from Nova Scotia south to Bermuda, the Bahamas, Barbados, Antigua, and Jamaica. This organisation foresaw the need to deal not only with the American Navy and privateers in the waters off the east coast of the United States and Canada but also the need to protect Britain's economically valuable and politically significant sugar islands in the West Indies as well as British trade in the Caribbean.[44] With the outbreak of war, Britain's diplomatic representatives in the United States would necessarily leave the county, so Warren also acquired an ambassadorial role with the Americans. Warren had practical experience to undertake these roles having had some naval success as a rear admiral during the blockade of France in 1799–1801, as well as having served as ambassador to Russia in 1802–4.[45]

Warren reached Halifax on 26 September. By his instructions, his first move was diplomatic. Four days after his arrival he wrote to the American Secretary of State, James Monroe, to offer an armistice on land and sea in light of the withdrawal of the Orders in Council with the condition that the United States would withdraw the Letters of Marque and Reprisal of the American privateers. Monroe replied on 27 October that to accept the offer Britain must stop all impressment. Warren forwarded the letter to London, where it arrived on Christmas Day. This condition

---

[43]   Lambert, *The Challenge*, p. 105.

[44]   Arthur, 'Warren and the Royal Navy's Blockade', pp. 209–10.

[45]   On Warren, see P. Le Fevre, 'Sir John Borlase Warren 1753–1822', in P. Le Fevre and R. Harding (eds), *British Admirals of the Napoleonic Wars: The Contemporaries of Nelson* (London: Chatham, 2005), pp. 219–44.

was unacceptable, given the critical manpower shortage that Britain faced in the war against Napoleon.[46] Anticipating the American rejection, the Privy Council authorised on 13 October general reprisals against American ships, goods, and citizens. On 21 November, Lord Bathurst, the Secretary of State for War and the Colonies, issued orders to begin an immediate commercial blockade of the United States should the Americans decline the armistice offer.

On his arrival, Warren faced immediate problems that hindered following orders to implement a commercial blockade. With the outbreak of war, American privateers proved an immediate problem as they attacked with considerable success British shipping that was heading to and from Halifax. The shortage of a sufficient number of ships that could blockade all American ports forced Warren to use his ships to convoy and patrol in Nova Scotian waters rather than to blockade the privateers' home ports. Warren faced a further complication in creating a blockade that would strangle the American economy. The British system of licensing merchant shipping to trade allowed a significant number of New England shipping companies to carry grain, flour, and timber in support of the Duke of Wellington's army for the Peninsular War. At the same time, this practice has created an opportunity for forging licences, making enforcement nearly impossible. While the licensing provided vital supplies to British military operations with a higher priority than those in North America, this relief and favoritism for New England was also a political tool based on the region's strong opposition to the war and the predominance of Federalist voters. In addition to the ministry's support for the higher priority strategic goal of defeating Napoleon in the war in Europe, British leaders thought that it might be an opportunity to split the New England states away from the union.[47] Reflecting these priotities, Lord Bathurst, in the name of the Prince Regent, sent an order on 21 November 1812 to the admiralty to direct the naval officer commanding on the North American Station to

> forthwith institute a strict and rigorous Blockade of the Ports and Harbors of the Bay of Chesapeake, and of the River Delaware, in the United States of America and do maintain and enforce the same according to the Usages of War in similar Cases and in the Event of the Blockade of the said Ports and Harbors being de facto Instituted, that he do lose no time in reporting the same, that the usual Notification may be made to Neutral Powers.[48]

[46]   Knight, *Britain against Napoleon*, pp. 433–48; Arthur, 'Warren and the Royal Navy's Blockade', p. 211.

[47]   Lambert, *The Challenge*, p. 105.

[48]   Arthur, 'Warren and the Royal Navy's Blockade', p. 230: Bathurst to Admiralty, 21 Nov. 1812.

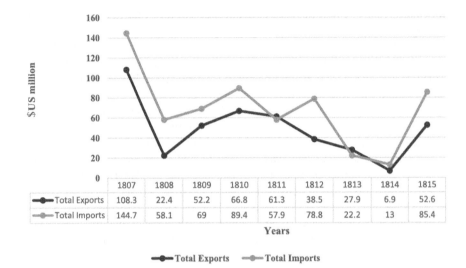

Figure 7.1: American imports and exports, 1807–1815
Source: Lance E. Davis and Stanley L. Engerman, *Naval Blockades in Peace and War* (Cambridge: Cambridge University Press, 2006), pp. 80–1.

By spring 1813, better weather, combined with the additional ships that the admiralty provided, enabled Warren to put an effective blockade into effect. This focus on Chesapeake and Delaware Bays specifically used an economic weapon to target the political centres that supported the war along with the capital in Washington, DC. This strategy had a direct crippling effect on the American economy (see Figure 7.1). The blockade of the Delaware River produced a 90 per cent decrease in Philadelphia's maritime revenue. At the same time, British success with its blockade also acted as a defence for British trade in the West Indies.[49]

In London, the admiralty sent orders to Warren that he widen the blockade to the principal ports south of Rhode Island, including New York, Charleston, Port Royal, Savannah, and the mouth of the Mississippi River.[50] The admiralty optimistically believed that Warren could efficiently blockade the American ports with his force of ten 74-gun ships of the line, 30 frigates, and some small vessels. In a private letter that elaborated on his admiralty orders, Lord Melville, the First Lord of the Admiralty, explained to Warren: 'We do not intend this as a mere paper blockade, but as a complete stop to all trade & intercourse by Sea with

---

[49]   Lambert, *The Challenge*, pp. 110–11.
[50]   Arthur, 'Warren and the Royal Navy's Blockade', p. 237: Admiralty to Warren, 26 Mar. 1813.

those Ports, as far as wind & weather & the continued presence of a sufficient armed force will permit & ensure'.[51]

By summer 1813, the blockade had nearly immobilised the small United States Navy, leaving only the American privateers to operate in place of a navy. Nevertheless, American privateers continued to slip through the blockade and capture 435 British merchant vessels.[52] Given the huge range and scale of British global shipping – comprising 20,951 ships totalling 2,349,000 tons, and rising[53] – the American response did not inflict any serious attrition. While the Americans demonstrated a continuing attempt to contest British command of the sea, their efforts had only a relatively small effect. As Mahan later pointed out, privateering was not a decisive weapon of war,[54] and was nowhere near the level of effect that the mid-twentieth-century theorist J.C. Wylie would term a successful 'cumulative' strategy.[55] The effect of the British commercial blockade forced American maritime trade to continue its sharp decline. In continuing its efforts to enforce the blockade the Royal Navy was not acting alone, but was supported by the independent work of the British privateers based mainly in New Brunswick and Nova Scotia. In contrast to the American situation, in which the US Navy was largely eliminated as an active opposition and the privateers were left in the navy's place as the country's maritime force, British privateers supplemented the results of the Royal Navy's attack on American trade. A significant number of seamen in the maritime provinces turned from smuggling to privateering, as it transformed from a dubious and sometimes lawless activity into a respectable and profitable business that supported the maritime defence of the British provinces and increased the pressure on the American economy.[56]

The combined effect of the Royal Navy's blockade and British privateers' attacks on American shipping placed a stranglehold on the American economy (Figure 7.2). These activities forced Madison's administration to increase the national debt to a point where the interest on the debt nearly equalled the payments necessary to service the increasing amounts of loan needed to maintain the government (Figure 7.3). At this point the American government could

---

[51] Arthur, 'Warren and the Royal Navy's Blockade', p. 236: Melville to Warren, 26 Mar. 1813.

[52] W.G. Dudley, 'The Flawed British Blockade 1812–1815', in B.A. Elleman and S.C.M. Paine (eds), *Naval Blockades and Seapower: Strategies and Counter-Strategies, 1805–2005* (London: Routledge, 2006).

[53] Arthur, *How Britain Won the War of 1812*, p. 248.

[54] Alfred Thayer Mahan, *The Influence of Sea Power upon History, 1660–1783* (Boston: Little Brown, 1890; repr. 1918), p. 138.

[55] J.C. Wylie, *Military Strategy: A General Theory of Power Control*, ed. with an introduction by John B. Hattendorf, Classics of Sea Power series (Annapolis, MD: Naval Institute Press, 1989), Appendix A (pp. 117–21).

[56] Kert, *Prize and Prejudice*, pp. 155–8.

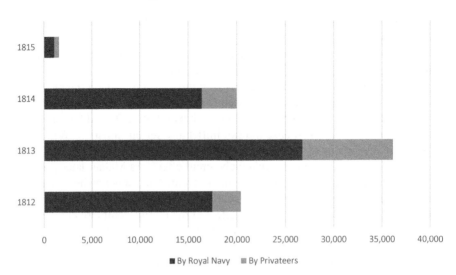

Figure 7.2: British captures of American merchant shipping tonnage, 1812–1815
Source: Faye M. Kert, *Prize and Prejudice: Privateering and Naval Prize in Atlantic Canada in the War of 1812* (St. John's, Newfoundland: International Maritime Economic History Association, 1997), p. 154.

neither obtain loans nor raise cash. This further economic pressure, in turn, led to unemployment, bankruptcy, and inflation in the United States. It seemed that Madison had little room to manoeuvre before political discontent would force the government to capitulate.[57] From mid-June 1813, North American supplies were no longer vital to Lord Wellington's Army and the war in the Peninsula. In November, Warren expanded the blockade northward to include Long Island Sound, effectively closing New York and Connecticut ports.[58] In consequence, the licensing for New England merchants was no longer useful.

Despite Warren's success with the blockade, the war in Europe and politics at home continued to interfere with his priorities. The blockade along the American coast had operated strategically to keep at bay American threats to the British West Indian trade, but this was not enough for the West Indian planters. Highly influential in London, the planters repeatedly complained that there was no visible naval presence for their local protection and more convoy protection for ships carrying specie. Political pressure from the West Indies forced the admiralty to recall Warren in November 1813 and to re-establish the separate

---

[57]   Bickham, *Weight of Vengeance*, pp. 132–3.
[58]   Arthur, 'Warren and the Royal Navy's Blockade', pp. 240–1: Warren to Croker, 20 Nov. 1813.

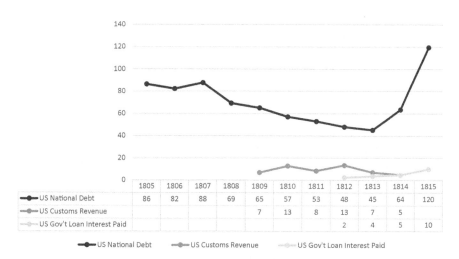

Figure 7.3: US national debt, loan interest, and customs revenue in $US million
Source: Brian Arthur, *How Britain Won the War of 1812: The Royal Navy's Blockades of the United States, 1812–1815* (Woodbridge: Boydell Press, 2011), pp. 230, 231, 249.

Leeward Island and Jamaica Stations under different flag officers who could be more attentive to local demands.

For both Scottish and West Indian political reasons, Lord Melville chose Vice-Admiral Sir Alexander Cochrane, governor of the West Indian island of Guadaloupe, to replace Warren.[59] Within six months, it was also clear that the British strategy of exempting New England from the blockade would not split the American states. As the weather improved in the summer of 1814, Cochrane extended the commercial blockade to the New England coast and even seized a part of Maine.

Meanwhile, in 1814, representatives of the United States and Britain had begun formal peace negotiations at Ghent, where discussions continued into December. The military and naval events of the latter part of 1814 remained unknown to the negotiators, but it was clear that the United States was now on the defensive. The Americans had failed in all of their attacks, and the issues of free trade, neutral rights, and impressment had disappeared as current issues with the defeat of Napoleon's regime in Europe. For the United States, the fundamental purpose of the war that lay behind all these issues remained: persuading Britain to treat the United States as an equal among nations. In the peace negotiations,

---

[59]   Arthur, 'Warren and the Royal Navy's Blockade', pp. 242–3; Melville to Warren, 24 Nov. 1813; Lambert, *The Challenge*, pp. 305–6.

Britain was apparently in control. In their opening position the British delegates made it clear that Britain had soundly defeated the United States on both land and sea. Until well into December 1814, Lord Liverpool's government persisted in its harsh view to leverage the effectiveness of the commercial blockade and British victories on land and sea to punish the obstreperous Americans. It began to appear that with the intransigence of the Americans, British forces would need to enforce the peace terms by invasion and occupation.[60]

The ministry suddenly backed away from its harsh peace terms and accepted American intransigence by agreeing to settle the war on the basis of the status quo ante. The ministry reversed its policy not because the blockade had been unsuccessful, nor because of any American military or naval victory. The determining factor was an internal political one. Just as the ministry was increasing diplomatic pressure on the Americans to accept their military, naval, and economic losses, the opposition in Parliament threatened to force the public revelation that the continuing high cost of the war in America was the only basis for high taxes. This point was one the ministry could not deny. There was also significant public opinion in Britain that continuation of the war only benefited the war contractors not the broader British economy, and increased taxes. At the same time on the international scene an extension to the war in America indirectly threatened the establishment of a balance of power for Europe in the delicate negotiations at the Congress of Vienna.[61]

As a case study of maritime strategic theory, recent historical scholarship has clearly demonstrated that Mahan's view of the War of 1812 is too narrowly focused. The Royal Navy soon eliminated the minor threat that the United States Navy and its large frigates posed at the outset of the war, giving Mahan evidence to make his main point that the US Navy was not significant enough to defend American commerce or to attack British trade. Mahan's central purpose in his historical study was to argue for the construction in his own time of a large American navy. Mahan correctly observed that attacks on commerce were effective only when supported by ships of the line and squadrons.[62] Mahan had declared:

> It is not the taking of individual ships or convoys, be they few or many, that strikes down the money power of a nation; it is the possession of that overpowering power on the sea which drives the enemy's flag from it, or allows it to appear as a fugitive; and which, by closing the great

---

[60]   Bickham, *Weight of Vengeance,* pp. 164–5, 241–2.

[61]   Norman Gash, *Lord Liverpool: The Life and Political Career of Robert Banks Jenkinson, Second Earl of Liverpool, 1770–1828* (Cambridge, MA: Harvard University Press, 1984), pp. 110–11, 113.

[62]   J.T. Sumida, *Inventing Grand Strategy and Teaching Command: The Classic Works of Alfred Thayer Mahan Reconsidered* (Baltimore, MD: Johns Hopkins University Press, 1999), pp. 45–7.

commons, closes the highways by which commerce moves to and from the enemy's shores.[63]

Recent scholars have demonstrated that Britain's maritime economic warfare against the United States during the War of 1812 was a powerful tool that had a great effect, particularly when naval operations paralleled and complemented the results produced by privateering operations that seriously damaged America's vital sea-based trade. Nevertheless, neither naval nor privateering operations was decisive in this war.

Although Britain had thwarted efficiently all American offensive operations both on land and at sea, and had pushed the United States to the brink of economic collapse through its commercial blockade, the Americans achieved their fundamental war aim to be fully recognised as a nation and develop the country as they wished. On both sides, offensive attacks and defensive counter-attacks had created a stalemate that had increased political resistance on both sides, rather than shattering it. The resulting impasse was resolved only by the domestic political forces inside Britain that demanded an end to the war and to the high wartime taxes in Britain that came with it.[64] The example of the War of 1812 provides a useful reminder that no matter how successfully Britain had utilised the tools of war, they were not necessarily the final determinants for the outcome of the war. In this case, the fruit of a victory of arms was too costly, inconvenient, and pointless for the victor to sustain. As soon as Britain lifted the commercial blockade, the American maritime-based economy recovered rapidly.

---

[63]   Mahan, *The Influence of Sea Power upon History*, p. 138.
[64]   Bickham, *Weight of Vengeance*, pp. 223–8, 251–61.

CHAPTER EIGHT

Protecting Neutrality at Sea in a Global Age, 1815–1914

Maartje Abbenhuis

When it comes to maritime warfare, the 'long' nineteenth century had a distinct character. Bookended by the Napoleonic Wars that concluded in 1815 and the outbreak of the First World War in 1914, this century witnessed exceptional levels of global change, not least in the expansion of industrial empires and the extensive use of the world's seas and oceans as highways of commerce, migration, investment, and ideas. Historians estimate that more than 100 million people migrated across the planet between 1815 and 1914.[1] Many did so by moving across the Pacific, Atlantic, and Indian Oceans. The invention and development of steamships sped up the movement of these peoples as much as it mobilised the sinews of global industrial capitalism. The laying of transoceanic telegraph cables, which by the turn of the century traversed the planet, also globalised communications. The nineteenth century, then, was an age in which the seas played a pivotal role. Perhaps surprisingly, it was also a century in which not a lot of formal naval warfare occurred. Rather, the relationship between the naval powers and the seas seemed to be less about asserting military dominance over the world's salty waters (even if in practice and by dint of its sheer size, the British Royal Navy dwarfed all others and did just that)[2] and more about opening up the highways of trade and exchange that crossed on and under the open seas.

As Stephen Neff argues, most economic warfare contends with two questions: who controls the seas and who owns goods captured outside of sovereign territory?[3] Seas can act as barriers between states as much as they are conduits of commerce, migration, communication, and state power. They are also sources

---

[1] Adam McKeown, 'Global Migration, 1846–1940', *Journal of World History* 15.2 (2004), pp. 155–89.

[2] Cf. Olive Anderson, *A Liberal State at War: English Politics and Economics during the Crimean War* (New York: St Martin's Press, 1967); M. Abbenhuis, 'A Most Useful Tool of Diplomacy and Statecraft: Neutrality and the "Long" Nineteenth Century, 1815–1914', *International History Review* 35.1 (2013), pp. 1–22.

[3] Stephen Neff, *The Rights and Duties of Neutrals: A General History* (Manchester: Manchester University Press, 2000), pp. 23–4.

for foodstuffs and commodities. The act of claiming authority and rights to access and utilise the seas as channels for the movement of ships, people, goods, money, and ideas as well as to extract resources has preoccupied human society for centuries. In the nineteenth-century age of industrial imperialism, these waterways were particularly essential. Without easy access to the 'open seas', the mechanisms of a globalising industrial capitalist economic system with its heart in the Atlantic world could not have formed.

This chapter contends with the shift away from the 'might makes right' premises that sat at the heart of much early modern maritime warfare, particularly when conducted by the British in the Atlantic world. It focuses on the shift to what the American Rear Admiral C.H. Stockton described in 1920 as the anomalous separation between a 'commercial peace' and a 'military war' that evolved in the wake of the Congress of Vienna of 1815.[4] Stockton's conceptualisation of 'limited warfare' as a situation in which the global economy operates almost uninhibited by the occurrence of a military or naval conflict is essential to understanding the nature of the 'concert of Europe' system that operated among the nineteenth-century great powers.[5] In this system of 'limited warfare', wars occurred frequently but were almost always constrained geographically and economically by the neutrality declarations of other states, many of which were great powers. In this era of limited warfare, neutrals were almost always in the majority and, thus, the assertion of neutral rights (as opposed to belligerent rights) came to predominate. Much of the shift to limited warfare depended on Britain's willingness to forego some of its traditional naval warfare strategies – not least the 'rule of 1756' – in favour of protecting its own neutral maritime rights when other states were at war.[6]

As the previous chapters in this collection show, across the centuries, the principle of neutrality formed the fulcrum of the regulation of maritime and economic warfare: who could trade with whom in time of war and what they could legitimately carry across the seas without fearing interference from a belligerent power. The policing of economic warfare was almost always done at sea and relied on the interpretation of key principles of belligerency and neutrality in international law.[7] Throughout the early modern period, neutral states proclaimed the right to trade unhindered ('free ships make free goods'). Some even suggested that private property should be free from belligerent capture altogether. They certainly demanded that contraband should be defined and that belligerent blockades were only binding on neutrals when they were effectively sustained at the entrance of

---

    [4] C.H. Stockton, 'The Declaration of Paris', *American Journal of International Law* 14.3 (1920), p. 360.

    [5] M. Abbenhuis and G. Morrell, *The First Age of Industrial Globalization: An International History 1815–1914* (London: Bloomsbury, 2019), esp. chap. 2.

    [6] M. Abbenhuis, *An Age of Neutrals: Great Power Politics, 1815–1914* (Cambridge: Cambridge University Press, 2014), pp. 15, 35.

    [7] Neff, *Rights and Duties*, p. 24.

a port. In turn, and depending on circumstances, belligerent powers were keen to defend their right to capture enemy goods (even when carried on a neutral vessel), the right to issue letters of marque to privateers, and, at minimum, to itemise contraband, impose the principle of continuous voyage (that the ultimate destination of goods determined whether they could be captured even when carried by a neutral ship), and to sustain blockades by declaring them in name only. The British were particularly staunch enforcers of the concept of 'might makes right' and repeatedly ignored or overruled the proclaimed trading rights of neutrals in favour of interfering in the economic affairs of their enemies. Much early modern warfare involving Britain revolved around competing assertions of neutral and belligerent rights.[8] Even the War of 1812 was the product of the clash of interpretations between the erstwhile neutral United States and Britain regarding the principles of economic warfare conducted by the Royal Navy on the Atlantic Ocean during their Napoleonic campaigns.[9]

In 1815, at the conclusion of the Napoleonic Wars and the War of 1812, the British government certainly was unwilling to concede any of its jealously guarded belligerent rights. As victors in the Napoleonic Wars, it asserted that naval might continued to determine which rules applied in time of war.[10] Yet, in the wake of the Congress of Vienna, which redesigned the map of Europe and established the principle of war avoidance among the European great powers, British naval policies and practices shifted radically away from judiciously protecting belligerency. During the wars of independence which rocked Latin America in the immediate aftermath of the Napoleonic era, the British government quickly embraced the advantages of neutrality, not least to placate embattled Spain but also as a means to avoid losing commercial and diplomatic influence among the newly formed South American states.[11] In 1819, the British government designed

---

[8] Abbenhuis, *Age of Neutrals*, pp. 27–38.

[9] For a useful overview, see Kathleen Burk, *Old World, New World: Great Britain and America from the Beginning* (New York: Atlantic Monthly Press, 2007), pp. 212–21, 247–8. Also W.G. Dudley, 'The Flawed British Blockade 1812–1815', in B.A. Elleman and S.C.M. Paine (eds), *Naval Blockades and Seapower: Strategies and Counter-Strategies, 1805–2005* (London: Routledge, 2006), pp. 34–45; John Hattendorf, 'The US Navy and the "Freedom of the Seas" 1775–1918', in R. Hobson and T. Kristiansen (eds), *Navies in Northern Waters, 1721–2000* (Portland, OR: Frank Cass, 2004), *circa* p. 165; Robin F.A. Fabel, 'The Laws of War in the 1812 Conflict', *American Studies* 14.2 (1980), pp. 199–218.

[10] Abbenhuis, *Age of Neutrals*, p. 35.

[11] D.A.G. Waddell, 'British Neutrality and Spanish-American Independence: The Problem of Foreign Enlistment', *Journal of Latin American Studies* 19.1 (1987), pp. 1–18. Also Matthew McCarthy, '"A Delicate Question of a Political Nature": The *Corso Insurgente* and British Commercial Policy during the Spanish-American Wars of Independence 1810–1824', *International Journal of Maritime History* 23.1 (2011), pp. 277–92.

its own version of the US Foreign Enlistment Act, to prevent its citizens from signing up to fight for Latin America's revolutionary armies.[12]

Britain sustained its neutrality practices at sea for much of the ensuing century. Britain's position as the century's superpower, the principles of the *Pax Britannica*, and the country's phenomenal imperial and commercial expansionism after 1815 were built on neutrality: on its ability to avoid going to war with its great power and imperial rivals and to sustain its neutral rights (and thus also the neutral rights of other powers) at sea in times when its rivals went to war with each other.[13] Neutrality presented a most useful tool for the consolidation and expansion of British industrial, imperial, and commercial power, much of which moved across the global seas.[14] Thus the protection of neutral rights came to feature prominently in British diplomacy, foreign policy, and naval strategies alongside their assertion of a universal international law of war to define those rights. Clearly, British naval might still defined right, but that right now gave precedence to neutrality.

The best example of the centrality of these ideas to British practices has to be the one great power war that Britain did engage in, namely the Crimean War (1853–56). In keeping with the globalising nature of all geostrategic developments of the time, this war involved four major powers: Russia on one side; France, Britain, and the Ottoman empire on the other. It was also a global event.[15] Not only did military and naval operations occur across the planet, but the potential for the war to disrupt the global economy and to draw in even more belligerents was recognised by all. Tellingly, the French and British negotiated an agreement early on not only to renounce privateering but also to limit their economic warfare campaigns against the Russians to effective blockades of particular ports only.[16] By localising their naval engagements, they protected their relationship with key neutral powers (including Prussia, Austria, and the United States) and kept their own mercantile and passenger shipping routes open.[17]

---

[12]   Nir Arielli, Gabriela A. Frei, and Inge van Hulle, 'The Foreign Enlistment Act, International Law, and British Politics, 1819–2014', *International History Review* 38.4 (2016), pp. 636–56.

[13]   Cf. Quincy Wright, 'The Present Status of Neutrality', *American Journal of International Law* 34.3 (1940), pp. 411–13.

[14]   Abbenhuis, 'A Most Useful Tool'.

[15]   Ian R. Stone, 'The Crimean War in the Arctic', *Polar Record* 21.135 (1985), pp. 577–81; John D. Grainger, *The First Pacific War: Britain and Russia 1854–1856* (Woodbridge: Boydell Press, 2008); Jan Lemnitzer, *Power, Law and the End of Privateering* (London: Palgrave Macmillan, 2014); Andrew C. Rath, *The Crimean War in Imperial Context 1854–1856* (London: Palgrave Macmillan, 2015).

[16]   For an excellent study of Britain's economic and foreign policies during the Crimean War, see Anderson, *Liberal State*.

[17]   For a detailed study of the principles and diplomacy involved, see Lemnitzer, *Power, Law and the End of Privateering*.

Essentially, this Anglo-French move suspended the 'rule of 1756', which asserted that all enemy trade regardless of its point of origin or the neutrality of the flag of its carrier was liable for capture by a belligerent.[18] Thus, when it came to the movement of goods at sea during the Crimean War, neutral rights trumped belligerency and did so with full acquiescence of the warring powers.

That these same principles were subsequently codified at the end of the war in the Declaration of Paris (1856) is, therefore, highly significant. As Olav Riste describes it, the Declaration of Paris was 'the most remarkable of the milestones that marked the progress of neutrality' because it consecrated previously claimed neutral rights in a multilateral treaty ratified by all the great powers, excepting the United States. By 1914, all sovereign states in existence, bar the United States and Venezuela, had signed the treaty.[19] The Declaration, furthermore, highlighted how useful codification could be to enforcing compliance of contentious international laws.[20] In its wake, as Nicholas Tracy rightly asserts, Britain could no longer enforce its traditional maritime rights.[21] Stronger still, as Scott Andrew Keefer argues, after 1856, Britain used its global naval dominance to enforce the terms of the Declaration and of neutral rights more generally, much as it already did in administering the abolition of the slave trade on the open seas.[22]

The Declaration of Paris asserted four basic principles of maritime warfare, namely: the abolition of privateering, 'free ships make for free goods' (except for contraband), neutral goods in belligerent ships are free from capture (except for contraband), and blockades, in order to be binding, have to be effectively maintained.[23] The implications of these four precepts were revolutionary. They privileged the power of states with navies (and thus also confirmed the power of the Royal Navy) as it ended a government's right to buy foreign-owned ships and employ their crews to conduct economic warfare on their behalf.[24] Furthermore, the Declaration of Paris legitimated the right of neutral merchants (ships flying

---

[18]   Cf. Neff, *Rights and Duties*, pp. 111–12.

[19]   Olav Riste, *The Neutral's Ally: Norway's Relations with Belligerent Powers in the First World War* (Oslo: Universitetsforlaget, 1965), p. 18.

[20]   Amos S. Hershey, 'The History of International Law since the Peace of Westphalia', *American Journal of International Law* 6.1 (1912), p. 50.

[21]   Nicholas Tracy (ed.), *Sea Power and the Control of Trade: Belligerent Rights from the Russian War to the Beira Patrol, 1854–1970* (Aldershot, Ashgate and Navy Records Society, 2005), p. 3.

[22]   Scott Andrew Keefer, '"An Obstacle, Though Not a Barrier": The Role of International Law in Security Planning during the *Pax Britannica*', *International History Review* 35.5 (2013), p. 1043; Jonathan Chappell, 'Maritime Raiding, International Law and the Suppression of Piracy on the South China Coast 1842–1869', *International History Review* 40.3 (2018), pp. 473–92.

[23]   Francis R. Stark, *Abolition of Privateering and the Declaration of Paris* (New York: Columbia University Press, 1987), p. 141.

[24]   Pat O'Malley, 'The Discipline of Violence: State, Capital and the Regulation of Naval Warfare', *Sociology* 22.2 (1988), p. 265.

a neutral flag) to conduct their economic affairs almost unhindered. Except for contraband (which had to be declared and defined by a belligerent) and the imposition of a belligerent blockade at port (which restricted all neutral access to the port), an instance of warfare should not affect neutral-to-neutral commercial affairs at all and should only have a limited impact on the economic relationships between neutrals and belligerents.

In effect, the Declaration of Paris restricted the use of naval power to influence the commercial use of the open seas in wartime. While belligerent navies continued to have the right to 'search and visit' neutral vessels and to take them, their crew, and cargo into a belligerent port as a prize to be adjudicated by a belligerent prize court, the collective weight of the Declaration reduced the likelihood of that happening and localised such activities to the waters in and around a belligerent state. The age of the privateer had essentially ended, making the world's seas and oceans safer for all civilian traffic and, thus, enabling global commerce to thrive and industrial empires to expand their formal and informal networks across the world.[25]

Significantly, the neutralisation clause inserted in the Treaty of Paris,[26] which drew the Crimean War to a formal close in 1856, also aimed at the opening up of commercial opportunities in the Black Sea region. By neutralising the Black Sea, not only were Russian and Ottoman naval ships restricted from using the waters, but in so doing, the sea also opened up as an 'international space' (as opposed to a militarised frontier of Ottoman–Russian imperial rivalry). In a fascinating account, the historian Charles King describes the move as befitting the general tenor of the nineteenth century as an age in which the internationalisation of waterways for the attainment of geostrategic, commercial, and imperial goals was commonplace.[27] That the Danube river was also internationalised in the Treaty of Paris, enabling any merchant to navigate its entire length unopposed, spoke volumes about the commercial opportunism of the time.[28] When the Suez Canal opened in 1869, as the world's first inter-oceanic waterway with widespread commercial and military potential (the Panama Canal would open at the outbreak of the First World War in 1914), it too was neutralised by international agreement. For, as Stockton so astutely described in a 1904 commentary, 'there are certain things that rise above even

---

[25]   For a complementary argument showing that the opportunity costs of 'legitimate' shipping grew after 1815 making privateering less profitable, see Henning Hillman and Christina Gathmann, 'Overseas Trade and the Decline of Privateering', *Journal of Economic History* 71.3 (2011), pp. 730–61. See also Lemnitzer, *Power, Law and the End of Privateering*.

[26]   Not to be confused with the Declaration of Paris that was negotiated soon after.

[27]   Charles King, *The Black Sea: A History* (Oxford: Oxford University Press, 2004), esp. pp. 191–6.

[28]   The Rhine River was internationalised for similar reasons in 1815: Joseph Perkins Chamberlain, *The Regime of International Rivers: Danube and the Rhine* (New York: s.n., 1923), p. 5.

narrow territoriality, and the world's use of great straits connecting oceans is among such matters'.[29] In the 'long' nineteenth century, at least, the great powers recognised the peculiar advantages afforded to all of them by guaranteeing such access.

Both the neutralisation of the Black Sea and the Declaration of Paris occasioned resistance, particularly from naval strategists (as did the neutralisation of the Suez for that matter).[30] In 1856, the Russians were particularly incensed that the Black Sea decision reduced their imperial security. The Romanov government managed to abrogate the neutrality clause at the conclusion of the Franco-Prussian War in 1870.[31] By this stage, however, the commercial expansion of foreign interests in the region was in full swing: railway lines wound their way to the Black Sea ports and tourism expanded across the Caucasus.[32] By contrast, the only major resistance to the Declaration of Paris of 1856 came from the United States, whose government was unwilling to forego privateering unless a more universal declaration ending all warfare against private property was entered into.[33] Without privateers, so its argument went, the United States was left vulnerable to naval attack. Rather ironically, then, the United States attempted to sign up to the Declaration of Paris upon the secession of the southern states in 1861, hoping to prevent the Confederates from employing privateers during the American Civil War. Without privateers, the Confederacy would have difficulty conducting any naval operations, as it did not own a navy. As a result, it would either need to purchase naval ships from a neutral supplier (which neutral states were prohibited from doing by international law) or build them from scratch (which the Confederacy did not have the resources to do).

When during the course of the war the British government failed to prevent the sale of several naval vessels (of which the CSS *Alabama* remains the most famous) by private British firms to the Confederacy, the United States government exacted compensation.[34] Based on the claim that a neutral government should take responsibility for enforcing neutrality laws on its citizenry (as already occurred in terms of the Foreign Enlistment Act), the United States demanded that, as a neutral state, the British government was responsible for monitoring the anti-neutral activities of its citizens. Amidst much public furore,[35] the *Alabama* arbitration

---

[29]   C.H. Stockton, in T.J. Lawrence, 'Problems Connected with the Russo-Japanese War', *Royal United Services Institution Journal* 48.318 (1904), p. 932.

[30]   For the British debate, see Tracy, *Sea Power and the Control of Trade*, pp. xxiii, 4–5.

[31]   W.E. Moss, 'The End of the Crimean System: England, Russia and the Neutrality of the Black Sea, 1870–1', *Historical Journal* 4.2 (1961), pp. 164–90.

[32]   King, *Black Sea*, p. 200.

[33]   Lemnitzer describes the diplomacy involved in *Power, Law and the End of Privateering*.

[34]   For an excellent overview of the *Alabama* situation, see Frank J. Merli, *The* Alabama, *British Neutrality and the American Civil War* (Bloomington: Indiana University Press, 2004).

[35]   For which, see William Mulligan, 'Mobs and Diplomats: The *Alabama* Affair

settlement of 1872 and the preceding Treaty of Washington not only saw Britain pay sizable indemnities for the belligerent actions undertaken by the Confederate ships but also confirmed that the principle of neutral rights coexisted with that of neutral duties.[36] Neutral governments from this point on had to become more proactive in applying those duties to their subjects. That the British government signed the Treaty of Washington and accepted the arbitral settlement without much opposition reflects how well it understood that imposing rights and duties on neutrals and demanding due diligence in enforcing them offered advantages to stabilising the rules surrounding neutrality.

The American Civil War and subsequent *Alabama* arbitration settlement also confirmed the central importance of neutrality in the global economy. After all, the whole world remained formally neutral while the United States waged war upon itself. As neutrals, how much economic support they could give either belligerent side was paramount not only to the course of the war but also to these neutrals' profit margins. The operation of an 'effective' blockade of the Southern ports by the Northern navy, for example, determined the rights of the neutrals to trade with the Confederates. Needless to say, blockade running was rife throughout the war, particularly out of the neutral Caribbean islands. How to sustain effective blockading tactics without upsetting the great power neutrals plagued the North and enticed the South.[37]

The limits of neutral trading rights in time of war were repeatedly contested in the latter decades of the nineteenth century. Without a clear delineation of the rules (as had occurred with the Declaration of Paris), essential questions resurfaced repeatedly: could British merchants freely trade in arms with France and Germany and move them into their unblockaded ports as happened during the Franco-Prussian War of 1870–71? Could France declare rice contraband, as it did during the Sino-French War of 1884–85?[38] Could Japan legitimately sink a neutral British passenger liner carrying Chinese troops, as it did during the first Sino-Japanese War of 1894–95?[39] Could the United States cut neutral sub-oceanic telegraph cables leading to the Spanish territories of Cuba and the Philippines as it did during the Spanish-American War of 1898?[40] How might

---

and British Diplomacy 1865–1872', in Markus Mösslang and Torsten Riotte (eds), *The Diplomats' World: A Cultural History of Diplomacy 1815–1914* (Oxford: Oxford University Press, 2008), pp. 105–32.

[36]   Elizabeth Chadwick, *Traditional Neutrality Revisited: Law, Theory and Case Studies* (The Hague: Kluwer Law, 2002), pp. 22, 45, 56; Abbenhuis, 'A Most Useful Tool', pp. 10–11.

[37]   For some of the issues involved, see Samuel Negus, 'A Notorious Nest of Offence: Neutrals, Belligerents and Union Jails in Civil War Blockade Running', *Civil War History* 56.4 (2010), pp. 350–85.

[38]   Lawrence, 'Problems of Neutrality', p. 928.

[39]   Neff, *Rights and Duties*, p. 114.

[40]   For a list of requests for clarity on such rules by neutral merchants and shipowners

wireless telegraphy carried on neutral ships on the open seas be monitored and policed?[41] All these issues created serious diplomatic situations, with the potential to endanger the peaceful relationship between neutral and belligerent states.[42] Clarity on what neutral governments and their subjects could and could not do was paramount to sustaining the principles of the congress system and the balance of power between the great naval states.

Such clarity was also sought by merchants, financiers, insurance brokers, bankers, telegraph companies, and anyone involved in the shipping industry. Neutrality handbooks, often written by international lawyers, proliferated at any time a war erupted. These advised companies and interested citizens on how to conduct their business across the seas and outlined the legal requirements of neutrality and belligerency. Newspapers too filled with articles describing the rules of war and the expected code of conduct of neutrals and belligerents alike, often with a focus on commercial issues. When disputes arose (as they did in all the cases cited in the previous paragraph), lengthy editorials explained the legal and diplomatic complexities involved. Nineteenth-century newspaper readers were well informed about neutrality and understood well the stakes involved for the maintenance of their society's wealth, well-being, and national prestige.[43] Enterprising businesses, international law associations, and liberal internationalists lobbied their governments to adopt ever more lenient neutrality rules both to protect the global economy from the harsh impact of war and to advance the peace of the seas for the movement of private property.[44]

It is utterly unsurprising then that neutrality featured prominently in the negotiations at the first Hague Conference of 1899. The subsequent Hague Conventions extended humanitarian medical intervention by the Red Cross to warfare conducted at sea and created clearly defined laws for the internment of belligerent soldiers found in neutral sovereign territory, including on board neutral ships.[45] The international lawyer, J. Helenus Ferguson, described the

---

from the British Foreign Office during the Spanish-American War of 1898, see Foreign Office, 'General Correspondence before 1906, Spain', The National Archives, Kew (TNA), FO72/2092 and 2093. Also Elbert J. Benton, *International Law and Diplomacy of the Spanish-American War* (Gloucester, MA: Peter Smith, 1968), pp. 183–218.

[41] Amos S. Hershey, *International Law and Diplomacy of the Russo-Japanese War* (Indianapolis, IN: Hollenbeck Press, 1906), pp. 115–23.

[42] Abbenhuis, *Age of Neutrals*, pp. 180–7.

[43] Cf. Annalise Higgins, 'The Idea of Neutrality in British Newspapers at the Turn of the Twentieth Century, c.1898–1902', *New Zealand Journal of Research on Europe* 11 (2017), pp. 2–51.

[44] As an example, 'The International Law Association Exemption of Private Property at Sea', *Incorporated Chamber of Commerce of Liverpool Monthly Magazine* 5.10 (Oct. 1906), pp. 147–9.

[45] M. Abbenhuis, *The Hague Conferences and International Politics 1898–1915* (London: Bloomsbury, 2018), pp. 86, 90; Abbenhuis, *Age of Neutrals*, pp. 191–4.

importance of these conventions in 1899 in effusive terms, for not only were neutral vessels entrusted with a grave humanitarian duty to rescue the victims of a naval battle (much as the Red Cross did on land) but 'a vessel distinguished by the *neutral flag*' will also 'henceforth ... incur no danger ... She is now intrusted to the honour of the naval Commanders; her safeguard is the humanity of civilized nations. ... [a] sacred trust!'[46] Above all, the successful ratification of the conventions highlighted the potential of a future Hague meeting to regulate the law of neutrality at sea more fully. All the major powers were cognisant of what they stood to gain from neutrality if and when their neighbours went to war, and what they might lose if those same belligerent neighbours interfered with their economic affairs too greatly. The potential to localise a war economically remained a powerful incentive to regulate the rules of economic warfare.

Such issues became particularly important in the contexts of the Boer War (1899–1901) and the Russo-Japanese War (1904–5).[47] In the former conflict, Britain was particularly careful to monitor the movement of contraband into the Transvaal by neutral ships and from neutral territory, but it did very little to enforce its rights of capture for fear of alienating neutral Germany and the United States.[48] In the latter conflict, the Russian navy conducted itself with an eye to reasserting key belligerent rights. It sank several neutral merchant vessels carrying what it declared as contraband (including post and cotton) and did so without first taking the ship into a Russian prize court for adjudication.[49] It also tested the limits of a neutral's territorial neutrality when it sent several Russian merchant ships through the Dardanelles Straits before converting them into armed naval vessels on the open seas. This development not only risked the neutrality of the Ottoman empire (which controlled the straits) but was of particular concern to Britain as it proffered the possibility of the reintroduction of privateering, increasing the danger of a full-blown 'war' against all neutral shipping and a decline in the Royal Navy's dominance of the global seas.[50]

---

[46]   J. Helenus Ferguson, *The International Conference of The Hague: A Plea for Peace in Social Evolution* (The Hague: Martinus Nijhoff, 1899), pp. 72–3.

[47]   Abbenhuis, *Age of Neutrals*, pp. 194–209.

[48]   J.H.W. Verzijl, *International Law in Historical Perspective*, vol. 10 (Leiden: A.W. Stijhoff, 1968), pp. 136–7; Richard Langhorne, *The Collapse of the Concert of Europe: International Politics 1890–1914* (New York: St Martin's Press, 1981), pp. 79–80; J.W. Coogan, *The End of Neutrality: The United States, Britain and Maritime Rights 1899–1915* (Ithaca, NY: Cornell University Press, 1981), p. 30.

[49]   For the controversies with neutrals, see Foreign Office, 'Confidential Print: Japan & Russia', TNA, FO881/8512. Also Lance E. Davis and Stanley L. Engerman, *Naval Blockades in Peace and War: An Economic History since 1750* (Cambridge: Cambridge University Press, 2006), p. 10.

[50]   For more, see M.S. Seligmann, *The Royal Navy and the German Threat, 1901–1914: Admiralty Plans to Protect British Trade in a War against Germany* (Oxford: Oxford University Press, 2012), pp. 95–105.

When the American President Theodore Roosevelt called for a second Hague conference in the midst of the Russo-Japanese War, he did so in part to delineate more clearly the rules of neutrality as they applied at sea.[51] The second Hague conference met in 1907 and made considerable progress on issues that plagued maritime relations, and not least on the limits of territorial waters, the laying of sea mines, the exemption from capture of post at sea (even when destined for an enemy port), the conversion of merchant vessels into belligerent vessels on the open seas, the right to coal in neutral harbours, and the right to bombard undefended coastal towns by naval means.[52] Most importantly, the conference established the International Prize Court (IPC), a court of appeal to which private individuals and companies (mainly from neutral states) could take their grievances if and when a domestic belligerent prize court decision went awry. The IPC was revolutionary: it was the first international court of appeal in existence, staffed by judges from neutral and belligerent states. Furthermore, its creation confirmed the principle that private individuals (as opposed to states) had 'inalienable rights' in international law.[53] Above all, the IPC and the Hague conventions recognised the centrality of neutrality in the international environment and highlighted the contemporary expectation that neutrals would continue to feature prominently in modern warfare.[54] As the American lawyer James Brown Scott described it in 1909, the IPC confirmed that 'the interests of neutrals should be safeguarded by neutrals'.[55]

But the 1907 Hague conference failed to do what many had hoped it would, namely to codify more clearly the major laws of maritime warfare. Without a universally defined law, the IPC, for one, would not be able to operate. While the delegates at The Hague spent many weeks negotiating, no consensus was achieved on central issues relating to contraband, blockade, and the right of capture. Remarkably, a meeting of the great naval powers in London in 1909–10, however, managed to overcome many of these same obstacles. The Declaration of London (1910) was another extraordinary document: it offered a well-defined set of laws relating the rights and duties of maritime powers in time of war, including classifying which resources could be declared as contraband, the powers of

---

[51]   Abbenhuis, *Age of Neutrals*, p. 209. Also Douglas Howland, 'Contraband and Private Property in the Age of Imperialism', *Journal of the History of International Law* 13 (2011), pp. 117–53.

[52]   James Brown Scott, *The Hague Peace Conferences of 1899 and 1907*, vol. 1 (Baltimore, MD: Johns Hopkins University Press, 1909), p. 114.

[53]   Francis Anthony Boyle, *Foundations of World Order: The Legalist Approach to International Relations 1898–1922* (Durham, NC: Duke University Press, 1999), p. 61.

[54]   Cf. Clive Parry, 'Foreign Policy and International Law', in F.H. Hinsley (ed.), *British Foreign Policy under Sir Edward Grey* (Cambridge: Cambridge University Press, 1977), pp. 103–4.

[55]   Scott, *Hague Peace Conferences*, i. 466.

blockade, and the abolition of continuous voyage.[56] The Declaration of London was even more neutrality-friendly than its Parisian predecessor of 1856.

That Britain ultimately failed to ratify the Declaration signals something essential about the growth of neutral rights in the years leading up to the First World War, namely the fear that it would leave Britain, as the world's foremost naval power, without the means to maximise its one major military asset – the Royal Navy – when it went to war. The debate over neutral and belligerent rights consumed British newspapers after 1909.[57] The President of the British Chamber of Shipping ably summed up the debate in a speech given to the Penzance Chamber of Commerce in March 1911:

> The problem of the Declaration of London must be regarded from two entirely different standpoints. We must ask ourselves first: 'How will it affect us as a neutral Power when other nations are engaged in war; and, secondly, how will it affect us as belligerents when we ourselves are engaged in war?' I fear that the underlying assumption in this country is that we shall never again be engaged in serious warfare. ... [If this proves to be the case, then] ... we need not trouble further about the Declaration of London, the main object of which ... is to safeguard the interests of a neutral Power. While other nations were engaged in war, this country, with its vast mercantile marine, would be busily employed in carrying on the ocean the trade of the world at enormous profit to ourselves, and should any of our ships be captured by a belligerent Power we should not resent it, nor declare war, but merely ... refer the matter in dispute to the Hague Tribunal [the IPC] ... [but if it proves not to be the case then] ... No Declaration, no code of international law can do anything for our protection when we are engaged in war. Our only hope then rests on the supremacy of the British fleet.[58]

In the end in 1912, the House of Lords voted in favour of protecting the Royal Navy's belligerency and refused to ratify the Declaration or the accompanying Naval Prize Bill. Without Britain's adhesion to the Declaration of London, the IPC could not be formed either.

Above all, the British debate around the Declaration of London highlights how the contestation between neutrality and belligerency was never straightforward.

---

[56]   Coogan, *End of Neutrality*, pp. 117–22.

[57]   As an example, see N.C. Fleming, 'The Imperial Maritime League: British Navalism, Conflict and the Radical Right *c.*1907–1920', *War in History* 23.3 (2016), pp. 296–322. Also Bernard Semmel, *Liberalism and Naval Strategy: Ideology, Interest and Sea Power during the* Pax Britannica (Boston: Allen & Unwin, 1986), pp. 109–18.

[58]   Edward Hain, President of Shipping, 3 Mar. 1911, in Board of Trade and Ministry of Transport, 'Consultative Marine Files: Hague Peace Conference: International Prize Court – Declaration of London, 1908–1911', TNA, MT15/135.

Between 1815 and 1914, naval strategists in every major state had to plan for a potential wartime future in which their country was either a belligerent or a neutral. Given that most neutral rights impeded belligerent rights (and vice versa), how to advocate for a suitable balance between them was never easy and always contentious. Between 1815 and 1914, however, the odds almost always favoured the neutrals, where before 1815, and certainly after 1914, they favoured the belligerents. Yet for most naval powers in 1910, neutrality continued to win out: Germany, the United States, and France all ratified the Declaration of London, for example. At the outbreak of the First World War, Britain also declared it would adhere to its terms, if only to stabilise the global economy in the short term. Neutrality then still defined the global landscape of maritime warfare at the outbreak of this global war.

Of course, the dominance of neutrality did not mean that all belligerent rights were signed away between 1815 and 1914. Nor did it mean that states did not invest in expanding their naval power. The naval races of the turn-of-the-century period highlight just how significant they considered that power to be both in sustaining their relative global power and in planning for a future war as a belligerent.[59] The key to their naval planning revolved around understanding the need to prepare for the possibility of going to war and the likelihood of remaining neutral.[60] As had been the case for centuries, the balance between peace and war thus lay in the details of their diplomatic negotiations. Thus, Britain's advocacy at The Hague in 1907 for an end to contraband but the continuation of the right of blockade spoke volumes about how the admiralty expected to maximise the size and strength of the Royal Navy (and not least the development of its *Dreadnought*-class battleships) to impose effective blockades. After all, if a blockade could be maintained, then neutrals could not reach an enemy port and thus the need for a defined list of contraband items disappeared alongside.[61]

By contrast, the German delegation at The Hague in 1907 presented a case for the right of any nation to lay sea mines in their own territorial waters to protect their borders.[62] The argument aimed squarely at avoiding a future

---

[59]   Cf. J. Glete, 'Naval Power and Warfare 1815–2000', in J. Black (ed.), *War in the Modern World since 1815* (London: Routledge, 2003), pp. 217–27.

[60]   Cf. Gabriela A. Frei, 'Great Britain, Contraband and Future Maritime Conflict (1885–1916)', *Francia* 40 (2013), pp. 409–18.

[61]   Cf. Tracy, *Sea Power and the Control of Trade*, pp. 96–7; Neff, *Rights and Duties*, p. 97; M.S. Seligmann, 'Failing to Prepare for the Great War? The Absence of Grand Strategy in British War Planning before 1914', *War in History* 24.4 (2017), pp. 414–37. Cf. Andrew Lambert, 'Great Britain and Maritime Law from the Declaration of Paris to the Era of Total War', in Hobson and Kristiansen, *Navies in Northern Waters, 1721–2000*, pp. 23–31.

[62]   For an excellent review of the subject of sea mines at the 1907 Hague conference, see Richard Dunley, *Britain and the Mine, 1900–1915: Culture, Strategy and International Law* (Basingstoke: Palgrave Macmillan, 2018), pp. 97–130.

blockade of its own ports. Another way in which German naval planners expected to make the most of neutrality, even if they were belligerents, was to protect the right of neutrals to trade with belligerents. After all, several small but powerful trading nations like the Netherlands and the Scandinavian states (all of which were renowned long-term neutrals) bordered Germany. From a German perspective, a future British blockade may be surmounted if enough goods could be funnelled by river or rail through these neutral territories. To that end, the German delegation supported the creation of the International Prize Court. Protecting a neutral's right to trade, then, was as much part of the German naval platform as building up a sizable fleet of warships.

Similarly, where France's *Jeune École* strategists considered that a fleet of torpedo boats could keep the ships of a future enemy (including Britain) in port, they also supported the widening of neutral merchant rights which could easily coexist with the expansion of its naval power that would ultimately target belligerent commerce.[63] Meanwhile, the United States' ongoing demand to declare all private property free of capture on the open seas reflected its government's expectation that it was unlikely to go to war with its major industrial and imperial rivals. Here, too, Alfred Mahan's promotion of building up the United States as a strong naval power capable of asserting its 'grip on the sea' when necessary sat comfortably alongside its reputation as a staunchly neutral power willing and able to protect those neutral rights.[64]

Some historians argue for a continuity in naval warfare strategies employed by Britain and the United States between 1815 and the First World War. Granted, the expansion of belligerent rights was phenomenal during the total industrial war that erupted in 1914 and the justification for many of them harked back to the pre-1815 period. But to suggest those rules stayed static across the century that separated these global wars is to fail to recognise how central a role nineteenth-century globalisation played in affecting the naval policies of the great powers. For between 1815 and 1914, as John Coogan also argues, Britain came to rely increasingly on neutrality and the rights of neutrals to trade and freely access the open seas.[65] Everyone's foreign and economic policies as well as their naval strategies had to accommodate this essential shift.

---

[63]   Tracy, *Sea Power and the Control of Trade*, p. xxiii; B. Ranft, 'Restraints on War at Sea before 1945', in Michael Howard (ed.), *Restraints on War* (Oxford: Oxford University Press, 1979), p. 51. Cf. L. Paine, *The Sea and Civilization: A Maritime History of the World* (New York: Vintage, 2013), p. 555.

[64]   John B. Hattendorf, 'Maritime Conflict', in Michael Howard, George G. Andreopoulos, and Mark R. Shulman (eds), *The Laws of War: Constraints on Warfare in the Western World* (New Haven, CT: Yale University Press, 1994), p. 110.

[65]   Coogan, *End of Neutrality, passim.*

CHAPTER NINE

Postal Censorship and the Alchemy
of Victory at Sea during the First World War

John Ferris

Historians agree that economic warfare shaped the First World War, but neither its execution nor its effect has been studied thoroughly.[1] Economic warfare usually is seen just as a matter of warships; in fact, it stemmed from and was waged through a combination of diplomacy, intelligence, law, and sea power. The Royal Navy might have exercised blockade simply through strength, but Britannia did not wish to rule the waves by force alone. Britain cared about being, and seeming to be, lawful in action, which also eased the diplomacy of economic warfare. Intelligence, especially intercepted cables, wireless, and, above all, the international post, unified sea power and sea law. Sea power enabled diplomacy, intelligence, and law. They executed sea power. In particular, only overwhelming sea power could make every neutral at once tolerate British interception of their sea-mail, which was inconvenient and, neutrals thought, illegal. Postal censorship, Britain's most feminised department of state, matched the most manly of arms as a tool of sea power. This relationship between economic warfare and letter intercepts was unique. Historically, only ship's papers – documents about the

---

[1] ADM, CAB, CO, DEFE, DO, HO, FO, T, TS and WO files are held at The National Archives, Kew. L/MIL files are held at the India Office Record Library, The British Library. Useful accounts of the blockade include Archibald C. Bell, *A History of the Blockade of Germany and of the Countries Associated with Her in the Great War, Austria, Bulgaria and Turkey, 1914–1918* (London: HMSO, 1937); M.C. Siney, *The Allied Blockade of Germany, 1914–1916* (Ann Arbor: University of Michigan Press, 1957); N.A. Lambert, *Planning Armageddon: British Economic Warfare and the First World War* (Cambridge, MA: Harvard University Press, 2012), pp. 60–100; A. Offer, *The First World War: An Agrarian Interpretation* (Oxford: Clarendon Press); E.W. Osborne, *Britain's Economic Blockade of Germany, 1914–1919* (London: Frank Cass, 2004), pp. 34–41. For my interpretation, see J.R. Ferris, 'The Origins of the Hunger Blockade: Irony, Intelligence and International Law, 1914–15', in M. Epkenhans and S. Huck, *Der Erste Weltkrieg zur See* (Munich: De Gruyter Oldenbourg, 2017), pp. 83–98. For parallel views, cf. M.S. Seligmann, 'Failing to Prepare for the Great War? The Absence of Grand Strategy in British War Planning before 1914', *War in History* 24.2 (2017), pp. 414–37.

ownership of goods carried on board merchantmen – aided economic warfare, while the interception of communications affected diplomacy, or internal politics, but not war at sea. Nor did this relationship continue: though Britain applied this same system during the Second World War, its effect was different, because the naval war was different. That instance of economic warfare was unique, but it illuminates the alchemy of victory at sea between 1914 and 1918.

Intelligence and Policy for Economic Warfare before 1914

After 1815, governments feared threats from the lower orders at home, against which they turned to spies and the interception of mail. Postmasters regularly intercepted the mail of specific people. Under general warrants, sometimes they were told to 'detain and open any letters' sent to and from towns thought 'to be of a suspicious nature, and likely to convey seditious or treasonable information, or to contain money likely to be applied for the purpose of promoting seditious or other Disturbances'. From 1830, however, far fewer intercepts were ordered, but which were more specific and less political than before.[2] Statesmen who wished to reform politics feared internal threat less than those who resisted any change. After the Mazzini scandal of 1844 sparked public outrage about state interception of mail, authorities reduced even these practices. Letters and cables were intercepted only under special warrants signed by a Secretary of State, enabling collection simply of messages to or from specific individuals or addresses, and overwhelmingly for criminal matters. In 1910, as Home Secretary, Winston Churchill loosened these restrictions, to ease the interception of letters thought linked to espionage, which was a matter of concern at the time. Even so, until August 1914, mail rarely was intercepted within Britain.

Meanwhile, for the first time, postal censorship became linked to British preparations for war. During the 1890s, British authorities debated the value of cable censorship in war. Between 1899 and 1902, in the South African war, Britain practised every possible form of signals intelligence, primarily to prevent communication abroad by Boer authorities, but also to gather military intelligence, defeat espionage, and aid blockade. The main focus, with the greatest effect, was on cables, but British authorities in South Africa also intercepted letters to and from businessmen, especially in firms suspected of or known to be working with the Boer republics. A black chamber of the General Post Office (GPO) also secretly opened mail within Britain, probably involving individuals suspected of espionage or shipping contraband to the Boers. The Colonial Office warned that 'to avoid unpleasantness all letters to and from foreign countries which are opened by censors should be treated carefully by experts and before being sent

---

[2]   Home Office to General Post Office, 7 Apr. 1818, *passim*, HO 79/3; cf. HO 32/20, HO 151/7.

on or returned to sender should be closed up etc so to remove trace of opening so far as possible'. Letters 'should not be forwarded if they bear marks of having been opened'.[3] This expertise appears to have existed. At least no foreigners complained of Britain opening their mail.

The Boer war illustrated the power of censorship in war, and its requirements. Over the next decade, Britain generalised these experiences into a policy and a structure. It created a mature policy for communications power in war, which combined an aggressive attack on enemy-owned cables and censorship of British cables.[4] It aimed to prevent any enemy from communicating in war, and to control and exploit messages sent on British cables. Several interdepartmental committees formulated policy on censoring cable, wireless, and mail. Thinking globally and acting locally, they integrated complex technical issues into every decision. These committees defined protocols, applied them across the empire, and established a structure to embody these ideas. Its skeleton was strong, but this structure needed much more muscle than Whitehall realised.[5] This planning emphasised military security and intelligence, and contra-espionage. Planners barely considered what would become the greatest role of censorship – economic warfare – although they knew they would handle commercial traffic, as censors had done during 1899 and 1900. While the cause for this indifference is uncertain, perhaps it reflected the view that control of contraband would be tertiary in economic warfare. Planners did not understand how far they must collect and use personal communications: in 1914, this capacity rapidly became the greatest form of surveillance ever known, but that effect was unintended.

This policy spurred preparations for communications intelligence, focused on cables, and, to a lesser degree, wireless, both of which centred on communications between foreigners in other countries. Assessments of postal censorship, conversely, addressed mail sent within Britain, reflecting the assumption that international mail was sacred. This assumption was written into international law. Article XI of The Hague Convention for the Pacific Settlement of International Disputes of 1907 declared 'inviolable' the private or official correspondence of neutrals and belligerents, on neutral or belligerent vessels. Only mail being carried to or from an area under effective blockade was exempt from this rule. Similarly, Article 4 (1) of the Universal Postal Convention of 1906 guaranteed

---

[3]   High Commissioner for South Africa to Governor, Natal, telegram No. 2, 24 Feb. 1900, DO 119/513; War Office to Kitchener, 6 July 1901, WO 108/306; Milner to Colonial Office, 16 May 1900, FO 2/441.

[4]   115th CID meeting, 14 Dec. 1911, CAB 2/2.

[5]   Report of Censorship of Cable and Wireless Committee, 7 July 1908, CAB 17/92; Admiralty Intelligence Division, 8.11, 'Regulations for the Censorship of Radio-Telegraphy in Time of War', FO 371/ 1283, 38965; GS 1906, A 1039, 'Censorship of Submarine Cables in Times of War', and 1907 version, and Second Meeting, 'Censorship of Submarine Cables', 26 Oct. 1910, L/MIL 7/13581.

'the right of transit ... throughout the entire territory of the Union'.[6] Nothing better reflects the naive legalism underlying British strategy than the assumption that belligerents at war all would follow such statutes.

Legally, however, British authorities believed that they could censor any mail (belligerent, British, or neutral) within British territory, and any traffic on British-owned cables. The War Office was assigned responsibility to censor cable and letters, and the Royal Navy to do the same with wireless. Though procedures for cable and wireless censorship were well considered, these organisations existed only on paper. Procedures for postal censorship were not even defined. The GPO's secret section, a seasoned if shadowy body which executed all warrants for letter intercepts during peacetime, was excluded from this planning for wartime censorship, because that suited institutional interests. The GPO wished to move mail, not to stop and open it, while the War Office thought that only military agencies could control espionage. In these discussions on censorship, the GPO denied that 'the systematic examination of all letters passing through the post' was possible, and probably thought the same about cables. The GPO, however, noted that it 'had an excellent organisation for putting the censorship of letters into force when authorised to do so'. Churchill advocated the creation of a list of suspected spies whose correspondence would be read secretly during a period of strained relations, though 'this authority should be used sparingly, in order that public confidence in the Post Office should not be shaken'.[7] The GPO's special section applied this power against suspected spies in Britain during peacetime. MI5's successful campaign against German intelligence in Britain rested on these letter intercepts.[8]

Meanwhile, between 1900 and 1914, British authorities were divided over policy for economic warfare. Most wished to avoid having to seize contraband, property bound to enemy destinations. Authorities believed such a policy would make neutral powers into foes. Seizing contraband would be hard, and condemning it almost impossible, because of problems of proving the status of any item in a Prize Court, a national court enforcing the international law of the sea. Authorities hoped to avoid these problems through effective blockade of enemy ports or coasts. Legally, Britain could seize any items bound to such places, but this policy was increasingly difficult to execute. In 1912, leading members of the Liberal government defined a policy for economic warfare based around the recommendations of the Desart committee. That policy broke almost as soon as it was adopted, without anyone knowing the fact. It assumed that the Declaration of

---

[6]   Chapter 1, Article XI of The Hague Convention for the Pacific Settlement of International Disputes, 1907; Convention of Rome, 6th Universal Postal Union Conference, 1906 (London, 1907).

[7]   First meeting of the Censorship of Cables Committee, 11 July 1910, 'Observations by the Secretary of the Post Office on the Memorandum by the General Staff', CAB 17/92.

[8]   C. Andrew, *The Defence of the Realm: The Authorised History of MI5* (Toronto: Penguin Canada, 2009), pp. 37–48.

London, which aimed to rewrite maritime law, and close blockade, would govern economic war with Germany. The Royal Navy would effectively close German ports in the North Sea, preventing imports or exports of any items through them. Few items would be termed contraband, so minimising the need to seize neutral goods, especially those of American firms, thus angering their governments. Between 1912 and 1914, however, the Declaration of London became stalemated, and lacked legal standing. Sailors abandoned close blockade, without informing landsmen, and adopted a policy of distant blockade. The latter could prevent any trade by German ships and reduce neutral exports to Germany, but was not legally effective. In war, no German ports would be blockaded effectively, leaving neutrals free to trade with them. Contraband, conversely, was not legally redefined. It could cover any item which any belligerent cared to name. In August 1914, no British decision makers knew what contraband would be, nor how to control it.[9]

Before 1914, British authorities understood the strategic value of censorship, but not how to handle the mass of material it must collect. When war began the censorship was overwhelmed by traffic. Personnel were unready to do their job of censorship, let alone to exploit the intelligence it offered. These failures were inevitable, but not fatal, and outweighed by the successes. Decision makers had much to learn about censorship as the war began, but they did so well enough to guide key decisions during its opening phase, and to act effectively on the possibilities which opened by February 1915. Pre-war British authorities misunderstood the nature and value of economic warfare, and scarcely recognised the idea of intelligence for it. Yet Britain possessed powerful means to conduct economic warfare. In practice, though not in law, the Royal Navy could control all merchant shipping to and from Europe. Intelligence was preadapted to the problem of contraband, not because a problem was recognised, and solutions were followed, but through Britain's position in world communications and information processing. British authorities had intelligence fit for economic warfare without realising it – for a campaign most of them did not expect or want to fight.

Censorship and the First Months of War

When war began, the secret section of the GPO enabled MI5 to strangle Germany's spy networks, and monitored suspicious letters in Britain. It only ceased such work in spring 1915, once military censorship became efficient. Cable, postal, and wireless censorship were unprepared for their tasks. None was staffed even as skeletons. Each had a head, assigned just days before war began, and only then shown the documents to guide his organisation, but no staff. Every censorship agency faced great problems in organisation, and of procuring staff with the appropriate technical and intellectual skills. None thought economic warfare a main

---

[9]   Ferris, 'The Origins of the Hunger Blockade'.

task. Neither consumers nor intelligence authorities drove the initial development of communications intelligence for economic warfare because they did not know what they wanted nor what it could deliver. The drivers were junior personnel who censored messages, grasped what could be done with commercial communications, drew attention to the fact, and began to compile, index, distribute, and, finally, analyse that material. This process occurred independently almost on the same day within the cable censorship and the Naval Intelligence Department (NID), and within the same week that Churchill, now First Lord of the Admiralty, wrote the charter for Room 40, the newly established codebreaking agency of the Royal Navy. The product of censorship quickly impressed intelligence authorities who supported that work and, against some indifference and opposition from politicians and officials, tried to expand it. Success came only between February and June 1915, as desperate consumers pursued the intelligence and evidence necessary to control contraband successfully, only to discover it already was at hand.

In this process, postal censorship and the GPO's special section were less active than the cable or wireless censorships and the Military Intelligence Department (MID) or NID. Although postal censorship proved the greatest source of intelligence for economic warfare, its strength was the hardest to unleash. Among the censorship agencies, it faced the greatest problems of basic organisation, and the largest volume of traffic. As the postal censorship wrote after the war:

> The Chief Cable Censor had a precise, elaborate and skilful system worked out in the light of the experience of the South African War, and enjoyed almost an absolutism over the world's ocean cable communications; The Postal Censor (in the absence of any thought-out procedure), had to break down departmental opposition, and while devising temporary expedients, was never certain of the length he would be permitted to go or of the authority he was justified in exercising.[10]

The postal censorship began on 3 August, in a 'dark and crowded basement'; by 1918, it colonised much of the salubrious area around the Science Museum in South Kensington. Initially, its head, Colonel A.C. Pearson, 'had to improvise everything, policy and practice, at once', and create a staff.[11] Meanwhile, all mail from Britain abroad was simply delayed for a few days, which, allegedly, would render innocuous any espionage letters that passed, because the danger feared was accurate reports of immediate military or naval movements.

The censorship confronted 'terminal' mails, involving Britain and any other country, and 'transit' mails, letters between any two other countries but carried

---

[10]   The National Archives, Kew, Records of the Ministry of Defence, *Report on Postal Censorship, 1914–1919* (1921), DEFE 1/131, p. 37.

[11]   *Report on Postal Censorship, 1914–1919*, DEFE 1/131, pp. 6–10.

on ships which entered British ports, which often were transhipped from vessels in Liverpool to those in east coast ports, and vice versa, by special trains. The first warrant to censor mail, of 3 August 1914, enabled censors to open any letters to and from people in enemy states, and some specific neutral addresses, with persons in the United Kingdom. By September, this power was extended to all terminal mails between Britain and the Netherlands, Denmark, and Norway; and, by December 1914, those with every European neutral except Bulgaria and Greece, which were added in May 1915.[12] The intellectual requirements for staff, expected to determine the nuances of meaning for espionage traffic, in several languages, were high. By 10 August, its staff consisted of seven officers and some 25 clerks, whose 'duties are those of interpreters or translators, the 2nd grade clerks having to be fully conversant in two foreign languages, while the 1st grade clerks know from 3 to 5 foreign languages'. By September, with the extension of countries covered, the numbers of censors grew 'as a matter of urgency' to 130 clerks and 39 officers.[13] While this staff was large, it focused on preventing leaks of military secrets and espionage, and also censored letters to and from prisoners of war and the British Expeditionary Force. Systematic censoring of all mails would require a far greater organisation. Many authorities disliked the idea of censoring mail, and few supported an aggressive expansion of the practice. The postal censorship itself noted that censors felt

> an involuntary and deep-seated disgust when (they) first broke the seal of an unknown person's correspondence, a natural sentiment which indicates how profound was the confidence men had in the posts and how inconsistent with liberty and with good faith it appeared to invade 'the rights of private correspondence which ought to be held peculiarly sacred'.[14]

This situation changed through an incident famous from the memoirs of the Director of Naval Intelligence, Admiral Reginald Hall, though many other accounts support his description of a seminal moment in British intelligence, and economic warfare.[15] In November 1914, Hall heard inaccurate stories that espionage messages were passing the postal censorship, and accurate ones that only 5 per cent of mail bags were even examined. Hence, he met Colonel George Cockerill, who had just weeks before become the Director of Special Intelligence,

---

[12] W.H.W. Carless Davis, *Emergency Departments* (London: HMSO, 1920), pp. 55–60.

[13] WO to Treasury, 10 Aug. 1914, 17 Sept. 1914, 26 Oct. 1914, T 1/11971.

[14] *Report on Postal Censorship, 1914–1919*, DEFE 1/131, p. 77.

[15] Chapter regarding 1914–15, Reginald Hall Papers, 3/2, Churchill College, Cambridge; Carless Davis, *History of the Blockade: Emergency Departments*, p. 21; *Report on Postal Censorship, 1914–1919*, DEFE 1/131, pp. 19–20; Stephen McKenna, *While I Remember* (London: Butterworth, 1921), p. 165; Stephen McKenna, *Tex, A Chapter in the Life of Alexander Teixiera de Mattos* (New York: Dodd, Mead and Co., 1922), pp. 7, 10.

in charge of censoring the press, mail, and cables. Cockerill correctly noted that his men were overwhelmed: 'It was necessary to create something out of nothing and at the very beginning opposition from more than one quarter had to be faced. The available staff remained lamentably small'. On a recent visit to the Returned Letter Office of the GPO at Mount Pleasant, where postal censorship was based, 'Col. Cockerill admitted that he had found something like chaos: piles of letters awaiting special attention, cheques strewn about the floor, dozens of bags which had not yet been examined at all'. Hall offered Cockerill a small staff, at no cost, to examine 'all the foreign mails', as he characterised his words years later. In fact, Hall referred just to all of the terminal mails between Britain and European neutrals, which the GPO processed. Britain was free to censor these letters, so long as a Secretary of State authorised the action, as Reginald McKenna, the Home Secretary, had done for Cockerill. Hall did not refer to transit mails, or those on ships at sea, neither of which entered GPO control, and were inviolable under conventional views of international law. Cockerill, replying, 'In other words a little private censorship of your own!' accepted the offer for a two-month trial. He appreciated the assistance of an 'Admiralty super-censorship', and preferred working with Hall to fighting him.

Hall convinced Churchill to authorise £1,600 for an unspecified project and organised staff through a friend, Colonel 'Freddy' Browning, later Deputy Chief of the British secret service. Browning delegated the task to friends at the National Service League, a pressure group which favoured conscription and had, as Stephen McKenna, later associated with these efforts, stated, 'premises, furniture, a staff', and no functions. 'A press gang of two, working the clubs of London and the colleges of Oxford, established the nucleus of a staff; and the first recruits were given, as their earliest duty, the task of bringing in more recruits', the men being paid 'stealthily each week, like a member of some criminal association, with a furtive bundle of notes'. This 'press gang' included Browning and probably Henry Penson, a well-known economist and statistician, or else William Carless Davis, a medieval historian, from Worcester College and Balliol College respectively. These men pressed not drunken sailors, but scholars and businessmen. Only educated gentlemen, sober or otherwise, could be trusted to find and keep secret the letters of spies.

This inspection found no signs of espionage traffic, but massive amounts of information valuable to the economic war, about how German and neutral businesses handled contraband. Penson and Carless Davis wrote a report on these issues. This document, fundamental to the development of British economic warfare between 1914 and 1918, and one of the most influential analytic reports in the history of British intelligence, apparently no longer survives. Given their academic background, and the nature of their reports on economic intelligence between 1915 and 1918, however, probably it used communications intelligence to illuminate economic warfare with unprecedented power. The chief wireless censor, Admiral Brownrigg, the head of cable censorship, Major Phillips, and Frederick Leverton Harris, an intelligence officer in the NID, and the world's first analyst of communications intelligence, already had reported that intelligence about contraband was best found in traffic between neutrals, like transit letters. The Penson and Davis report finally made senior intelligence officers appreciate the general significance of communications intelligence to economic warfare, precisely as Whitehall emphasised the value to victory of those activities.

These censors, unlike the GPO's secret section, lacked the skill to hide their work. In mid-December, Reginald McKenna, discovering the development, called Cockerill and Hall on the carpet. He declared Hall's actions illegal, which the Admiral and his subordinates thought was true. Both were mistaken. Cockerill rightly defended Hall as operating under his delegated authority, and the two persuaded McKenna that the work on commercial letters should continue officially, under the War Office. The wording of the warrants for interception justified this enhanced effort. McKenna's liberalism pulled him in different directions: his faith in economic warfare overrode his distaste for reading other people's mail. McKenna was much more willing to coerce neutral citizens and states over shipping food to Germany, and to starve enemy civilians, than he was to intercept the private mail of enemies or neutrals.

Communications Intelligence and Economic Warfare, 1914–1915

Like the butterfly effect in chaos theory, this incident disproportionately drove the outcome of a complex process. When war began, British policy for economic warfare collapsed. Britain needed a new one. That endeavour involved Britain's relations with every neutral state and national firm, and its own subjects. These issues caused controversy between decision makers. Initially, they wished to respect international law and neutral rights, and to minimise control over contraband. This approach would ease Germany's ability to import and export goods through neutrals, and blunt a blade against it. Many British authorities resisted changes in this approach. The diplomats, intelligence officials, and naval officers who managed economic warfare, however, soon developed a different view, as the cost of the war and its likely length grew. Britain must prevent Germany from importing any item useful to its war economy, including any raw material used by industries, and food. Some said openly that Britain must starve the German economy and people into surrender.[16]

Communications intelligence shaped these views. During autumn 1914, public information and intercepted messages of neutral and enemy firms increasingly illuminated the problems with contraband. German and American firms cooperated to send vast supplies to the Central Powers via neutral countries. Chicago meatpackers sent 23 million pounds of meat and lard to Denmark. American firms boosted their shipments of copper to Italy, Scandinavia, and

---

[16]   Ferris, 'The Origins of the Hunger Blockade'.

Switzerland from eight million tons per year to 33 million tons in four months. At that rate, Germany easily could maintain its pre-war imports of copper through neutrals, avoid any shortage of food, and escape economic pressure. Britain stopped most of these shipments before they reached their destination.[17] Sporadic intercepts of cables drove these actions, and also of letters, some taken by the postal censorship from British mails, but most provided by Britons or friendly neutrals from abroad. Cables provided effervescent, through sometimes powerful, intelligence on specific cases. Letters illuminated the structural problem, and how firms and the foe operated. In October 1914, for example, an intercepted letter regarding the smuggling of contraband noted:

> For greater security our friends in America would retain proprietary rights over the goods until they are out of danger and have passed into the possession of the purchasers or their goods. Should anything happen to the goods, there would in the first place be the insurance and in the second place I should represent the interests of the purchaser in my name and in the name of my American friends under the protection of the American Government and of the Swiss authorities, without thereby incurring any risk.[18]

In October 1914, several Danish firms reminded exporters abroad 'that the war will considerably increase the importance of Copenhagen as a transit port', absorbing pre-war trade to Hamburg. 'Denmark is likely to work up a good transit-trade with the belligerent powers under shelter of its neutrality; there is hardly any reasonable supposition why the neutrality of Denmark should be offended; on the contrary, the belligerent powers are interested in Denmark's continued neutrality'.[19] In markets across the world, German firms advertised their ability to import contraband to Germany, while announcing auctions of such material at home.[20] From 19 October, Britain's main blockade agency, The Restriction of Enemy Supplies Committee (RESC), cited intercepted letters or cables when discussing contraband and suspects. These messages were the largest category of intelligence outside of open or official sources, and trusted. More than any other source, they drove specific suspicions and cases, provoking two of the greatest events in Britain's campaign of economic warfare: its first use of intelligence to guide a major seizure of cargo, over copper, and proof that American meat packers were shifting shipments to Germany from Rotterdam to Copenhagen.[21]

---

[17]   ADM 186/603, memorandum by W.E. Arnold-Forster, Lt Commander RNVR, undated, 'The Economic Blockade, 1914–1919', p. 36. ADM 183/603.

[18]   Slade to Mellor, 16 Oct. 1914, TS 13/234 B.

[19]   Foreign Office to Copenhagen Legation, Despatch NO 53 Commercial, 11 Nov. 1914, FO 211/300; cf. memorandum by Trade Clearing House, 25 Apr. 1915, FO 211/311.

[20]   Despatch, Legation Christiana to Foreign Office, 13 Feb. 1915, FO 382/304, 19264.

[21]   Ferris, 'The Origins of the Hunger Blockade'.

British policy on economic warfare slowly radicalised, against much resistance. It was far from radical when Germany declared unrestricted submarine warfare in February 1915. In response, Britain and France declared a blockade of Germany, which did not fit legal definitions of the practice. They sought to prevent Germany from importing or exporting anything at all. They called contraband all items which neutrals imported or exported to the Central Powers. The international law of the sea became a battleground. Britain actually could enforce the most effective blockade in history, but that power did not fit prevailing ideas of law, and centred on matters which authorities thought dynamite: control of contraband, and proof of its status. Britain could conduct economic warfare only through means it had thought impossible before the war.

Unrestricted submarine warfare created a crisis in sea power. So did British retaliation, in a form combining intelligence and law. During March 1915, reflecting its respect for the power of law and the United States in economic warfare, the admiralty believed that inadequate intelligence would cripple expanded control of contraband. The only solution was more and better spies. No one mentioned communications intelligence, not even those who had used it for economic warfare. Churchill relied on communications intelligence for naval operations, but did not refer to it regarding economic warfare. He did not know about its use there and thought the limits to intelligence forced a cautious start to blockade, noting 'Prima facie, I do not favour the wholesale arrest of ships. We should proceed selectively where we have clear cases, and operate by a deterrent effect on others who may think they come near the line'. The Royal Navy should seize 'at least a dozen ships sailing within the next week, for arrest on grounds outside the penalties of international law' while rapidly developing spy networks 'so that we get really good information of cargoes which are German tainted, and do not have to rummage ships unnecessarily'. His subordinates shared this view. Before the Prize Court, the Trade Division claimed, real evidence of contraband status, such as proving 'probable enemy destination', was essential. 'These cases should for the present be comparatively few in number, and vessels should not be sent to discharge cargo unless the information is considered sufficiently definite to ensure condemnation in the Prize Court'.[22] The Procurator General's Department (PGD), which prosecuted contraband cases, and the Foreign Office, routinely bemoaned how lack of evidence wrecked arguments before the Prize Court.

The managers of economic warfare saw law and intelligence as a problem matching submarines, which would rise the more contraband was seized. Hence, retaliation must be restrained. These views were widespread, but misconstrued the power of intelligence in economic warfare. Three great sources were effective: cable and wireless censorship and official reports; while another was starting:

---

[22]   Minute by Churchill, 3 Mar. 1915, memoranda by DTD, 4 Mar. 1915 and Landan, 6 Mar. 1915, FO 800/909.

postal censorship; with a minor source: espionage. Good intelligence and the means to cross-check it were increasingly available. Modes of data processing were about to improve. The combination of communications intelligence and the Trade Clearing House (TCH), later renamed the War Trade Intelligence Department (WTID), solved the problem of blockade intelligence at the microeconomic level with astonishing speed. They enabled the 'evidentiary system', which proved the provenance of contraband, through evidence the Prize Courts accepted as proof – communications intelligence, so long as the material demonstrably was real, and entered as public evidence. Until the court showed it would treat intelligence as evidence, British authorities were unsure whether their new system of blockade would work. Then, this system became easy to execute, and to radicalise.

Evidence and Economic Warfare, 1915–1916

Following the Penson and Davis report, Cockerill and Hall immediately advocated what became two core elements of economic warfare: aggressive postal censorship and the coordination of intelligence on economic warfare, leading to the creation of the TCH, 'a central bureau of war trade and economic information which would receive intelligence from all available sources; would examine, collate, and re-issue it in a convenient form to all the departments interested'.[23] Their joint power re-engineered British machinery for economic warfare. By March 1915, the TCH occasionally provided reports from several sources on suspects. By April, that practice was routine. Its intelligence was good enough to help destroy major contraband organisations. By May, the TCH's familiarity with firms, their communications, and trade, and the increasing power of data retrieval, enabled it to offer powerful analyses regularly. By June 1915, the TCH settled into a pattern which continued through the war. Its basic sources were official reports, newspapers, and intercepted cables, with some wireless messages between neutral and German firms, which, 'edited and studied by a special staff, proved to be specially constructive'. Agents' reports were small in number, and used only with care, after their accuracy was cross-checked with other sources.[24] Intercepted transit letters were the prize source, regularly receiving detailed analyses. The head of the Report Section wrote in early 1916,

> The Transit Letters are the most valuable evidence that the Censors bring to light. Cables are usually ambiguous; letters to and from this country are written in the expectation that they will be censored. It is only in the Transit Letter, which he thinks may not be opened, that the

23   Carless Davis, *History of the Blockade: Emergency Departments*, pp. 58–9.
24   Carless Davis, *History of the Blockade: Emergency Departments*, pp. 143–8.

enemy frankly and fully reveals the details of his business. The evidence derived from Transit Letters is invaluable to the Contraband Committee, and the Procurator General. The Transit Letters are also of good value to the War Trade Department, both in dealing with licences and in answering the inquiries of British firms about their neutral customers.[25]

Meanwhile, Cockerill and Hall also pressed to censor transit mail, challenging prevailing concepts of law on grounds of necessity. On 23 December 1914, an inter-departmental meeting rejected that idea. As Carless Davis later noted, 'The opinion still prevailed in many quarters that the Postal Censors should devote their energies mainly, if not exclusively, to the detection of espionage and to the prevention of trade between the enemy and persons resident in the United Kingdom'. The Foreign Secretary, Edward Grey, rejected Cockerill's request for a warrant authorising the opening of transit mail moving through Britain from Latin America. In January 1915, however, Cockerill reinterpreted existing warrants to enable that interception. That reinterpretation, on the extreme end of exegesis and contradicting the assumptions of the original instrument, if not its wording, carried the interception of international mail into new legal waters. The censors examined the politically most innocuous category of transit mail. A 'batch of business letters' from South America to Holland, via Britain, had correspondence for Germany: 'A strong impression was left by the examination that, if these letters were a fair sample of the correspondence between Germany and America which passes through this country under neutral cover, a rigorous censorship' was needed. These letters 'contained much valuable information as to the destination of cargoes and other matters' like cover names and arrangements for secret communications, which 'materially assisted the enemy in obtaining supplies and credit'. The censors thereafter censored all transit mails from Latin America which reached British ports, and aimed to stop all letters which 'would materially assist the enemy in any manner, except those to United States'.[26] In fact, initially they just opened letters between businesses, having no intelligence to guide their efforts, though that situation rapidly improved. Cockerill took these steps without requesting a new warrant, or even informing the Home Office, but with the tacit consent of the Foreign Office. Several neutral states claimed that the Universal Postal Convention of 1906 prevented such actions. The War Office replied that this convention merely prevented states from stopping, as against opening, mail. The Law Officers supported this position on 31 March 1915: 'it

---

[25]   'Memorandum on the Censoring of Transit Mails', HMSO, By Trade Clearing House and Trade Branch, Postal Censorship, n.d., *c*.1 Jan. 1916, in D. Stephenson, K. Bourne, and D. Cameron Watt (gen. eds), *British Documents on Foreign Affairs, Part II, Series H, The First World War, 1914–1918*, vol. 6, *Blockade and Economic Warfare*, II, *July 1915– January 1916* (Frederick, MD: University Publications of America, 1989), pp. 385–94.
[26]   Director of Military Operations to Foreign Office, 18 Jan. 1915, FO 371/2515, 8184; *Report on Postal Censorship, 1914–1919*, DEFE 1/131, pp. 20–32.

would indeed be a ridiculous construction' of the Universal Postal Convention 'to suppose that it bound a belligerent state to make itself the conduit-pipe for communications intended to defeat its own measures of war'. This decision killed some old rules regarding the censorship of neutral mail, strengthening Cockerill against McKenna and Grey.[27]

As unrestricted submarine warfare loomed, postal censorship finally became efficient, but it was far behind the competence of those censoring cable and wireless. The postal censors opened only 40 per cent of letters on the terminal mails which it censored, and few transit letters. By March 1915, the MID and the War Trade Department (WTD) concluded that Britain needed 'a continuous and thorough examination of letters'. The postal censors must read all terminal mails between Britain and European neutrals. This task would require an increase in staff from 300 to 750, which would enable 'an extra provision for special examination in other cases, although it may be later on advisable to censor the whole of certain extra European mails'. The censorship was divided into a 'private section', with 25 censors, and a 'trade section' of 45 censors, including four 'with special commercial experience'. The postal censorship emphasised the 'special trade experience required from those in the trade branch and also the highly confidential nature of the work', but placed it under the head of the private section, to keep proper focus on counterespionage, which was 'the primary object of the censorship'.[28] The services and the Foreign Office agreed that transit letters passing through Britain and carried on British ships should be censored, but not those on board neutral vessels, even when they were in British ports. That would be a breach too far of the inviolability of mail. Ministers authorised this proposal, but with misgivings. Reading the letters of innocents was wrong, illegal, and would outrage neutral people and states. When Reginald McKenna received a draft copy of the warrant for this purpose, his reply to Grey personalised the formal terms of interdepartmental correspondence:

> Before, however, he signs this warrant, Mr. McKenna would be glad to know more precisely what Sir Edward Grey's views in the matter are. It should be clearly understood that if the Warrant is issued in this form, it will be inevitable that large masses of innocent correspondence will be opened and sent on marked 'opened by the Censor'. The secret opening of letters is possible only on a very small scale, and, where any particular person or address is suspected, the correspondence to that person or address can be, and is now, examined.[29]

---

[27]   *Report on Postal Censorship, 1914–1919*, DEFE 1/131, pp. 20–2.
[28]   War Office to Treasury, 4 Mar. 1915, 63/2343 (F.2), T 1/11971/5894; War Office to Treasury 12 Mar. 1915, T1/11971/6745; War Office to Treasury, 8 Aug. 1915, T1/11971/19074.
[29]   Home Office to Foreign Office, 254,721/157 23 Apr. 1915, FO 371/2522, 49058.

The warrant was signed as drafted but enacted slowly. Initial searches in May to June 1915 were limited to transit mails carried across Britain. 'As it was not thought possible to censor all the transit letters without unduly delaying them', the censors simply opened the letters of suspects identified by the TCH, 'subsequently modified to enable censorship examiners who were convinced from previous experience that certain other trade letters should be examined'. The number of opened letters rose with the TCH's power. This censorship discovered much intelligence about contraband, especially 'large dealings in cotton and other commodities' with Sweden, and more suspects.[30] After this haul became public, however, rapidly neutrals sent their mail by neutral vessels, beyond Britain's reach.[31] Despite this limit, over coming months, letter intercepts proved the key source of intelligence for every function of economic warfare, teaching consumers the problem at hand, and particularly of evidence in the Prize Courts. One letter, intercepted from the German consul in Haiti, provided a list of hundreds of names of neutral firms which aided German contraband. In 1916, the PGD concluded that without the aid of transit letters, virtually all of its cases would collapse, and economic warfare too. A senior member of the PGD, R.W. Wood, wrote:

> in practically every contraband case, in which there is any prospect of success in the Prize Court, the information upon which the Crown rely is derived from letters intercepted in the post ... and, if the censorship of the through steamers is to be discontinued, it is to be anticipated that the chances of success in the Prize Court will be so diminished as practically to cease to exist.[32]

By October 1915, 'test examinations of portions' of the transit mails between the United States and European neutrals showed much correspondence about contraband. The MID and the NID agreed that censoring all of 'the American mails' was 'absolutely necessary in order to ensure the effectiveness of the censorship as a whole'. The Foreign Office agreed, 'on the understanding that measures will be taken to interfere as little as possible with correspondence of an innocent character'.[33] The postal censors and the TCH emphasised that smugglers would exploit any gap in the censorship, so compromising the whole.[34] Yet such

---

[30]   Pearson, 'Memo': FO 371/2536, 84884; memorandum by Armstrong, Strand House, 2 July 1918, TS 13/138 A.

[31]   Minute by Montgomery, 22 June 1915, FO 371/2536, 84884.

[32]   R.W. Wood, PG Department, 24 June 1916, 'The Value of the Censorship in Prize Work', TS 13/138 A.

[33]   War Office to Treasury, 22 Oct. 1915, T 1/11971/25555; War Office to Treasury, 11 Oct. 1915, T1/11971/23900.

[34]   Stephenson, Bourne, and Cameron Watt, *British Documents on Foreign Affairs*, vol. 6, *Blockade and Economic Warfare*, pp. 385–94,

actions would repudiate all pre-war liberal legalist efforts to rewrite international law, including the Hague Convention, which Germany already had denounced. In a last stand for legalism, many ministers questioned the concomitants, that all neutral shipping must be forced into British ports for examination, and all transit mails censored. During December 1915, blockading cruisers and blockade runners played cat and mouse. Politicians were split over policy, while a belief that 'for political reasons it is of the utmost importance that the Christmas parcel mail should reach Scandinavia' caused a demand for a 'Very Urgent' telegram to be sent 'at once'. In this confusing environment, sailors forced neutral ships and their mails into British ports, which censors searched before the Cabinet formally authorised these actions. Even the Minister for Blockade, Robert Cecil, who thought it 'very important' to seize 'all mails', believed 'to stop mails on neutral vessels on the high seas will cause great wrath'. However, these searches demonstrated the necessity of these actions even to Grey, who months before had written, 'I have always doubted whether interference with *letters* was worth while'.[35] The parcel post carried contraband in quantity, and strengthened Britain's public case for these actions.

The government announced this policy in January 1916, after it had happened, using as precedent the fact that German cruisers destroyed transit mails bound for Britain across the Baltic.[36] The neutrals complained, but concurred. Sweden protested but the United States agreed that Britain could search the parcel post for contraband and sea-mail for letters on that topic, and seize any parcels or financial instruments bound for Germany. This position, the PGD noted, 'opens the door to a fairly extensive exercise of the right of search, which, after all, is the point of vital importance to retain, inasmuch as letters in indefinite numbers can be photographed and the photograph is quite adequate evidence for the purpose of Prize Court Proceedings'.[37] The State Department, however, denounced any practical delay and damage that censorship, or the blockade, caused to American trade.[38] Thus, speed and efficiency shaped the political value of postal censorship.

---

[35] *Report on Postal Censorship, 1914–1919*, DEFE 1/131, p. 31.

[36] Minutes by Cecil, 27.12.15, 30.12.15, FO 382/383, 107641; minute, no author, 'Parcel Mails in the "HELLVIG OLAV" and from Java', FO 382/382, 194626; K. Bourne and D. Cameron Watt (gen. eds), *British Documents on Foreign Affairs, Part II, Series H, The First World War, 1914–1918*, ed. David Stephenson, vol. 7, *Blockade and Economic Warfare*, III, *Jan.–Oct. 1916* (Frederick, MD: University Publications of America, 1989), pp. 1–2.

[37] C.D. Webb, 19 June 1916, 'The Censorship of Mails', TS13/138A.

[38] Stephenson, *British Documents on Foreign Affairs*, vol. 7, *Blockade and Economic Warfare*, pp. 219–21.

The Process of Postal Censorship

Postal censorship was sporadic during 1914–15. Though it produced significant intelligence, most letters from neutral and enemy countries crossed the Atlantic Ocean unscathed until January 1916. Then, the censorship caught most such letters, but three mouseholes remained. Britain did not normally censor letters carried by Spanish ships to and from Latin America, a sailor might be suborned to carry mail, while, until late 1917, the Swedish government carried some neutral and enemy letters in its diplomatic bags.

In order to appease neutral states and the American public, the postal censorship must work quickly. Bags of neutral transit mail were sent from ports by fast train to the censorship offices in London and Liverpool, processed immediately, and then returned to their destinations. All mail bags were opened, but not all letters were read. During the war, the postal censors touched 650 million postal packets, although commercial letters often included ten or more letters. With 2,000 to 4,000 examiners reading about 100 letters per day, the postal censors might have scanned perhaps 300 million letters during the war, but read few of them thoroughly. Astonishingly few of these letters actually were used. From hundreds of millions of envelopes touched between June 1915 to September 1917, the censors detained only 1,300,000 while the WTID generated only 15,516 Transit Letters.[39] Once the United States entered the war, censorship became harsher. By August 1918, another series of letter intercepts, for the War Trade Department, had reached the number WTD 81,431.[40] These numbers would have been far greater had postal censorship been more ruthless and powerful before 1916. Skilled staff and the list of names provided by the WTID were essential to this system, but even so, no doubt many unknown unknowns slipped through. In order to inspect the mail bags of every neutral vessel in British ports, Britain added another 62 censors and 1,015 examiners to its present staff of 79 and 953.[41] At its peak, the postal censorship involved 4,800 people, 75 per cent of whom were women. Women proved 'especially useful' in censoring private letters, the censorship held, 'as they have an excellent memory for detail, and can recognise peculiarities of handwriting, form and expression, which they have once remarked in a letter'. This large female staff marked the organisation of the postal censorship, including the maintenance of a nurse with volunteer assistants at the offices in London, the appointment of a senior female censor (ranking within the top four officials in the agency, the highest post reached by any woman in British signals intelligence during the twentieth century) to oversee all women

---

[39]   TL 15516, German Post Office Shanghai to Denmark, 6 Aug. 1917, FO 382/1770; *Report on Postal Censorship, 1914–1919*, DEFE 1/131, pp. 71, 328, 331.

[40]   Minutes of meeting on Swedish Black List Cases, 16 Aug. 1918, TS 13/689.

[41]   War Office to Treasury, 22 Oct. 1915, T 1/11971/25555; War Office to Treasury, 11 Oct. 1915, T1/11971/23900.

employees, and the decision not to adopt shift work at Liverpool, where the offices stood in the docks, and trams closed at 23.30. In both offices, Girl Guides carried messages and staffed the lifts.[42]

Postal censors were instructed above all else to prevent the passage of military information and espionage letters. Letters were to be treated with respect, opened cleanly, one at a time so contents never became confused or lost, and resealed with a censors' label, without any effort to hide the fact they had been read. An elaborate and labour-intensive routine, which absorbed more effort than the actual acquisition of intelligence, aimed to return every innocent letter to the same bag from which it came, with all of its contents and nothing more. Among the lower staff, 'sorters', working solely from 'Suspect Lists' (addresses and categories of mail, and names of senders or addressees), assigned the letters to various groups of 'examiners', who read them, identified suspicious letters, and passed them for detailed inspection by a censor, or specialised departments. Suspicious features included concealed writing, strange phrases, indications of plain language codes, or any references to contraband. Examiners sat at specific tables, supervised by a Deputy Assistant Censor charged to enforce regulations, read any letter deemed suspicious, and forward to experts those which really were so. Some experts, veterans of specific trades, knew their jargon and practices. Assigned the correspondence from specific localities and using local trade directories, they acquired expertise in its normal patterns of business, and the ability to sense unusual developments. The censors maintained many card indexes, including a 'Watching List', which coordinated and simplified the rankings from other handbooks, and a Central Registry with a staff of 65 at its peak. The postal censors advertised publicly for applicants, interviewed them twice, and vetted them through MI5. They created a formal system of instruction into procedures, 'the subterfuges used by enemy agents to evade censorship, the signs of secret writing, and the ordinary tests for detecting its presence'. The censors even offered language classes.[43] The postal censorship and the army provided far more thorough training for signals intelligence than any other such agencies in Britain.

The censors maintained a 'Testing Department', with 'a few technical experts' in photography, invisible inks, chemistry, and cryptography, including one 'in musical codes', which were used with espionage messages, and thirty women assistants.[44] A specific staff handled all registered letters. 'Special Sorters' were the only people allowed to open the letters of celebrities, so to prevent loose talk.[45] No personnel were to discuss the contents of any letter with unauthorised

---

[42]   *Report on Postal Censorship, 1914–1919*, DEFE 1/131, pp. 321, 329, 330.

[43]   *Report on Postal Censorship, 1914–1919*, DEFE 1/131, pp. 179, 313, 320–1, 352.

[44]   War Office to Treasury, 8 Aug. 1915, T1/11971/19074; *Report on Postal Censorship, 1914–1919*, DEFE 1/131, pp. 407–8.

[45]   *History of the Postal and Telegraph Censorship Department, 1938–1946*, vol. 1, DEFE 1/333, p. 168.

people.[46] Significant letters, especially those which might have to be entered as evidence in court, were photographed. Usually they went on to their final address, though some were held to the end of the war, and then delivered, with only a few destroyed in 1919. Censors required high standards of 'general education, reliability and personal fitness', and often unusual linguistic skills.[47] Female examiners, one demi-official account of 1916 noted, 'read only the letters that do not require a business education. They pass on to a competent male censor all letters dealing with commercial or financial matters'. In fact, around this time, after receiving a month's training in specific trades, women censored business letters, where they proved 'efficient'.[48] The censorship maintained experts in 47 languages (from Albanian to Zulu, including Braille and Esperanto), and regularly consulted experts in London regarding another nine languages. Sometimes, they approached Edward Dennison Ross, the director of the School for Oriental and African Studies, who read 49 languages himself. An 'Uncommon Languages Section' tackled those topics. It was carefully watched, as many of its staff were not British subjects, raising concerns over loyalty.[49] Thus, Russian and Polish Jews censored letters written in Yiddish, Ladino, and Hebrew. For political reasons, in 1918, the postal censorship ended this practice as 'undesirable'. Given 'the great difficulty in replacing these persons by reliable British subjects, capable of reading the languages in question but not liable to be influenced by racial sympathy', the censors blocked all such letter unless written in Roman letters, and 'without the use of Hebrew characters which formed the chief difficulty to the censors'.[50] The 'Uncommon Languages Section' provided fortnightly reports on political developments, especially among Czechoslovak, Polish, Yugoslav, and Zionist political organisations. 'Correspondence with bookmakers, literary agencies, fortune-tellers, pseudo-scientific instructions and the like, is not forwarded by the British Censor as the volume and obscurity of such correspondence make it difficult to control'.[51]

Postal censorship involved constant 'quick decisions on subjects of great variety and difficulty'. The censorship emphasised the need to maintain 'the interest of the girls employed. If that flagged, they might let through some important information. As no one girl acts as supervisor to another so for passing letters each girl is a final voice'.[52] Similarly, in 1945, the authorities in Cable and

---

[46]　'Instructions to the Postal Censorship Staff', War Office, 1 Oct. 1916, 'Instructions to the Trade Branch of the Postal Censorship', 20 May 1915, T 1/11971/28061.

[47]　War Office to Treasury, 22 Sept. 1915, T 1/11971/19679.

[48]　Sydney Brooks, 'Censorship Seeks to Balk German Use of Mails', *New York Times*, 22 Nov. 1916, p. 5; *Report on Postal Censorship, 1914–1919*, DEFE 1/131, p. 321.

[49]　*Report on Postal Censorship, 1914–1919*, DEFE 1/131, p. 126.

[50]　MID to FO, 19 May 1918, FO 211/456; *Report on Postal Censorship, 1914–1919*, DEFE 1/131, pp. 409–10.

[51]　War Office to Colonial Office, 10 Oct 1918, 63/2/10 (M.I.9) CO 323/782, 49383.

[52]　Unsigned memo, Treasury, Dec. 1915, 'Postal Censorship', T1/11971/28061, *Report on Postal Censorship, 1914–1919*, DEFE 1/131, p. 79.

Letter censorship concluded that 'Censorship is, as it were, an inverted pyramid, dependent more upon the intelligence of the junior than the skill of the senior'. Censorship did not follow 'the normal functioning of a normal administrative or executive organisation', where brains were needed more at the top than the bottom. Instead, censorship

> rests on the intellectual effort of each individual examiner. The most important members of the staff are the most junior. Ninety-nine per cent. of the communications read are seen by no one else. If an examiner lets slip a piece of information of value to the enemy the slip cannot be recovered. If the examiner misses a piece of intelligence it is lost for ever. The examiners work in large numbers and in large aggregations, but the essential work of censorship remains personal and individual. The reading of each communication is a separate operation requiring application of thought, judgment and discretion on the part of the examiners. The maintenance of an alert active interest in the minds of the examiners is always of more vital concern than the creation of a neat uniform system. The perfect, though unattainable, ideal is that the whole *apparatus criticus* of the Censorship machine and the using departments should be at the disposal of each examiner as he or she reads each communication.[53]

The Alchemy of Victory at Sea

Between 1916 and 1918, far-distant weather-beaten ships, and British dominance in communications, coal depots, and insurance, commanded neutral merchantmen in the Atlantic Ocean. Neutrals accepted British rules for international commerce, and the authority of its Prize Court, with reluctance. Only fair judgments, based on compelling and publicly available evidence about the provenance of contraband, could sustain this consent. Thus, communications intelligence and law became a sharp end of sea power. At its peak, the TCH received roughly 1,000 intercepted cables per day and 700 letters per month. Material on firms was constantly updated, with intelligence collated on cards. The Traders' Index had 250,000 cards by the end of 1915, and grew by 4,000 to 5,000 cards each week, probably reaching 1 million by the end of the war. Remarkable data processing allowed thorough and rapid analysis and retrieval of this material, and dissemination to consumers, within hours of a request for information.

Intelligence on contraband was assessed in two ways, by distinct groups: to justify holding a shipment, and to win a case condemning it. Through a

---

[53] *History of the Postal and Telegraph Censorship Department, 1938–1946*, vol. 1, DEFE 1/333, pp. 8, 33.

quick essay of evidence, the interdepartmental specialists on the Contraband Committee, the first line of defence, held all suspicious items for investigation, and rapidly passed the innocent, which was essential in diplomatic terms. The Contraband Committee became ever more confident and discriminate. It received manifests for every ship calling in British ports, cross-checking each item of cargo against all compiled and indexed information. Each day intelligence or suspicion held cargoes for further investigation, by collecting material from consuls, or shippers. With more hard material than before, on innocents and suspects, the Contraband Committee instructed customs on what cargoes to search, and which to ignore, though random investigations remained an important tool. Behind this line, the Naval Trade Division and the Contraband Department of the Foreign Office used intelligence in detail to address strategies for attack or defence. Finally, after a thorough examination of all the evidence available, the PGD determined whether a successful case could be made in Prize Court, and specific items should be held or released. The Contraband Department usually accepted this recommendation.

O.R.A. Simpkin, a Chancery barrister and leader in blockade intelligence, later wrote: 'the ultimate sanction of the Blockade is the Prize Court'. Seizure of goods required 'evidence, or at least a strong presumption, of intended enemy destination or origin such as would satisfy a Court of Law'. Intelligence confronted 'a two-fold problem: namely (a) the collection of facts constituting evidence or creating a presumption, of enemy destination or origin; and (b) the application of these facts accurately and rapidly' to ships and cargoes.[54] Whenever a cargo was assigned to the Prize Court, the PGD continued to pursue evidence on it. Here, intelligence became evidence, almost all of it communications intelligence, with letter intercepts the crown jewels. This evidence, publicly available, convinced both the court and neutrals that however inconvenient the British system of economic warfare might be it was fair. One member of the PGD wrote:

> I have hardly a case in which I can see any prospect of condemnation without the help of intercepted letters, and certainly not without the help of intercepted cables. Further, the latter, even when very good ones from this point of view, are seldom definitively damning, but usually only raise suspicion or a prima facie case, which is confirmed by a letter. The letter may even be necessary before the cables can be understood – not only where the letter gives a code, but because the letter (though possibly not referring to the particular transaction in question) shown [*sic*] the relationship or course of dealing between the parties, or it may be simply because the heading or signature of the letter shown [*sic*] the members of a firm. I don't think it is too much to say that without intercepted letters, we shall not secure condemnation of half the goods

---

[54] Carless Davis, *History of the Blockade: Emergency Departments*, pp. 23–32.

in the Prize Court, even if cables etc. are intercepted as heretofore: and if the interception of cables is also given up, not one per cent will be condemned.[55]

This system provided the highest possible marginal balance of the factors involved in economic warfare. It minimised the movement of contraband to and from Germany, British interference with legitimate neutral trade, and neutral anger at British interference with their interests. Blockade was all too easily employed as a battleaxe, rather than a scalpel. It could damage relations with firms and states in a counterproductive fashion. Intelligence helped Britain to wield it with some accuracy. The effect of blockade remains a vexed question, but intelligence was a fundamental factor on the margin, minimising the damage to Britain while maximising that on the enemy. Otherwise, the blockade might have failed, or else caused more damage to Britain, than to Germany.

---

[55]   Memorandum by R.H.S., undated, but June 1916 by internal evidence, 'Censorship', TS 13/138 A.

CHAPTER TEN

Britain and Economic Warfare in German Naval Thinking in the Era of the Great War

Matthew S. Seligmann

On two occasions in the twentieth century German naval forces waged what amounted to the most destructive campaigns of commerce warfare ever undertaken by any nation. The scale of the losses inflicted upon Britain, its Empire, and its allies, and the comparatively short timeframes in which these results were achieved, marked the apex of commerce warfare and remain to this day the prime expression of the industrialisation of destruction at sea. Across the two world wars over 11,000 allied merchant vessels, amounting to some 34 million gross registered tons of shipping, were sunk by German forces. Yet, remarkably, this was achieved by a naval establishment that is largely seen by historians as having possessed little or no interest in commerce warfare and a state which placed scant emphasis on economic campaigns in its national strategy and war-planning. Indeed, prior to both world wars, it was the development of a battle fleet that absorbed the vast majority of the material and intellectual efforts expended in the creation of German naval power; *guerre de course* was seemingly rejected.

These subsequent wartime successes, achieved in spite of the apparent lack of interest in such methods of applying naval pressure, have tended to be explained as unexpected developments and the products of adaptation, geopolitical gains, and technological innovations that changed the means available to the navy with which to prosecute commerce warfare. The use and evolution of the U-boat as a commerce warfare tool ultimately was the key factor; yet it was at odds with pre-war thinking in both wars. The historiography tends to focus on the individual wars or operational, technological, and human aspects rather than providing a broad overview of commerce warfare in German naval thinking across the first half of the twentieth century. This chapter will offer an alternative to this explanatory framework. It will provide a view of how this thinking evolved, the political and strategic factors that shaped the navy's relationship with commerce warfare. The focus lies on Britain as it bore the brunt of the German attack and was the lynchpin of the alliances against Germany. It addresses aspects such as how Britain's economy was understood, and whether and how any economic vulnerability should be exploited.

I

If ever there was a figure that epitomised the rejection of the *guerre de course* in favour of an emphasis on the decisive role of the battle fleet in a traditional *guerre d'escadre* it was Alfred von Tirpitz. Appointed State Secretary at the Imperial Navy Office in 1897, he guided German naval development across the succeeding two decades, only bowing out in the middle of the First World War. There is a common and widespread assumption that under his prolonged leadership the German navy not only had no interest in economic warfare, but actually found the concept 'repulsive'.[1] Two factors underpin this belief. First, Tirpitz was dismissive of the prospects for economic warfare in his famous memorandum of 15 June 1897 in which he outlined to Kaiser Wilhelm II the rationale for his proposed naval programme. As he explained, although for 'Germany the most dangerous naval enemy … is England', Germany could not realistically hope to use naval force to apply economic pressure to Britain. His reasoning:

> Commerce raiding and transatlantic war against England is so hopeless, because of the shortage of bases on our side and the superfluity on England's side, that we must ignore this type of war against England in our plans for the constitution of our fleet.

Accordingly, he maintained that what was needed was a traditional battle fleet that could be used to weaken British maritime power in a sea engagement. 'The military situation against England', he proclaimed, 'demands battleships in as great a number as possible'.[2] That this was not mere rhetoric seemed apparent from what happened next. As he had promised, Tirpitz built a fleet strong in battleships – the type of warship required for a sea battle – and weak in cruisers – the vessels most needed for a campaign of economic warfare.[3]

That was not all. Although Tirpitz's battleship-centred programme for German naval power development coincided with the view made prevalent by

---

[1]   See, for example, Holger H. Herwig, 'The Failure of German Sea Power, 1914–1945: Mahan, Tirpitz, and Raeder Reconsidered', *International History Review* 10 (1988), p. 73; Gary E. Weir, *Building the Kaiser's Navy: The Imperial Navy Office and German Industry in the von Tirpitz Era, 1890–1919* (Annapolis, MD: Naval Institute Press, 1992), p. 77.

[2]   Tirpitz, 'General Considerations on the Constitution of our Fleet according to Ship Classes and Designs', 15 June 1897. Quoted in J. Steinberg, *Yesterday's Deterrent: Tirpitz and the Birth of the German Battle Fleet* (New York: Macdonald, 1965), pp. 208–11.

[3]   Key texts on Tirpitz and his battleship-building programme, include V.R. Berghahn, *Der Tirpitz-Plan: Genesis und Verfall einer innenpolitischen Kreisenstrategie unter Wilhelm II* (Düsseldorf: Droste Verlag, 1971); M. Epkenhans, *Tirpitz: Architect of the German High Seas Fleet* (Washington, DC: Potomac Books, 2008); P.J. Kelly, *Tirpitz and the Imperial German Navy* (Bloomington: Indiana University Press, 2011).

the American naval officer and writer on naval affairs Alfred Thayer Mahan that commerce warfare was the 'strategy of the weak', and thus to be avoided, whereas seeking command of the sea after a decisive battle was the correct historically validated approach for a serious naval power, it was not universally accepted by Germany's naval establishment. There were a number of naval theorists, mostly retired naval officers, such as Captain Lothar Persius and Lieutenant Commander Franz Rust, who wrote about naval affairs, who believed that far from economic warfare against Britain being hopeless it was the attempt to build a fleet to rival the Royal Navy that was destined to inevitable and catastrophic failure. Many of them – for example, Vizeadmiral Karl Galster – were ultimately ready to say so publicly in the popular press. Tirpitz came down hard on such critics. Not only did the government's News Bureau commission special articles criticising the ideas of such people, but orders were also issued for Galster to be ostracised from the naval officer corps. A similar fate awaited any other former officer turned commentator who had the temerity to dissent from the Tirpitz plan in public. As for serving officers, they were strictly forbidden from publishing anything that ran counter to Tirpitz's battleship doctrine.[4] All of this seemed to underline the idea that under Tirpitz the battle fleet strategy was the sole orthodoxy and that any emphasis on economic warfare would be treated as a form of heresy.

Reinforcing this idea was what happened at the outbreak of the First World War. The German naval command had always intended to use submarines against the Royal Navy's warships and, indeed, achieved some notable early successes in this regard, most famously the sinking by U-9 of the three armoured cruisers *Aboukir*, *Cressy*, and *Hogue* on the Broad Fourteens on 22 September 1914. However, prior to 1914 there were no plans to wage economic warfare against merchant vessels by such means, although the idea had been advocated by several of Tirpitz's critics, Persius in particular, for many years.[5] The apparent discovery early in the war that U-boats might actually be highly effective in this capacity serves to underscore the fact that this had not been appreciated previously, an oversight that says much about the seeming inadequacy of the planning for a campaign against British trade and, hence, the lack of interest in economic warfare.

While neither of these points is in itself wrong, they can be fundamentally misleading. Once again, two factors underpin this. First, Tirpitz was not as opposed to economic warfare as his 1897 memorandum might lead one to suppose. As Jost Dülffer has shown, in interdepartmental battles over the German stance at the second Hague Conference, he argued strongly, repeatedly, and on

---

[4]   C.-A. Gemzell, *Organization, Conflict, and Innovation: A Study of German Naval Strategic Planning, 1888–1940* (Lund: Esselte Studium, 1973), pp. 56–61; W. Deist, *Flottenpolitik und Flottenpropaganda: Das Nachrichtenbureau des Reichsmarineamtes, 1897–1914* (Stuttgart: Deutsche Verlags-Anstalt, 1976), pp. 250–7.

[5]   G.E. Weir, 'Tirpitz, Technology and U-Boat Building, 1897–1916', *International History Review* 6 (1984), pp. 180–1.

his own for maintenance of the right of capture of private property on the high seas;[6] and he did so because he wanted to apply economic pressure to Britain. As he explained in one memorandum from February 1907:

> Given the fact that English policy is almost completely determined by the interests of the 'City', decisions on war and peace also depend essentially on the wishes of the big traders. As long as the right to take prizes exists, the interests of English trade, and that means the interest of the 'City', will be exposed to considerable danger in the event of war.[7]

He repeated the point a few months later:

> The only course of action open to us in order to make England feel the effects of war would be to apply in the most reckless way the right of capture ... We do not have any other means and England knows this. A blockade of England imposed by us is out of the question for the foreseeable future. Landings in England in the manner of Napoleon are a pipedream as long as the English fleet exists ... But ... as the right of capture interferes with commerce and the City of London is very sensitive in this matter, the maintenance of the right of capture constitutes a peacekeeping factor for us ... The English city man must consider that, as long as the right of capture persists, a war against Germany requires a considerable investment.[8]

Some historians have expressed incredulity about whether Tirpitz really meant what he said in such memoranda.[9] However, given the mercantilism that underpinned much of his thinking, it seems likely that he was sincere, that he genuinely thought that the City dictated British policy, and that he therefore regarded a credible economic warfare threat as essential for browbeating Britain.

---

[6]   Jost Dülffer, 'Limitations on Naval Warfare and Germany's Future as a World Power: A German Debate, 1904–6', *War and Society* 3 (1985), pp. 23–43; the same point, made more briefly, can be found in I.N. Lambi, *The Navy and German Power Politics, 1862–1914* (Boston: Allen & Unwin, 1984), p. 235.

[7]   Dülffer, 'Limitations', p. 35.

[8]   Tirpitz, 'Memorandum concerning the Significance of the Right of Capture', 20 Apr. 1907. Johannes Lepsius et al. (eds), *Die Große Politik der Europäischen Kabinette, 1871–1914*, 40 vols (Berlin: Deutsche Verlagsgesellschaft für Politik und Geschichte, 1922–7), xxiii, pt. 2, pp. 361–7. A full translation can be found in M.S. Seligmann, Michael Epkenhans, and Frank Nägler (eds), *The Naval Route to the Abyss: The Anglo-German Naval Race, 1895–1914* (Farnham: Ashgate for the Navy Records Society, 2015), pp. 222–6.

[9]   Rolf Hobson, *Imperialism at Sea: Naval Strategic Thought, the Ideology of Sea Power and the Tirpitz Plan, 1875–1914* (Leiden: Brill, 2002), p. 282.

The second factor that militates against the idea of a total lack of interest in economic warfare in Tirpitz's navy was the organisation of Germany's naval administration. On its foundation in 1871, the German Empire had adopted a unitary structure for control of German naval affairs, with responsibility for all aspects of this resting with a single body, the *Kaiserliche Admiralität* or Imperial Admiralty. However, in 1889, barely a year after ascending the throne, Kaiser Wilhelm II decided to fragment this structure, dividing responsibility for naval affairs among three main agencies: the *Reichsmarineamt* (Imperial Navy Office), the *Kaiserliches Marinekabinett* (Imperial Naval Cabinet), and the *Oberkommando der Marine* (Naval High Command), the last of which was transformed in 1899 into the *Admiralstab* (Admiralty Staff). Each of these organisations had unique and distinct areas of competence. The consequence of this was that even if Tirpitz had been totally opposed to economic warfare – and, as we have seen, he was not – the office he headed – the *Reichsmarineamt* – was not the one responsible for war planning. That task lay with a different agency – the *Admiralstab* – and the heads of this body consistently saw value in economic warfare or, as they termed it, *Handelskrieg*. There were two reasons for this.

First of all, they recognised Britain's vulnerability to a disruption of seaborne trade. It should be said that no great acumen or analytical skill was required in order to draw this conclusion. Ever since the middle of the nineteenth century, Britain had ceased to be agriculturally self-sufficient and needed to import the bulk of its food supplies. This was no secret and, as a result, at frequent intervals, the question was ventilated in the public arena as to what this implied for national security. The late nineteenth and early twentieth centuries saw a recrudescence of debate on this issue. For example, in 1897, George Clarke, the future secretary to the Committee of Imperial Defence, published an article in the *National Review* warning of the dangers of war-induced shortages.[10] More striking was the spectre of outright famine conjured by retired army officer Stewart Lygon Murray.[11] His 1901 paper, 'Our Food Supply in Time of War and Imperial Defence', was just an initial step in a comprehensive campaign to alert his countrymen to the fact that unless active steps were taken war could mean starvation for the British people.[12] Such was the anxiety that these and similar works created that in the end the government felt obliged to investigate the matter and to do so multiple times. Thus, in 1903, the administration of Arthur James Balfour appointed a Royal Commission on the Supply of Food and Raw Materials in Time of War, the report of which was published in 1905. Hard on its heels was the 1908 report of the Committee on a National Guarantee for the

---

[10]   Clarke, 'War, Trade and Food Supply', *National Review* 29 (1897), pp. 756–69.
[11]   A. Offer, *The First World War: An Agrarian Interpretation* (Oxford: Clarendon Press, 1989), pp. 223–5.
[12]   Stewart L. Murray, 'Our Food Supply in Time of War and Imperial Defence', *Journal of the Royal United Services Institution* 45 (1901), pp. 656–729.

War Risks of Shipping. Both of these documents were testament to the anxiety that conflict might bring serious economic dislocation and that this in turn might lead to domestic unrest. All of this was played out in the public domain and, as a result, it is hardly surprising that the German Navy, which closely monitored such developments, was aware of all the details. For example, Carl von Coeper, the German naval attaché in London, assiduously reported on the activities of the Royal Commission on the Supply of Food and Raw Materials in Time of War.[13] He also notified Berlin when a year later a major British newspaper, *The Standard*, ran a series of articles on the threat that war might pose to essential imports.[14] For such reasons, it was more or less impossible for the authorities in Berlin not to recognise that in the interdiction of the flow of food and raw material supplies to Britain lay a weakness capable of being exploited. Certainly, it did not escape public notice, with articles on the topic by the economist and close Tirpitz ally Ernest von Halle, appearing in the key navalist journal, *Marine Rundschau*, in both 1906 and 1908.[15]

On the basis of the information received from London, the *Admiralstab* conducted its own analysis of British vulnerability to economic warfare and the implications of this for German war strategy.[16] Certain definite conclusions were quickly reached. To begin with, the officers on the *Admiralstab* were clear that the battle fleet remained Germany's primary maritime weapon and that economic warfare would not be the decisive instrument of war. However, they were equally clear that acceptance of the primacy of Tirpitz's battle fleet strategy did not preclude an important role for *Handelskrieg*. As they explained, a German attack on British merchant shipping launched vigorously from the very outset of any conflict and pursued across multiple seas and oceans across the globe must inevitably divert some Royal Navy assets away from the main theatre in northern Europe towards more distant trade routes, and that this weakening of British forces in the North Sea must consequently improve the odds for the German Navy in home waters. Moreover, they were certain that such a result could be achieved at minimal cost to Germany; a few overseas cruisers augmented with some suitable civilian vessels that were transformed into armed merchant cruisers would be more than sufficient to achieve the desired end. In effect, therefore, in their assessment, a carefully planned and determined

[13]   See, for example, Coerper to Tirpitz, 'Nahrungszufuhr im Kriege', 5 Feb. 1905 and 11 Aug. 1905. NHB: GFM 26/91.

[14]   P. Overlack, 'The Function of Commerce Warfare in an Anglo-German Conflict to 1914', *Journal of Strategic Studies* 20.4 (1997), p. 97.

[15]   E. von Halle, 'Die englische Seemachtpolitik und die Versorgung Großbritanniens in Kriegszeiten', *Marine Rundschau* 17 (1906), pp. 911–27 and 19 (1908), pp. 804–15. Cited in Dirk Bönker, *Militarism in a Global Age: Naval Ambitions in Germany and the United States before World War I* (Ithaca, NY: Cornell University Press, 2012), p. 156.

[16]   Admiralstab, 'Denkschrift betreffend die Lebensmittelzufuhr Großbritanniens in Kriegszeiten', 27 Apr. 1911. NHB: GFM 26/82.

Handelskrieg was a highly effective force multiplier for Germany. By distracting the Royal Navy and forcing a dispersion of resource it would enhance Germany's relative strength where that mattered. This conclusion, which regularly featured in German planning documents, became ever more prevalent with the passage of time as the argument was refined, schemes were developed, and 'All-Highest' sanction obtained. Seven snapshots, taken from key planning documents spread across an 11-year period, allow us to trace the inception and development of this idea in the German naval planning system.

One of the earliest detailed assessments of the value to Germany of an attack on British trade was written by Fregattenkapitän Max Grapow, the head of A4 Section of the *Admiralstab*, the division that oversaw operational planning against Britain. Dated 20 September 1902, and running to 23 closely typed pages, the Grapow memorandum took as its primary goal an assessment of the prospects and value of cruiser warfare against Britain. Grapow was clear that with careful planning and prior preparation of such essentials as hidden supply points a campaign could be launched in distant waters that targeted British cargo vessels. Furthermore, he had no doubt that it should be done. His reasoning was clear:

> Our fleet at home shall take every possible opportunity to demolish the enemy, but it must not expose itself to total destruction. In this difficult task the fleet can be significantly supported if our overseas and auxiliary cruisers undertake as quickly as possible a ruthless attack on British trade. By such means we can ensure that our opponent strengthens the forces in overseas waters and on his coast.[17]

Here was one of the earliest expressions of the idea that, through economic warfare, the odds of a successful battle fleet action could be materially improved and that, given the German fleet's inferiority in battleships to the Royal Navy, this was an essential component of a fully rounded strategy against Britain.

Grapow's conclusions found a distinct echo in a major 1906 evaluation. This was another substantial document and it was no less clear than its predecessor on why a campaign needed to be mounted against British floating trade.

> The goal of commerce warfare is through the energetic damaging of English seaborne trade and English shipping to bring before the eyes of all sections of the English population that, in spite of their strong fleet, war causes losses.[18]

---

[17] Grapow, 'Welche Aussichten bietet für uns die Führung eines Kreuzerkrieges gegen England?' 20 Sept. 1902. BA-MA: RM5/1610.
[18] A.102.II, 'Nebenkriegführung: Handelskrieg von der Heimat aus', 4 Apr. 1906. BA-MA: RM 5/1611.

As with Grapow's 1902 memorandum, this assessment proposed to use a mix of warships and armed merchant cruisers in order to create the global disruption that would lead to a dispersal of British force. However, it differed from its predecessor in one crucial respect. Whereas weakening the Royal Navy in home waters was the sole objective of the 1902 proposals, this document incorporated a strong element of psychological warfare. The aim was not just to force the Royal Navy to send cruisers to distant waters, it was also to undermine 'the feeling of absolute trade security' that the British public held owing to their belief in British maritime supremacy.

Three years later, in an audience with the Kaiser, the head of the *Admiralstab*, Friedrich von Baudissin, outlined the latest version of his organisation's ideas for economic warfare. The plan for attacking British trade was now more substantial and had been articulated in greater detail, with important points like rendezvous areas for the cruisers to meet colliers for refuelling being specified in advance. Consideration was also given to the best theatres for pressing the attack. As Baudissin informed the Kaiser:

> In a war against England operations against the steamers on the Argentine or Uruguay to England route are in the first instance recommended, since England can most effectively be harmed in cruiser warfare by disruption of the import of necessities. Great success can in the first instance be expected from such an energetic attack.[19]

Baudissin's successor as Chief of the *Admiralstab*, Max von Fischel, espoused similar views. On 24 January 1911, he outlined in an audience to the Kaiser his proposals for utilising *Handelskrieg* as a crucial component in a war against Britain. Recognising that at the start of a war the British high command would seek either to contain Germany's overseas naval assets or overwhelm them with superior force, he proposed to evade this and engage in a dispersed campaign against British commerce. As he explained:

> [Our] attack on trade should be so developed that our opponent is compelled to divide his sea forces as much as possible. In pursuit of this objective the auxiliary cruisers and the slowest of the small cruisers from our Cruiser Squadron will be dispatched to different areas, while armoured cruisers and the faster small cruisers shall together scour the most navigated trade routes.

After a discussion of the best deployment strategy and the advantages and disadvantages of several possible theatres of action, he went on to consider the likely results and benefits:

---

[19]   Baudissin, 'Zum Immediatvortrag', 2 Feb. 1909. BA-MA: RM 5/895.

A simultaneous disturbance in many areas would compel England to a costly division of its forces. The English cruiser squadrons must, in order to fulfil their task, be stronger in all places than the hostile German forces in each area. Or England must decide to introduce a convoy service on the trade routes.[20]

The latter might be an effective form of defence, but it would tie down considerable forces and make British floating trade slower and more expensive. As such it would achieve the objective of detaching British forces from the main theatre and undermining the British economy.

The issue was re-examined three months later by the head of the German East Asian Cruiser Squadron, Vizeadmiral Günther von Krosigk. Unsurprisingly for a naval leader whose forces were based upon the German concession of Tsingtau in China, Krosigk placed considerable emphasis on the German ability to cause depredations to British trade in the Pacific and Australasia. However, his fundamental analysis as to the useful role that economic warfare could play in support of Germany's battle fleet strategy reinforced the prevailing analysis. As he wrote:

It lies completely within the realm of the possible through cruiser warfare to bring about panic, thus leading to the detaching of considerable forces to distant regions of the world, which will contribute considerably to the relief of our operations in the home theatre and in conjunction with other operations make the English government more inclined to a peace favourable to us.[21]

As before, the goal was not just to produce a dispersal of force, important though this obviously was, but also to create a psychological impact in Britain, although the goal has now advanced from undermining the feeling of security to creating panic.

With the planners in the *Admiralstab* in broad harmony with the leaders of Germany's forces overseas over the role and function of *Handelskrieg*, it is little wonder that the process gained further momentum. This culminated in an important meeting between the Kaiser and the new Chief of the *Admiralstab*, Hugo von Pohl, in April 1913. In an imperial audience focusing on war planning against Britain, Pohl outlined the navy's latest strategic thinking and asked for and received 'All-Highest' approval to make economic warfare a major element in it. As he explained:

---

[20]   Fischel, 'Zum Immediatvortrag', 24 Jan. 1911. BA-MA: RM5/897.
[21]   Krosigk, 'Denkschrift über den Kreuzerkrieg im Kriege gegen England', Apr. 1911. Quoted in Peter Overlack, 'German Commerce Warfare Planning for the Australia Station, 1900–1914', *War and Society* 14 (1996), p. 44.

> Your Majesty has deigned to command that in the event of war our ships stationed abroad are to engage in cruiser warfare. In a cruiser war against England most targets will be found in the Atlantic Ocean across which all of England's imports of food and raw materials are transported.[22]

In proposing this, Pohl did not believe that such operations would be easy. However, he argued that with proper and detailed prior preparation a force of warships and auxiliary cruisers could be placed in the Atlantic and used to target British trade. The Kaiser's explicit endorsement, which came five days later, leaves no doubt that this was not just the recommended course of action; it was now official policy. Accordingly, it was reflected in the 1913 war orders, which, for ships stationed overseas, stated:

> In the event of war [against Britain] the damaging of English trade and of English shipping to England and to and from its colonies, especially the halting of imports of necessities to England, stands in the forefront. Such damage, will through its indirect effect on marine insurance and commercial undertakings among English trading circles have a greater impact the sooner to the start of the war it is successfully undertaken.[23]

Eight months later the point was further reinforced in an *Admiralstab* paper looking at commerce warfare in the light of international law. The author of the paper reasserted the now widely held orthodox view that attacking British trade could bring huge benefits. As the memorandum argued:

> The goal of the war against commerce is to create anxiety among the English people through damaging English shipping and interference with trade, to induce price hikes – especially for food stuffs, and to bring about the transfer of a part of the trade and shipping to neutrals.[24]

Once again, in addition to the psychological effect of anxiety and the prospect of shortages, there was now a new emphasis on the detrimental impact on Britain's all important carrying trade. Shipping, it was asserted, would fly to neutral flags, with a consequent financial cost.

As can be seen, across the space of 11 years, the German naval authorities considered in detail their options for a concerted campaign against British

---

[22] Pohl, 'Zum Immediatvortrag', 24 Apr. 1913. Quoted in J.C.G. Röhl, *Wilhelm II: Into the Abyss of War and Exile, 1900–1941* (Cambridge: Cambridge University Press, 2014), p. 998.

[23] 'Kriegsführung der Auslandsschiffe', 29 Apr. 1913. BA-MA: RM5/899, f. 211.

[24] 'Handelskrieg: Fragen des Seekriegsrechtes', 4 Dec. 1913. Quoted in Bönker, *Militarism in a Global Age*, p. 157.

commerce using surface warships. The shortage of available cruisers meant that the planning often revolved around fast merchant ships that were to be armed and commissioned as auxiliary cruisers at the outset of conflict. Nevertheless, whatever the limitations in the materiel available, these plans assumed that such a campaign would have significant impact. First, and in absolute conformity to the prevailing orthodoxy of the Tirpitz plan, it was assumed that *Handelskrieg* would be of principal value as a means of diverting Royal Navy assets away from the North Sea to distant waters and thereby contributing to an evening of the balance of force between the German and British fleets. Second, it was hoped that by attacking British trade, German action would affect the morale of the British people by undermining belief in their naval supremacy. Krosigk even spoke of panic. Finally, there was the prospect of undermining the British economy.

II

How did these plans work in practice when the test of war came? The reality of war would certainly endorse some aspects of the *Admiralstab*'s pre-war thinking. To begin with, several of the German light cruisers on overseas stations, namely the *Dresden, Emden, Leipzig, Königsberg,* and *Karlsruhe* – were used in a commerce raiding capacity. Some of them were more successful in this role than others, but, as will be shown, all had an impact.

Least effective, at least in terms of merchantmen captured, was the light cruiser *Königsberg*. Based at the outset of the conflict in Dar-es-Salem in German East Africa, the vessel's wartime instructions were to harass enemy shipping in the nearby Gulf of Aden, normally a well-travelled route not without importance. Nevertheless, this did not prove an easy task. For one thing, there were several allied warships in the area and as the *Königsberg* was intending to avoid battle in order to concentrate on the anti-commerce role it was necessary to evade them. More pressing still was the need to obtain supplies of fuel and the difficulties of so doing. Not only were coal stocks in German East Africa low, but stores in the Portuguese colonies were unavailable having been deliberately bought up by the British government as a means of hampering German operations. Furthermore, because it was known that the *Königsberg* was present in the area, the colliers and cargo ships of allied nations, whose coal might have been seized as a prize, were deliberately avoiding the region. As a result, the *Königsberg*'s efforts were severely hampered and only one merchant vessel, the *City of Winchester* (6,600 tons), was actually captured, although the *Königsberg* was also able to attack the elderly Pelorus class protected cruiser HMS *Pegasus* while the latter was undergoing repairs in Zanzibar harbour. However, want of fuel eventually forced the German cruiser to seek refuge in the Rufiji delta, where she was blockaded and eventually destroyed. Despite this meagre return, as a subsequent British Admiralty assessment of the impact of German cruiser warfare later admitted, the *Königsberg* caused the allies a lot of trouble. Shipping in the region was immobilised for several months for fear of her depredations, and her eventual destruction, although successfully achieved, tied down resources badly needed elsewhere.[25]

If the essentially unfruitful activities of the *Königsberg* validated the *Admiralstab*'s belief that even an unproductive *Handelskrieg* operation could be both disruptive and the cause of a forced dispersal of resource, the much more lucrative commerce-raiding activities of the *Karlsruhe* and the *Emden* showed just how effective such a method of warfare could be when skilfully undertaken. Like the *Königsberg*, the *Karlsruhe* and the *Emden* were fast, modern light cruisers. However, unlike the *Königsberg*, at the outset of the conflict they found themselves placed in much more promising theatres of operation, where greater freedom of action was possible.

In the case of the *Karlsruhe*, the start of the war found this vessel at anchor in the Florida Straits. After coaling she proceeded to the northern coast of Latin America, where she captured and sank 16 British vessels (and one Dutch one) totalling over 70,000 tons and with an insurance value of £1,500,000.[26] Her depredations only ended because on 4 November 1914, while returning to the West Indies, the *Karlsruhe* suffered a massive spontaneous internal explosion and sank. This was, however, unknown to the allied authorities, who continued to devote scarce assets to the hunt for Captain Köhler's evasive and troublesome vessel for several more months to come. This did not prove cheap in resource terms. More than 26 different cruisers and auxiliary cruisers were employed in the hunt for the *Karlsruhe*, searches being undertaken in the West Indies, South America, and the waters around the Azores and the Canary Islands. Even the West African coast was at one point scoured for signs of this elusive raider. In short, Captain Köhler and his crew caused considerable damage during their three-month cruise and managed to draw away many warships from other theatres. Indeed, they performed the extraordinary feat of still drawing off resources even after their ship had actually gone to the bottom.

The cruise of the *Emden* was, if anything, even more successful. Between 31 July, when the *Emden* first departed Qingdao, and 9 November, when the vessel was destroyed by HMAS *Sydney* off the Cocos Island, a period of some nine weeks, the cruiser first captured the Russian volunteer fleet steamer *Rjasan*, and then proceeded to the Indian Ocean, where she sank the Russian cruiser *Zhemchug* and the French destroyer *Mousquet*, set fire to the oil tanks in Madras, in the process destroying 346,000 tons of precious fuel, and captured more than 82,000 tons of British merchant shipping. Her depredations caused enormous disruption to civilian traffic in the Bay of Bengal, led to a resource-heavy multinational effort to locate her – many French, Russian, Japanese, and

---

25   The National Archives, Kew (TNA), ADM 275/22, OU 6337 (40), 'Review of German Cruiser Warfare 1914–1918', p. 9.

26   Admiralty, Naval Staff Monograph, vol. 9, 'The Atlantic Ocean, 1914–1915', p. 151.

British warships being required for this enterprise – and delayed the departure of British troopship convoys from Australia and India, which, when they eventually sailed, did so under the protection of strong escorts. Summing up the impact of the *Emden*'s exploits, Sir Julian Corbett, the author of the British official history of the war at sea, was unambiguous that 'Captain von Müller obtained a high measure of success both in the actual damage he had done and the strategical and economic disturbance he had caused'.[27] This, of course, was always the intent.

However, while the German overseas cruisers were able to inflict considerable disruption, owing first to the fact that they were so quickly hunted down – or, in the *Karlsruhe*'s case, lost – and second to the fact that they could not be replaced, their impact as a means of undertaking commerce warfare did not extend beyond the first few months of the war. More significant, both in terms of tonnage destroyed and the duration and chronology of their operations, were the activities of Germany's auxiliary cruisers, that is to say German merchant vessels converted into impromptu men-of-war.

As the British naval authorities had long suspected and greatly feared, the *Admiralstab* had long been interested in supplementing their regular overseas warships with auxiliary cruisers in the event of a conflict with Britain.[28] Moreover, as the German official history notes, the vessels most suitable for that role had already been identified long before 1914 and guns and ammunition had been set aside to enable their rapid conversion to take place, be that from storehouses in home ports or from the holds of German cruisers operating in distant waters or from the decks of elderly gunboats of little fighting value that were disarmed for this purpose and then abandoned.[29] As a result, it was hardly to be wondered that when the fighting between Britain and Germany commenced in August 1914 the German authorities promptly equipped and dispatched a number of converted merchantmen, the explicit role of which was to act as dedicated commerce raiders. Indeed, the only real surprise was the small number used; only five auxiliary cruisers were sent to sea when the war began. In part this was due to misfortune. For example, one of the designated vessels, the liner *Viktoria Luise*, had defective boilers that rendered her incapable of fulfilling her intended belligerent role. Others, such as the *Kaiser Wilhelm II*, *George Washington*, and the *Prinz Friedrich Wilhelm*, were at sea performing their regular peacetime duties as passenger vessels when war began. With no prospect of their making

---

[27]   Sir Julian S. Corbett, *Naval Operations*, i. 385.

[28]   M.S. Seligmann, *The Royal Navy and the German Threat, 1901–1914: Admiralty Plans to Protect British Trade in a War against Germany* (Oxford: Oxford University Press, 2012); M.S. Seligmann, 'Germany's Ocean Greyhounds and the Royal Navy's First Battle Cruisers: An Historiographical Problem', *Diplomacy and Statecraft* 27 (2016), pp. 162–82.

[29]   Paul Köppen, *Die Überwasserstreitkräfte und ihre Technik* (Berlin: Mittler, 1930), pp. 141–5; Eberhard von Mantey, *Der Kreuzerkrieg in den ausländischen Gewässern*, vol. 3, *Die deutschen Hilfskreuzer* (Berlin: Mittler, 1937), pp. 4–6.

it home or of discharging their passengers and rendezvousing with a ship that could supply them with armaments, they fled instead to neutral harbours. Finally, the early departure of auxiliary cruisers to their war stations, an expedient that the *Admiralstab* had long planned, conflicted with the diplomatic strategy the German government adopted in early August 1914. Desirous of securing British neutrality and believing that this might be secured by the semblance of peaceful intent, instructions were issued that all naval activities that might provoke a response in London were forbidden. As the order explained: 'It is considered of the utmost importance to avoid, under all circumstances any suspicious movements of ships which might lead to an unintended outbreak of hostilities with England'. Among the actions denied by this was the planned arming and sailing of auxiliary cruisers, which was thereby delayed for several days.[30]

Of the five that did make it to sea in 1914, namely the *Kaiser Wilhelm der Grosse*, *Cap Trafalgar*, *Kronprinz Wilhelm*, *Prinz Eitel Friedrich*, and the captured Russian vessel *Rjasan*, now renamed *Cormoran*, success was varied. Neither the *Cormoran* nor the *Cap Trafalgar* made any captures, the former being interned in Guam after running low on fuel, the latter being sunk upon encountering and then coming off second best in an engagement with the British armed merchant cruiser *Carmania*. A similar fate befell the *Kaiser Wilhelm der Grosse*, which was sent to the bottom after being caught coaling at the Rio de Oro by the cruiser HMS *Highflyer*, albeit not before she had herself captured two British merchantmen with a cumulative displacement of 10,400 tons. Much more successful was the *Prinz Eitel Friedrich*. Operating off both the east and west coasts of Latin America, this former Norddeutsche Lloyd liner managed, during a seven month cruise, to sink 11 vessels with a total displacement of 33,423 tons. However, most successful of all of the initial German merchant raiders was the *Kronprinz Wilhelm*, which made 15 captures, totalling 60,522 tons. Her 250-day commerce raiding operation was only brought to an end by the state of the ship, which needed urgent boiler repairs after eight months of near continuous hard steaming, and by the health of the crew, which was deteriorating as a result of an outbreak of beri-beri. Nevertheless, if both the *Prinz Eitel Friedrich* and the *Kronprinz Wilhelm* had ceased to operate by April 1915, it is worth recording that considerable efforts had been made by the allies to hunt them down, a fact that more than justified the *Admiralstab*'s pre-war belief that, at minimal cost to Germany, *Handelskrieg* could pay dividends in terms of nuisance value and dispersal of force.

Despite achieving some success with those surface raiders deployed at the very start of the war, little effort was made to build on the potential revealed by their activities until the very end of 1915, when a former fruit-carrying freighter, the *Pungo*, was commissioned as the auxiliary cruiser *Möwe*. She promptly embarked upon a two-month operation in the north and south Atlantic

---

[30]   Otto Groos, *Der Krieg in der Nordsee*, vol. 1 (Berlin: Mittler, 1922), pp. 26–7.

during which time she captured 15 vessels totalling 57,776 tons. She also laid a minefield off Cape Wrath that accounted for the pre-dreadnought battleship HMS *King Edward VII*. Eight months later, the *Möwe* tried her luck again. During her second cruise, which lasted some four months, she sank or captured over 120,000 tons of allied shipping. More significantly still, 24 British cruisers and auxiliary cruisers were diverted from other duties to hunt for the *Möwe*, fear of whose activities also delayed the departure of several troopships. The success of the *Möwe* spurred further efforts. In March 1916, the freighter *Wachtfels* was commissioned as the auxiliary cruiser *Wolf.* Her cruise lasted some 15 months and involved the capture of 14 ships and the laying of nine minefields that accounted for a further 13 vessels. In total she destroyed 114,279 tons of shipping. As with the *Möwe*, that was the least of her value. The unsuccessful allied efforts to hunt her down eventually involved 31 cruisers, 14 destroyers, and nine sloops, a huge dispersal of scarce allied assets.

When one adds in the 27,923 tons of shipping that was sunk by the converted sailing vessel *Seeadler*, the tally from Germany's auxiliary cruisers topped 427,000 tons. In combination with the captures of Germany's regular warships, *Handelskrieg* from surface units accounted for 623,406 tons. Given the limited number of vessels allocated to this role and the disproportionate diversion of resource this caused, it was not an insignificant result.

III

As can be seen, the reality of war would endorse certain aspects of the *Admiralstab*'s pre-war planning. The German navy did use surface raiders in the First World War and these raiders destroyed a not inconsiderable quantity of British and allied shipping. Some of the vessels engaged – for example, the cruiser *Emden* and the auxiliary cruiser *Kronprinz Wilhelm* – made such a nuisance of themselves that considerable allied naval forces were deployed in the effort to end their depredations. Nevertheless, it must also be recognised that German efforts at *Handelskrieg* with surface raiders never threatened to have more than a marginal impact upon the course of the war. This was not because warfare by such means was inherently ineffectual, as Mahan had argued, but because the German naval authorities never pursued it with real vigour. Very few regular cruisers were at sea in August 1914 and they were augmented by only five auxiliary cruisers. With such meagre numbers there was a definite limit to what could be achieved. Similarly, the efforts made from late 1915 onwards to revive this approach, despite the very large number of laid up German merchant vessels that might have been deployed in this capacity, only ever involved a handful of ships. To the relief of the British authorities, no large-scale or coordinated undertaking was ever attempted. Unsurprisingly, the results were once again only commensurate with the effort made. It would be no more than speculation to consider what impact greater numbers and better organisation and coordination might have made, but it is notable that the official British history of the war at sea thought that such efforts might have been genuinely damaging. As Sir Henry Newbolt explained:

> A German success could only have been attained if the attacks had been more prolonged and more intense: if, for example, our trade routes in each ocean had been threatened by six raiders instead of one ... It is noteworthy that the German Admiralty, who can hardly have failed to consider the point, did not instruct the captains of the *Moewe*, the Seeadler and the *Wolf* to act upon any combined plan, but sent them out as isolated and independent marauders ... The motive may have been the anxiety to preserve secrecy or to avoid misunderstandings ... If they only caused us inconvenience when they might have done real damage, the fault lies not with [the captains of the raiders] but with the High Command, which sent them on a vast enterprise with hopelessly inadequate means.[31]

If the reality of conflict proved that economic warfare had considerable potential, but only if exploited with energy and intent, it would also show that the mechanism for waging such a campaign rested not with the tools previously envisaged, namely surface raiders, but with a new weapon, the submarine. However, it is an irony of this clear wartime experience that in the run up to the Second World War the German naval leadership would once again focus its efforts on creating a battle fleet and would play down, until circumstances forced the lesson to be learnt anew, that *Handelskrieg*, both by surface raiders and by U-boats, had a major part to play.

---

[31]   Henry Newbolt, *Naval Operations* (London, 1928), iv. 227–8.

CHAPTER ELEVEN

Maritime Power and Economic Warfare in the Far East, 1937–1941

Greg Kennedy

The relationship between Anglo-American maritime power and economic warfare in the Far East, prior to the attack on Pearl Harbor on 7 December 1941, was a complex one.[1] It is also a somewhat surprising story. Given the history of the Royal Navy's role in waging economic warfare, and the United States Navy's exposure to that process,[2] it is understandable that many scholars continue to assume that those two nations were well prepared to conduct a campaign of economic coercion against Japan in the Pacific. However, this chapter will demonstrate that a lack of naval power, political will, and a culture of the offensive in both the Royal Navy and United States Navy (much like that demonstrated by French generals prior to the First World War),[3] allowed Japan, not Great Britain or the United States, to be strategically more proficient in utilising maritime power to wage economic warfare. That effectiveness on Japan's part created substantial strategic effect during the pre-war period, achieving many significant

---

[1] For an introduction to some of the issues, see G. Kennedy, 'Anglo-American Economic Warfare and the Deterrence of Japan, 1933–1941', in Anastasia Filippidou (ed.), *Deterrence: Concepts and Approaches for Current and Emerging Threats* (Bristol: Bristol University Press, 2020). For a complete analysis of Anglo-American Economic Warfare in the period from 1933 to 1941, see my forthcoming, 'Anglo-American Economic Warfare and the Deterrence of Japan, 1933–1941'.

[2] G. Kennedy, 'The British Strategic Assessment of the United States as a Maritime Power, 1900–1917', in G. Kennedy (ed.), *Britain's War at Sea, 1914–1918: The War They Thought and the War They Fought* (London: Routledge, 2016); G. Kennedy, 'Intelligence, Strategic Command and the Blockade: Britain's War Winning Strategy, 1914–1917', *Intelligence and National Security*, 5.22 (2007), pp. 699–721; N.A. Lambert, *Planning Armageddon: British Economic Warfare and the First World War* (Cambridge, MA: Harvard University Press, 2012).

[3] Jack Snyder, *The Ideology of the Offensive: Military Decision Making and the Disasters of 1914* (Ithaca, NY: Cornell University Press, 1982); Stephen Van Evera, 'The Cult of the Offensive and the Origins of the First World War,' *International Security* 9.1 (1984), pp. 58–107.

strategic ends, such as the diminishment of China's military power. By contrast, the Anglo-American strategy of deterring Japanese aggression in China and the Pacific did not, on the other hand, accomplish their desired outcomes. Instead, economic warfare escalated the tensions between the Western nations and Japan and acted as a catalyst for Japan's final decision to use military force to secure its aims on 7 December 1941.

Throughout the interwar period Great Britain and the United States faced the challenge of an expansionist Japan seeking to destabilise the established balance-of-power system that governed the international order in the Far East.[4] The growth of the perception, and eventual reality, of the Japanese threat to both Western nations' national security interests in the region provoked a number of responses.[5] International arms agreements, naval fortification agreements, declaration of spheres of influences in China, appeasement, coercion, persuasion, appeals to moderation, sanctions, embargoes, proxy war, and economic warfare were all utilised in the quest to arrest Japan's drive to dominate Pacific international affairs.[6] However, until the actual threat of war in Europe initiated real

[4]   Dorothy Borg and Shumpei Okamoto (eds), with the assistance of Dale K.A. Finlayson, *Pearl Harbor* as *History: Japanese-American Relations, 1931–1941* (New York: Columbia University Press, 1973); Ann Trotter, *Britain and East Asia, 1933–1937* (New York: Cambridge University Press, 1975); K. Neilson, '"Unbroken Threat": Japan, Maritime Power and British Imperial Defence, 1920–32', in G. Kennedy (ed.), *British Naval Strategy East of Suez, 1900–2000: Influences and Actions* (London: Routledge, 2005), pp. 62–72; Stephen L. Endicott, *Diplomacy and Enterprise: British China Policy, 1933–1937* (Vancouver: University of British Columbia Press, 1975); Jonathan Haslam, *The Soviet Union and the Threat from the East, 1933–41*, vol. 3, *Moscow, Tokyo, and the Prelude to the Pacific War* (Pittsburgh: University of Pittsburgh Press, 1992).

[5]   Ian Nish, 'Britain and Japan: Long-Range Images, 1900–52', *Diplomacy and Statecraft*, 15.1 (2004), pp. 149–61; Tae Jin Park, 'Guiding Public Opinion on the Far Eastern Crisis, 1931–1941: The American State Department and Propaganda on the Sino-Japanese Conflict', *Diplomacy and Statecraft*, 22.3 (2011), pp. 388–407.

[6]   E.O. Goldman, *Sunken Treaties: Naval Arms Control between the Wars* (University Park: Penn State Press, 1994); C.D. Hall, *Britain, America, and Arms Control* (New York Palgrave Macmillan, 1987); M.H. Murfett, 'Look Back in Anger: The Western Powers and the Washington Conference of 1921–1922', in B.J.C. McKercher (ed.), *Arms Limitation and Disarmament* (New York: Praeger, 1992), pp. 83–104; S.W. Roskill, *Naval Policy between the Wars*, vol. 1 (London: Walker, 1968), pp. 70–5; M.A. Barnhart, *Japan Prepares for Total War: The Search for Economic Security, 1919–1941* (Ithaca, NY: Cornell University Press, 1987); Ian Nish (ed.), *Anglo-Japanese Alienation, 1919–1952* (New York: Cambridge University Press, 1982); E.S. Miller, *War Plan Orange: The US Strategy to Defeat Japan, 1897–1945* (Annapolis, MD: Naval Institute Press, 1991); J.K. McDonald, 'The Washington Conference and the Naval Balance of Power, 1921–22', in John B. Hattendorf and Robert S. Jordan (eds), *Maritime Strategy and the Balance of Power* (London: Springer, 1989), pp. 189–213; I.H. Nish, *Alliance in Decline* (London: Athlone Press, 1972).

rearmament programmes for the Western powers, one of the few effective means of deterring Japan's challenge to Anglo-American strategic interests in the Far East was the use of economic and fiscal power.

Economic warfare was the weapon of choice for Great Britain. The British were the world's masters in its dark arts, having waged it successfully from the Napoleonic Wars through to the successful conclusion of the First World War. From 1914 to 1918 the British Empire's main strategic weapon had been the ability to deny Germany and its allies access to the necessary tools of modern industrial warfare. Access to credit, overseas markets, raw materials, and finished goods had all been disrupted or denied, and patterns of international trade significantly altered in the process. America had participated in that process during the First World War[7] and proved itself a quick and avid student of the concept in the interwar period.[8] Therefore, in the summer of 1937, Great Britain and the United States were the world's undisputed experts in utilising a strategy of economic warfare. It was no surprise, then, that following Japan's open use of military force in China in July 1937, the two Western nations immediately looked to economic warfare as a solution to the threat presented to their Far Eastern national security interests.

However, conditions in the Pacific Ocean off the coast of China were not the same as those present in 1914–18 in the North Sea and Atlantic Ocean. Viable, meaningful, immediate, and, most importantly, decisive naval power able to support maritime operations concerned with economic warfare, such as blockade, sanctions, and embargoes, was lacking. Furthermore, this strategic weakness was a condition which had been agreed to by the United States and Great Britain. The 1922 Washington Naval Treaty had ceded maritime control of the Western Pacific to the Japanese through its three-layered approach to security in the region, a critical part of which was the inability of either Western power effectively to power project maritime force into the Far East in a sustained and persistent manner.[9] The newest technology available for patrolling and providing

---

[7]   Lambert, *Planning Armageddon*; M. Simpson (ed.), *Anglo-American Naval Relations, 1917–1919* (Basingstoke: Navy Records Society, 1991); G. Kennedy, 'Anglo-American Strategic Relations and the Blockade, 1914–1916', *Journal of Transatlantic Studies* 6.1 (2008), pp. 22–33; G. Kennedy, 'Strategy and Power: The Royal Navy, The Foreign Office and the Blockade, 1914–1917', *Journal of Defence Studies* 8.2 (2008), pp. 190–206.

[8]   This assessment is at odds with the American Executive Director of the Board of Economic Warfare, Milo Perkins's estimation of the American situation in June 1943, which he depicted as small in size and suffering from years of neglect in the face of Germany and Japanese activity. He is, however, predominately focused on the stockpiling of strategic raw materials in his statement regarding how economic warfare is seen by his office. Letter from Perkins to Mariner Eccles (Chairman, Board of Governors, Federal Reserve System), 16 June 1943, Federal Reserve Bank of St Louis.

[9]   Zara Steiner, *The Triumph of the Dark: European International History, 1933–1939* (Oxford: Oxford University Press, 2011), chap. 9, pp. 474–551; Erik Goldstein and John

situational awareness with regard to ship movements, long-range maritime patrol aircraft, were also not available. Insufficient suitable aircraft, inadequate command and control infrastructure, inchoate communications arrangements, and the absence of an Anglo-American-controlled network of protected airfields meant that the vast waters of the Pacific remained an easy place for ships and shipping to avoid detection. Without the ability to conduct sustained and persistent patrolling, intercepting, and deterrence operations, Anglo-American naval power was not a significant factor during peacetime, making maritime economic warfare in anything other than war a non-starter operationally.[10] Such shortcomings in operational capability for waging economic warfare did not, however, keep Great Britain and the United States from contemplating the use of such a weapon for deterring Japan in peacetime.

In the immediate aftermath of the Japanese attack on China in July 1937, British planners began to consider whether or not economic warfare could be used as an effective tool to deter and contain, as well as perhaps coerce, Japan into ceasing military operations in China. Uppermost in the British planners' minds was a specific word used by President Franklin D. Roosevelt in his 5 October speech in Chicago on the state of international affairs. During the speech, FDR denounced the 'outlaw' states who were using military force to change the international system and called for these states to be 'quarantined'. Secretary of State for Foreign Affairs Anthony Eden asked the British embassy in Washington to get clarification on what the US President really meant in using that term, as London was concerned

> lest public opinion here and in the United States should too hastily and too easily assume that quarantine means economic sanctions without the risk of war ... It would surely be unjustifiable and therefore in her

---

Maurer (eds), 'Special Issue on the Washington Conference, 1921–22: Naval Rivalry, East Asian Stability and the Road to Pearl Harbor', *Diplomacy and Statecraft* 4 (1993); Roger Dingman, *Power in the Pacific: The Origins of Naval Arms Limitation* (Chicago: University of Chicago Press, 1976); W.R. Braisted, *The United States Navy in the Pacific, 1909–1922* (Austin: University of Texas Press, 1971).

[10]   On the requirements for the operational and legal needs of such actions, see A.C. Bell, *A History of the Blockade of Germany and of the Countries Associated with Her in the Great War: Austria–Hungary, Bulgaria, and Turkey, 1914–1918* (London: HMSO, 1937), pp. 33–189; W.N. Medlicott, *History of the Second World War: The Economic Blockade*, vol. 1 (London, HMSO, 1959); W.H.W. Carless Davis, *History of the Blockade: Emergency Departments* (London: HMSO, 1920); Arthur Marsden, 'The Blockade', in F.H. Hinsley (ed.), *British Foreign Policy under Sir Edward Grey* (Cambridge: Cambridge University Press, 1977), pp. 466–87; M.C. Siney, 'The Allied Blockade Committee and the Inter-Allied Trade Committees: The Machinery of Economic Warfare, 1917–1918', in K. Bourne and D.C. Watt (eds), *Studies in International History: Essays Presented to W. Norton Medlicott* (London: Archon Books, 1967), pp. 330–44.

present mood highly dangerous to assume that Japan would submit to a boycott or to economic sanctions that were effective without seeking to rectify the consequences.[11]

Immediate discussions with senior officials in the American State Department confirmed that the President's use of the word was seen as being a vague and imprecise description of how those states were to be seen, not as an indication of an intention to mount a blockade. The State Department was clear in its declaration that there was no prospect of economic actions being taken against Japan, or, indeed, any provocative acts being contemplated in Washington at all.[12] Yet, despite a lack of American interest in using economic warfare against Japan, the British Foreign Office (FO) had already begun to investigate the possibility of using economic pressure on Japan. Unbeknownst to the Americans, the United States played a prominent part in their appreciations.

The FO study highlighted four possible courses of action that could be thought of as economic warfare against Japan: (a) limiting the export of materials essential for Japan's war effort; (b) actions against Japanese imports that would cripple its trade and thus ability to generate financial power required to continue the war effort; (c) blockade against Japanese shipping; (d) and, finally, the withdrawal of the financial facilities required for Japan to find credit or maintain a stable currency.[13] This suite of options was a reasonably conceptual summary of Britain's strategic options; however, it betrayed a fundamental discrepancy between thinking at the FO and within the Royal Navy. Much like the United States Navy at the same time, the leadership of the Royal Navy were not culturally attuned to the idea of using naval power to conduct maritime economic warfare in the Pacific. In both cases the cult of the offensive, and total war, were the only conditions envisioned in either navy's planning for war in the Pacific.[14]

The Royal Navy's main concerns when it came to war in the Far East was how to relieve the great naval base of Singapore[15] and assessing how long exactly the besieged fortress would have to withstand Japan's combined air, land, and naval domination. Any idea of limited war, or of contesting Japanese regional naval primacy in peacetime, were a thing of the past, if they had ever played a major role in Royal Navy thought. Britain had chosen Japan as a naval ally at the turn of the century specifically so the Royal Navy would not have

---

[11]   Foreign Office (FO) 837, Ministry of Economic Warfare, Butler Committee, Economic Pressure – Japan/528, telegram from FO to Mallett (Washington), 12 Oct. 1937.

[12]   FO 837/528, telegram from Washington to FO, 12 October 1937.

[13]   FO 837/528, 'Economic Measures against Japan', undated, summer 1937.

[14]   W.N. Medlicott, *The Economic Blockade*, vol. 1 (London: HMSO, 1952), pp. 384–5; CAB 2/7, Minutes of CID Meetings, Nov.1937–July 1938, '301st meeting, Nov.18, 1937'.

[15]   B.P. Farrell and S. Hunter (eds), *A Great Betrayal? The Fall of Singapore Revisited* (Singapore: Marshall Cavendish, 2010).

to extend itself to protecting Far Eastern possessions in the face of a rising continental power: the Kaiser's Germany. Post-war financial restraints and the naval arms limitation system continued to condition Royal Navy strategic thinking to avoid the idea of utilising economic warfare in a limited or total war in the Pacific.[16] A defensive stance, able to limit Japanese expansion while at the same time drawing in international support for some sort of peace agreement, was the best that could be hoped for in a scenario in which Great Britain found itself facing German, Italian, and Japanese naval power. No one in the Admiralty ever really dared answer the question of 'what next' if Singapore was successfully relieved. The assumption, it appears, is that the losses the Japanese would inevitably suffer in attacking Singapore would allow Royal Navy forces to prevent an invasion of Australia or New Zealand. The position at Singapore would then be used as a base from which to wage offensive economic warfare across the Pacific. Submarines, aircraft, and fast surface units would attempt to disrupt Japan's sea lines of communication with China and Korea, and also her longer-range maritime trade with South America.[17] A long war of attrition would wear down the Japanese economy and war weariness would at some point cause the Japanese government to agree to some sort of peace agreement.[18]

These intentions were severely disrupted by the outbreak of war in Europe in 1939. The loss of France as an ally, Italy's decision to join the Axis powers, and the subsequent weakening of the Royal Navy during operations off Norway, in the Atlantic, and in the Mediterranean combined to mean that the Admiralty revised its plans to relieve Singapore, and could not commit to providing meaningful support for at least three months after a potential attack. The ambition of subsequent operations from the base were also downgraded.[19] The submarines which were to have provided the bulk of the offensive action against Japanese naval units and merchant shipping had been redeployed to Malta in an effort to interdict Italian and German shipping supporting the Axis war effort in

---

[16]   See C.M. Bell, *The Royal Navy, Seapower and Strategy between the Wars* (London: Palgrave Macmillan, 2000); I. Cowman, *Dominion or Decline: Anglo-American Naval Relations in the Pacific, 1937–1941* (Oxford: Berg, 1996); Andrew Boyd, *The Royal Navy in Eastern Waters: Linchpin of Victory, 1935–1942* (London: Seaforth Publishing, 2017).

[17]   CAB 53/44, COS Papers, No. 826-839, 18 Jan.–8 Feb. 1939, COS 826(JP), 20 Jan. 1939; CAB 53/49, COS Papers, No. 902-911, 10 May–23 May 1939, COS 910, 22 May 1939.

[18]   G. Kennedy, 'The Royal Navy and Imperial Defence, 1919–1945', in G. Kennedy (ed.), *British Imperial Defence: The Old World Order, 1856–1956* (London: Routledge, 2008), pp. 133–52.

[19]   Russell Grenfell, *Main Fleet to Singapore* (London: Faber and Faber, 1951); C.M. Bell, 'The "Singapore Strategy" and the Deterrence of Japan: Winston Churchill, the Admiralty and the Dispatch of Force Z', *English Historical Review* 16.2 (2001), pp. 604–34; C.M. Bell, '"Our Most Exposed Outpost": Hong Kong and the British Far Eastern Strategy, 1921–1941', *Journal of Military History* 60.3 (1996), pp. 61–88.

North Africa.[20] Modern long-ranged aircraft, vital to reconnaissance and attack missions against those same Japanese maritime targets, were simply too few in number and too vital to the war against the German U-boat threat in the Atlantic to be sent to the Far East. Finally, Royal Navy fast surface raiders, such as light and heavy cruisers, let alone battlecruisers, were in far too short supply to offer any hope of making any economic warfare threat to the Japanese a viable option.[21] By the spring of 1940, British national interests in the Far East were reliant on American political will and naval power for their protection from any Japanese aggression. British naval power was simply not strong enough to be able to conduct economic warfare on a global scale, given the strategic demands and centrality of the war in Europe. It would be up to America and her navy to provide any real deterrent to Japanese actions, including providing the bulk of the naval units required for blockade and interdiction if the maritime domain was to be used in the growing Anglo-American economic warfare effort in the Pacific.[22]

The United States Navy's strategic thinking was equally wedded to an operational plan, War Plan Orange, predicated upon forcing a decisive naval action on the Imperial Japanese Navy. American planners anticipated concentrating the majority of American naval power in the Pacific and undertaking offensive fleet operations against the Imperial Japanese Navy in the area around the Philippine Islands as soon as the appropriate concentration of force was prepared.[23] At no time were senior American naval officers ready or able to grasp the niceties of conducting naval economic warfare operations in peacetime. During considerations as to what sort of limited, or even greater, actions might be envisioned if Japan invaded the Dutch Netherlands East Indies for its valuable oil supply in the spring of 1940, the Chief of Naval Operations (CNO) Admiral 'Betty' Stark informed the Commander in Chief (CinC) United States Fleet Admiral J.O. Richardson that:

> I don't know and I think there is nobody on God's green earth who can tell you ... I would point out one thing and that is that even if ... the US to take no decisive action if the Japs should decide to go into the Dutch East Indies, we must not breathe it to a soul ... the Japs don't know what

[20]   G. Kennedy, '"To Throttle, Not Knock Out": The Role of Malta in the RN's Sea Denial, Interdiction and Naval Diplomacy Operations: 1939–43', in Ian Speller (ed.), *The Royal Navy and Maritime Power in the 20th Century* (London: Frank Cass, 2004).

[21]   Stephen Roskill, *The Navy at War, 1939–1945* (London: Wordsworth, 1998); H.P. Willmott, *Empires in the Balance: Japanese and Allied Pacific Strategies to April 1942* (Annapolis, MD: Naval Institute Press, 1982).

[22]   Ian Cowman, *Dominion or Decline*, chap. 4, pp. 165–206: 'Britain Cedes the Initiative: The Road to ABC-1'.

[23]   Edward S. Miller, *War Plan Orange: The US Strategy to Defeat Japan, 1897–1945* (Annapolis, MD: Naval Institute Press, 1991).

we are going to do and so long as they don't know they may hesitate, or be deterred.[24]

Richardson himself was sceptical of the worth of the Orange plan, pointing out the liability seizing necessary advanced anchorages such as Truk would entail in terms of development and investment, a condition that would severely limit its utility as an operational base from which to mount naval or blockade/strangulation operations.[25] By the autumn of 1940, as the United States moved from a posture of 'moral embargoes' following the passage of the Act of 2 July 1940, restrictions were imposed in the interests of national defence on an ever-increasing list of exports of strategic materials. These measures were intended as both a deterrent and an expression of America's opposition to Japan's continued aggression in the Far East.[26] Given the risk of provoking a war with Japan through the use of these economic measures, consideration to what kind of war the United States could wage in the Pacific, while building up its naval strength and assisting Great Britain in its war effort, needed to be taken.

The first real evidence of discussions by the United States Navy regarding the use of naval power to augment economic warfare as a means of limiting Japanese expansionism took place in the autumn of 1940. Concerned about possible Japanese retaliation against Great Britain for opening a land line of communication to supply Chiang Kai-shek's war effort, the so-called Burma Road,[27] President Roosevelt informed the Secretary of the Navy, Frank Knox that he was considering action against Japanese maritime trade.[28] Roosevelt wanted the Navy's assessment of three possible courses of action that would increase the economic pressure on Japan.[29] Admiral Richardson informed Admiral Thomas C. Hart, CinC Asiatic Fleet, that FDR was contemplating reinforcing the Asiatic Fleet with four

---

[24]   Pearl Harbor Investigation, Part 14, secret letter from Admiral Harold (Betty) R. Stark, Chief of Naval Operations to Admiral J.O. Richardson, CinC United States Fleet, 27 May 1940.

[25]   Pearl Harbor Investigation, Part 14, Exhibits, Exhibit No. 9, Selected letters between Admiral H.R. Stark and Admiral J.O. Richardson, Richardson to Stark, 26 Jan. 1940.

[26]   Pearl Harbor Investigation, Part 3, Testimony of Secretary of State Cordell Hull, p. 412.

[27]   The opening of the Burma Road as a supply line to China in its war against Japan was a serious threat to the balance of power relationship between Great Britain, America and Japan, and would remain so until the outbreak of war in December 1941: Pearl Harbor Investigation, Part 14, Exhibit No. 16, secret memo for the President, from the War and Navy Department, 5 Nov. 1941, p. 1061.

[28]   Pearl Harbor Investigation, Part 14, Exhibit No. 11, memo from CinC US Fleet to CinC US Asiatic Fleet, 6 Oct. 1940, p. 1006.

[29]   Pearl Harbor Investigation, Part 14, Exhibit No. 9, memo for CNO from Richardson, 'Points covered in talk with the President', 9 Oct. 1940; Pearl Harbor Investigation, Part 1, Testimony of Admiral Richardson, pp. 265, 268, 294, and 297.

older cruisers, which could be used for interdiction and the blockade of Japanese shipping if such actions were thought necessary in support of future sanctions and embargoes. The President was even contemplating declaring a complete embargo on shipments to and from Japan. Thirdly, any potential embargo might also entail stopping all trade between Japan and America.[30] In order to enforce the third measure, Roosevelt envisioned the 'establishment of patrol lines of light forces from Honolulu westward to the Philippines and a second line roughly from Samoa to Singapore, "in support of" the first line'.[31] While the Navy opposed the President's 'concepts' for employing the navy in an economic warfare capacity,[32] and had drafted a memorandum to that effect for Secretary Knox and FDR,[33] Richardson advised Hart that, nonetheless, 'it is believed that further study in the Department in the Fleet will result in some modifications in the operations proposed for the Fleet, particularly as regards distribution of patrol planes and cruisers'.[34] The movement of Fleet units in both the Atlantic and the Pacific to interdict Japanese trade with South America, particularly Brazil, Uruguay, and Argentina, was the main area of concern. Only South America presented a market

---

[30]   Pearl Harbor Investigation, Part 14, Exhibit No.11, memo from CinC US Fleet to CinC US Asiatic Fleet, 6 Oct. 1940, p. 1006.

[31]   Pearl Harbor Investigation, Part 14, Exhibit No.11, memo from CinC US Fleet to CinC US Asiatic Fleet, 6 Oct. 1940, p. 1006; Pearl Harbor Investigation, Part 1, Testimony of Admiral Richardson, pp. 305–34.

[32]   Pearl Harbor Investigation, Part 14, Exhibit No. 9, letter from Stark to Richardson, 24 Sept. 1940, p. 961. Shows the reluctance not only of the USN but also the State Department in escalating economic warfare as a deterrent to Japan:

> Spent over three hours in the State Department yesterday – something over two in the morning with Mr. Hull, Welles and Hornbeck, and then again in the afternoon over an hour with Mr. Welles. I believe had you been present you would have been in agreement with what I did and I pushed my thoughts home just as hard as I could. I may say that the same general picture so far as our attitude is concerned still holds, although I would not be surprised, confidentially between you and me, to see an embargo on scrap but this too would be along the lines State has been working on. I strongly opposed, and I believe carried my point, an embargo on fuel oil for reasons which are obvious to you and with which I may say I think the state Department is in concurrence.

[33]   Pearl Harbor Investigation, Part 1, Testimony of Admiral Richardson, pp. 334–40. Richardson had pointed out to Knox, as well as President Roosevelt, his lack of belief in the deterrent effect of the US Fleet acting as a Fleet in being at Pearl Harbor. The manpower and logistics need of a naval contribution to economic warfare would have exacerbated that unfavourable situation he told them both. Knox's reply to Richardson's critical assessment of the political handling of the USN was to be told, 'Richardson, we have never been ready, but we have always won'. FDR's response was to relieve Richardson as CinC US Fleet.

[34]   Pearl Harbor Investigation, Part 14, Exhibit No. 11, memo from CinC US Fleet to CinC US Asiatic Fleet, 6 Oct. 1940, p. 1006.

for Japanese goods to generate economic power, as Europe, the Soviet Union, the Middle East, and North America could not be accessed due to either fiscal and economic measures or Western naval dominance.[35]

Admiral Richardson viewed this political momentum for tightening economic warfare measures against Japan as a challenge to War Plan Orange and to the naval force posture-planning the navy had been moving towards over the previous decade.[36] He informed the CNO that, in his view, no serious officer could believe War Plan Orange was fit for purpose and now was the time for a new appreciation of the situation to be made. This new appreciation had to take into account the current state of international relations, the readiness and overall condition of the fleet, and the objectives to be achieved with regard to Japan involving the use of the United States Navy. Richardson called for a realistic new plan to be built around: '(1) Security and defense measures of the Western Hemisphere; (2) Long-range interdiction of enemy commerce; (3) Threats and raids against the enemy; (4) Extension of operations as the relative strength of the Naval Establishment ... is built up to support them'.[37] The CinC US Fleet remained an insightful critic of both the political decision to use the unprepared United States Navy as a 'fleet in being' at Pearl Harbor in an attempt to deter Japanese aggression, as well as a sceptic as to the ability of the United States Navy to operationalise any significant naval contributions to economic warfare against Japan. In late January 1941, President Roosevelt relieved him of his command due to their inability to reconcile their difference of opinion on the two topics.[38] Richardson viewed the episode as normal practice and elected to play no further role in pushing forward any ideas of how to get the United States Navy to prepare for large-scale conventional naval operations whilst simultaneously generating new operational capability to conduct economic warfare from nothing: 'The day I was made commander in chief I realised then and thereafter that the same power which made me commander in chief could unmake me at any time. When I arrive in Washington I shall keep my lips sealed and my eyes in the boat and put my weight on the oar in any duty assigned'.[39] As for the United States Navy's overall appreciation of the situation at the end of 1940, the CNO Admiral Stark made clear his own opinion to the CinC Asiatic Fleet Admiral Hart:

---

[35]   Pearl Harbor Investigation, Part 14, Exhibit No. 11, secret memo, 'Measures and Operations to be Undertaken by the US Fleet', 10 Nov. 1940, p. 1009.

[36]   Pearl Harbor Investigation, Part 14, 22 Oct. 1940, Secret memo, CinC US Fleet to CNO, 'War Plans – Status and Readiness of in View of the Current International Situation', pp. 963–70; Pearl Harbor Investigation, Part 3, Testimony of Secretary of State Cordell Hull, pp. 418–20.

[37]   Pearl Harbor Investigation, Part 3, Testimony of Secretary of State Cordell Hull, pp. 418–20.

[38]   Pearl Harbor Investigation, Part 1, Testimony of Admiral Richardson, p. 340; Pearl Harbor Investigation, Part 1, Testimony of Fleet Admiral William D. Leahy, pp. 342–4.

[39]   Pearl Harbor Investigation, Part 1, Testimony of Admiral Richardson, p. 340.

Ghormley tells me that [the] British expected us to be in the war within a few days after the re-election of the President – which is merely another evidence of their slack ways of thought, and of their non-realistic views of international political conditions, and of our own political system. They have been talking, in a large way, about the defense of the Malay Barrier, with an alliance between themselves, us, and the Dutch, without much thought as to what the effect would be in Europe. But we have no ideas as to whether they would at once begin to fight were the Dutch alone, or were we alone, to be attacked by the Japanese. Then again, the copy of the British Far Eastern War Plan which Thomas obtained at Singapore shows much evidence of their usual wishful thinking. Furthermore, though I believe the Dutch colonial authorities will resist an attempt to capture their islands, I question whether we would fight if only the Philippines, or only Singapore, were attacked ... I believe that the allied objective should be to reduce Japan's offensive power through economic starvation; the success of the blockade would surely depend upon allied ability to hold the major portion of the Malay Barrier. Your own action would, of course, be based upon your view as to the most effective method of contributing to the attainment of the ultimate objective.[40]

Throughout 1941, Anglo-American attempts to contain Japan through the use of economic warfare failed to incorporate the necessary naval power required to give legitimacy and authority to increasing economic warfare operations.[41] Staff talks occurred, and vague operational concepts moved around both the Admiralty and the United States Navy offices and planning departments. However, nowhere within the leadership or bureaucracy of either service had limited operations in support of long-term economic warfare aims been seriously addressed as a way to achieve the political ends desired in the Pacific.[42] In large part this was due to organisational cultures wedded to a particular view of what role naval power could and should play in a future war. A lack of political will in both nations to challenge Japan directly, naval opposition in both nations to politicians pushing the Western powers into a war in the Far East when both the Royal Navy and United States Navy believed they were not adequately prepared or positioned for such a fight, a lack of agreed strategic objectives, personality clashes, and

---

[40]   Pearl Harbor Investigation, Part 14, Exhibit No. 9, Secret Letter from CNO Stark to Admiral Hart CinC Asiatic Fleet, 12 Nov. 1940, pp. 972–3.

[41]   Pearl Harbor Investigation, Part 3, Testimony of Lt General Leonard T. Gerow, War Plans Division in the War Department in 1941, pp. 990–5; Ian Cowman, *Dominion or Decline*, chap. 5, pp. 207–79 ('Cooperation without Collaboration').

[42]   Pearl Harbor Investigation, Part 14, Exhibit No. 16, Secret, Minutes of Meeting of the Joint Board, 3 Nov 1941, pp. 1061–5; Pearl Harbor Investigation, Part 14, Exhibit No. 16, Memorandum for the Chief of Staff, 'The Far Eastern Situation', 3 Nov 1941, pp. 1066–7.

rapidly changing international conditions due to the war in Europe, all transpired to prevent a combination of naval power and economic power being forged into an effective deterrent.[43] Japan, however, had no hesitation or reluctance in using naval power to produce economic warfare effects through a strategy of blockade and denial of Anglo-American trade and access with China.

In early September 1937, the Imperial Japanese Navy issued warnings to all foreign vessels operating on the Yangtze and Whangpoo rivers that, for 'their own safety', it would be best if these ships would not operate in proximity to Japanese warships in the area.[44] By the 10th of that month Secretary Hull and the State Department considered that a Japanese blockade was in effect along the entire coast of China from Chinwangtao to Pakhoi.[45] While this blockade was notionally aimed only at Chinese shipping, the effects of that action were not limited to Chinese targets alone. On 14 September, President Roosevelt issued a statement announcing that US Government-owned merchant ships would not be permitted to transport to China, or Japan, any arms, ammunition, or implements of war as laid out in an earlier Presidential proclamation of May that year. Any other America-flagged vessels that were tempted to carry such cargoes would do so at their own risk.[46] The State Department issued instructions to the CinC of the Asiatic Fleet, Admiral H.E. Yarnell, for him to monitor American shipping traffic in the area, to be prepared to inform the Japanese naval authorities of any shipping he considered worthy of bringing to their attention in order to avoid unnecessary boardings, and in the case of where American vessels were stopped by Japanese naval vessels that the masters of those ships provide proof of nationality but no further information except under protest. Hull did not consider it necessary at this early stage to make any further statements of policy regarding cargoes given the President's declaration of 14 September.[47] British Foreign Office officials considered the American position to be a capitulation to the Japanese attempt to coerce any nations involved in trading with China to abandon that activity. Without Japan having officially declared their intention to interfere with the shipping of third parties, America's timidity was sure to encourage Japan to move towards a more complete blockade even more quickly.[48]

---

[43]   Pearl Harbor Investigation, Part 14, Exhibit No. 22B, Memorandum of Conversation, British–American Cooperation, Alexander Cadogan and Sumner Welles, 10 Aug. 1941 on board *Prince of Wales*.

[44]   Foreign Relations of the United States Diplomatic Papers (FRUS), Japan 1931–41, vol. 1, telegram from Navy Department to the State Department, 6 Sept. 1937, pp. 369–70.

[45]   Telegram from Navy Department to the State Department, 6 Sept. 1937, pp. 369–70; Press Release by the Department of State, 10 Sept. 1937, p. 371.

[46]   FO 371/20977, China 1937, F6557/130/10, telegram from Washington to FO, 15 Sept. 1937.

[47]   FRUS, Japan 1931–41, vol. 1, telegram from Hull to Consul General at Shanghai and CinC Asiatic Fleet, 22 Sept. 1937, pp. 371–2.

[48]   FO 371/20977, China 1937, F6557/130/10, Ronald minute, 17 Sept. 1937.

The Permanent Undersecretary in waiting, Sir Alexander Cadogan, pointed out how such actions made a combined Anglo-American response to the Japanese maritime strangulation of China unlikely:

> It is due to their blind fright of becoming involved in any incident. If there was only one chance in a hundred that the Japanese would interfere with this cargo, the US Government could be counted on to act as they have done. It is a poor performance and a warning to us, if such were needed, of what to expect from them.[49]

Under the pressures of a growing Far Eastern crisis, Anglo-American political collaboration to confront the Japanese use of maritime economic warfare was not trusting enough, robust enough, or coordinated enough to provide a reliable or legitimate resistance to such methods. While able to consider parallel approaches to the deteriorating situation, the two Western naval powers could not put real joint actions to deter Japan into operation.[50]

The Admiralty's assessment of conditions in the Far East at the time brought all of the various naval aspects of the situation into stark, if eloquently worded, relief. In discussions regarding the growing concerns about the stability of the situation in the Far East, the First Sea Lord, Admiral Alfred Ernle Montacute Chatfield, informed the British Cabinet of his, and therefore the Royal Navy's, perspective on how events could be expected to proceed. In his view, the Mediterranean and European situation continued to demand the presence of a large portion of the Royal Navy's strength and attention. If the Japanese were to tighten blockade controls and prevent contraband getting in to China, then the correct policy was for Great Britain to declare its neutrality, admit there was belligerency, and inform British shipping that they carried contraband to China at their own risk and without the protection of the state.[51] Most importantly, the First Sea Lord pointed out the differences between the European experiences for the Royal Navy in economic warfare and those that would be found in the Pacific. The case was simply one of mismatched strength: the Imperial Japanese Navy was too strong for the Royal Navy in those waters:

---

[49]   FO 371/20977, China 1937, F6557/130/10, Cadogan minute, 18 Sept. 1937.

[50]   FRUS, Japan 1931–41, vol. 1, telegram Hull to Minister in Switzerland (Harrison), 28 Sept. 1937, pp. 375–7; FO 371/20667, America 1937, A6896//448/45, despatch from Washington to FO, Cadogan minute, 28 Sept. 1937.

[51]   On the Royal Navy and the Spanish Civil War, see G. Kennedy, 'The Royal Navy, Intelligence and the Spanish Civil War: Lessons in Air Power, 1936–39', *Intelligence and National Security* 20.2 (2005), pp. 238–63; G. Kennedy, 'Anglo-American Strategic Relations and Intelligence Assessments of Japanese Air Power 1934–1941', *Journal of Military History* 74.3 (2010), pp. 737–73. FO 371/20979, China 1937, F8559/130/10, Most Secret FES(37)4 Cabinet Paper, Chatfield memo attached, 23 Sept. 1937.

Naval action would only be required if the Japanese greatly exceeded their rights under International Law or used unnecessary force to exact obedience from a merchant ship. It would be, in the opinion of the Naval Staff, a mistaken policy to use our naval forces in a manner comparable to that in which they have been used in Spanish waters, i.e. to give protection on the high seas against interference by the naval forces of a nation obviously actually at war whether or not a state of war had been admitted or declared to exist. Moreover, the situation in the Far East differs widely from that obtaining on the North Coast of Spain, where a single British battleship can exercise a dominating influence, since she is, in herself, more powerful than the whole of the naval forces by which British shipping in those waters might be threatened. In the Far East, however, any Japanese threat to British shipping which might arise could be backed by the strength of the Japanese fleet.[52]

Chatfield also raised the idea of sending two capital ships to act as a deterrent, but dismissed it as unwise. His recommendation was that sending such a token force would only aggravate the Japanese to escalate their own naval activities, not deter their maritime economic warfare. If the Japanese maritime economic warfare element of their war against China was to be nullified in order that Great Britain's trading rights and access to China were to be ensured then a great deal more effort, an effort resembling a full naval war in terms of its operational impact, would be required.[53] When this information was discussed in November 1937, neither the Foreign Office nor the Cabinet disagreed with the First Sea Lord's assessment or recommendation that no significant British naval reinforcements should be sent to the region.[54] As 1937 drew to a close, the Japanese revealed their growing confidence in and control of the maritime environment in China with deliberate attacks on the American and British gunboats USS *Panay* and HMS *Ladybird*. Their use of the maritime economic warfare weapon was set to increase in 1938.

Japanese naval and air activities in early 1938 began effectively to close off the Yangtse River to British and American naval and shipping activities.[55] By March of that year, similar Japanese actions were also denying those nations

---

[52]   FO 371/20979, China 1937, F8559/130/10, Most Secret FES(37)4 Cabinet Paper, Chatfield memo attached, 23 Sept. 1937.

[53]   FO 371/20979.

[54]   FO 371/20979, Vansittart minute, 4 Nov. 1937; FO 371/20979, F9710/130/10, Cabinet FES(37) 2nd Meeting, Cabinet Conclusions, 9 Nov. 1937.

[55]   FO 371/22119, China 1938, F1432/158/10, telegram from Howe (Shanghai) to FO, 1 Feb. 1938; FO 371/22119, China 1938, F2738/158/10, telegram from Shanghai to FO, 10 Mar. 1938; FO 371/22048, China 1938, in its entirety for Admiralty pressure on Jardine Matheson to not sail their ships on Yangtse in order to avoid creating tensions with Japanese naval authorities.

access to the Whangpoo River system, thus essentially cutting of China's ability to use the maritime domain for trade and the generation of economic power. Where some reasonable excuses regarding mines, military necessity, and danger of collateral damage were 'legitimate' for the Yangtse River situation, no such conditions existed on the Whangpoo. Japan's actions there were purely aimed at strangling Anglo-American trade with China.[56] Foreign Office attempts to convince even the British Board of Trade to allow some form of administrative reprisals against Japanese shipping (general obstructiveness, administrative delays, and an inability to grant even reasonable requests by Japanese shipping companies throughout the British Empire) met with no success.[57] Foreign Office officials were frustrated by both the Royal Navy's inability to consider even the most moderate of actions for fear of arousing Japan's ire and the Board of Trade's support for such an appeasing attitude: 'The FO are expected to protect British interests without weapons, to keep Chinese good-will without offering material assistance, and to uphold British prestige with nothing but ineffectual protests'.[58] The Admiralty had no intention of sending large reinforcements to the Far East, owing both to the situation in Europe and to fears of provoking an even greater Japanese response at a time when 'it would be strategically unsound to increase the number of ships that would be faced with the vastly superior forces of the Japanese'.[59] At best the Admiralty hoped that perhaps French and American support could be found for a convoy system on the Yangste if the Japanese continued to escalate their trade denial strategy, but failing that were content to let diplomacy take the lead on confronting Japan on the matter.[60] Throughout the year, continued Japanese restrictions to American and British shipping activities escalated the matter, tightening the maritime blockade of China through incremental measures.

By November of that year the British and American Governments had been driven closer in their appreciation of Japan's maritime economic warfare actions being a deliberate and long-term threat to those Western nations' trading interests in the region. The Foreign Office and State Department coordinated their diplomatic actions in November 1938 to bring joint pressure to bear on the

---

[56]   FO 371/22120, China 1938, F2971/158/10, telegram from Shanghai to FO, 15 Mar. 1938.

[57]   FO 371/22122, China 1938, F8213/158/10, letter from Board of Trade to FO, 28 July 1938.

[58]   FO 371/22122, China 1938, F8213/158/10, Thyne Henderson minute, 9 Aug. 1938, and private and personal letter from N.B. Ronald (FO) to Sir William Brown (BoT), 10 Aug. 1938.

[59]   FO 371/22123, China 1938, F8691/158/10, most secret letter from Jarrett (Admiralty) to Ronald (FO), 9 Aug. 1938.

[60]   FO 371/22123, China 1938, F8691/158/10, most secret letter from Jarrett (Admiralty) to Ronald (FO), 9 Aug. 1938; FO 371/22123, China 1938, F9185/158/10, secret Admiralty letter from Phillips to FO, 24 Aug. 1938.

Japanese Government to cease its continued blockade of China, particularly the Yangtse River.[61] Even more importantly, the by now obvious inability of either British or American naval power to influence Japan's maritime economic warfare in China prompted an Anglo-American search for other economic warfare ways to influence Japan's attitude toward their trade and economic rights in the region.

Without the necessary political will or naval power required to confront the Japanese blockade of China, Anglo-American strategic thinking turned to the idea of coordinated economic warfare at distance.[62] The two nations had come closer in their strategic assessments of the true danger and nature of the Japanese actions in China to their own, and international, interests. This common appreciation led to the consideration of more coordinated and linked actions to counter the Asiatic Power. As such, as 1938 drew to a close, both nations were not only considering crucial loans to help China continue to wear down Japan's war-making capabilities and avoid defeat on land but also contemplating the use of sanctions, embargoes, and freezing orders aimed at the Japanese economy.[63] The methods would not require naval power as they were directly linked to banking, trade, and commercial activities associated with economic power and its creation. The American denunciation of the 1911 Trade Treaty with Japan in July 1939, the coordination of large loans to China, continued forensic studies by both nations (with their information being shared between the State Department and Foreign Office, as well as between the two Treasuries), as well as the British provision of a Navicert (export shipping and licensing system) system for American preparations to restrict Japanese trade and shipping in 1940, were all precursors to the eventual embargoes and final freezing of Japanese assets in 1940 and 1941.[64] Those coordinated and escalating Anglo-American economic

---

[61]   FRUS, The Far East, 1938, vol. 3, telegram from Hull to Kennedy (US Ambassador in London), 2 Nov. 1938, p. 194; FRUS, The Far East, 1938, vol. 3, telegram from Kennedy to Hull, 3 Nov. 1938, p. 197; FO 371/22181, Japan 1938, F10611/71/23, telegram from FO to Craigie (British Ambassador in Tokyo), 10 Oct. 1938; FO 371/22144, China 1938, F12398/575/10, telegram from Craigie to FO, 22 Nov 1938 and Scott minute of 24 Nov.

[62]   FO 371/22092, China 1938, F7991/62/10, FO memo, 'Possibility of Retaliation against Action by Japan in the Far East Detrimental to British Interests', 27 July 1938.

[63]   FO 371/22125, China 1938, F11778/158/10, record of meeting between Ray Atherton (First Secretary of US Embassy in London) and Cadogan, and associated minutes, 3 Nov. 1938; FRUS, The Far East, 1938, vol. 3, memo of conversation by Hamilton, 5 Dec. 1938, pp. 577–81.

[64]   On the NAVICERT system's origins, see G. Kennedy, 'Strategy and Power: The Royal Navy, The Foreign Office and the Blockade, 1914–1917', for US and British coordination in 1940, see FO 371/24236, America 1940, A1618/86/45, confidential report from Lothian to FO, 20 Feb. 1940, and 1941 FO 371/28865, Co-ordination, W5376/150/49, entire file, 6 May 1941. On other elements of Anglo-American economic warfare at-a-distance cooperation, see FO 371/23551, China 1939, F8138/4027/61, record of meeting between Herschel Johnson (US Embassy London) and Rab Butler, Ministry of Economic Warfare,

warfare actions would fail in the end to provide a deterrent effect against Japan. Indeed, without the authority of convincing naval power to underpin the economic warfare at a distance strategy, the Anglo-American actions were more provocative and inflammatory than deterring.

The experience of the three powers engaged in economic warfare in the Pacific from 1937 until 1941 demonstrates the requirement for credible naval power to act as an enabler if economic coercion is to act as a realistic deterrent. Without the fear of an Anglo-American naval force to protect their maritime trade in the Far East, Japan was able to utilise the control of the sea to impose its will and favourable conditions for the prosecution of the war against China without interference. The necessity for Western powers to resort to supplying China's war effort over land rather than by sea severely weakened not only the Chinese war effort, but encouraged Japan in its pursuit of a dominant maritime economic warfare strategy. That confidence led to Japan's seizure of Indo-China in 1940, timed to coincide with the fall of France. Control of Indo-China's air and sea spaces threatened both Singapore and the Philippines to such a degree that the Western powers were forced to escalate their economic warfare at a distance, but this proved a weak and provocative choice which only further encouraged Japanese naval adventurism. Without sufficient naval power to support peacetime economic competition in the region, Great Britain and America could not gain the strategic initiative and remained reactive to Japanese actions. Only the step too far, the attack on Pearl Harbor and the transitioning of the strategic condition from one of a contested and tense peace to that of open warfare in 1941, allowed Anglo-American naval and economic power to wage a total war strategy instead of an economic warfare strategy.

---

29 July 1939; FO 371/25118, Co-ordination, W11802/8156/49, note from Minister of Ministry of Economic Warfare to Lord Lothian (British Ambassador in Washington), 15 Nov. 1940; FO 371/25137, Co-ordination, W1135/79/49, Cabinet Paper WP(G)(40)14, 'US Cooperation in Preventing Vital Commodities from Reaching Germany, Russia and Japan', 19 Jan. 1940; FO 371/23412, China 1939, F5848/11/10, letter from Bewley to Nixon and attached files, 22 June 1939; FO 837/534, 535. 536, 537 in their entirety on Anglo-American cooperation and freezing order of July 1941.

Epilogue:
Retrospect and Prospect

Daniel Moran

Mephisto's Trinity

All students of strategic theory are familiar with the 'remarkable trinity' put forward by Carl von Clausewitz at the start of *On War*, in which violence, chance, and reason are joined in a conceptual framework intended to encompass war in all its variety.[1] Less familiar is the trinity proposed by Clausewitz's great contemporary, Johann Wolfgang von Goethe, at the end of *Faust*, Part II, where Mephistopheles declares that he 'knows something of navigation', and proclaims 'war, trade, and piracy' to be 'a trinity indivisible'.

This volume has been concerned with how Mephisto's trinity came to be constructed, and how it developed during the centuries when economic warfare was a principal concern of powerful navies. This chapter surveys those developments from the perspective of more recent times, when economic, political, and military conditions have become unfriendly to economic warfare in its traditional forms, even as the economic stakes of war in general have continued to increase. It also aims to highlight some of the issues raised by the other contributors to this volume.

It is, admittedly, Satan who points us toward the trinity we are considering, which may not be an ideal recommendation; but to an audience in the 1830s his observation would have amounted to something like conventional wisdom. Navies of that era, as of ours, existed first of all to defend the coasts of the nations that created them. But they had long since been configured and operated to secure seaborne trade; which trade, managed and protected in the guise of empire, provided the wealth required to sustain the navies. And while no government at any time would have owned to piracy, the practice of seizing private property

---

The contents of this chapter are the responsibility of the author. They have not been reviewed or approved by any government agency, and do not represent the policy or official thinking of the United States Navy or Department of Defense.

[1] Carl von Clausewitz, *On War*, Book I, chap. 1.

at sea, and sharing out the proceeds among those responsible, could easily have deceived the Devil himself.

A number of chapters in this book consider the history on which Goethe looked back, and they show that he was right about how intimately violence and commerce were intertwined in what Stephen Conway calls the systemic underpinning of economic warfare. Even in the seventeenth century, economic warfare had already emerged as a normalised instrument of European diplomacy. Erik Odegard shows how its conduct could become directly aligned with the commercial interests concerned, while Silvia Marzagalli proposes that the capacity of merchants to reorganise trade flows in response to wartime conditions was crucial to strategic success.

Defence of friendly trade was no less critical, and it was in the expansion of such trade that strategic success often lay. Roger Knight concludes that, in terms of hulls if not of tonnage, the volume of trade convoyed by the Royal Navy in the Napoleonic period compares with that of the Second World War – a startling proposition in itself, and strong evidence of Britain's success in seizing control of colonial markets and carrying trade that had once belonged to France and its allies.

The agrarian economies of pre-industrial Europe provided few opportunities to deprive an enemy of material resources vital to war – though when such opportunity did arise there was ample will to seize them, as Britain tried to do during the politically induced food shortages experienced by Revolutionary France (see Silvia Mazargalli's discussion). In general, however, long-distance trade in the Age of Sail was dominated by goods of relatively high value compared to their weight or bulk, and the consequences of seizing them were primarily financial. Governments employed navies to protect trade (in whose proceeds they always had a share), and to prey upon it, because doing so promised to accumulate additional wealth under their control, while denying it to their enemies. It was with such wealth, accrued to the royal coffers by the Royal Navy, that Britain financed the armies that finally drove Napoleon from his throne.

'Mercantilism', as this general policy has come to be known, did not value economic connectivity and interdependence. Nor did it recognise them as instrumental to economic growth, a concept that was scarcely understood until the economic practices of pre-industrial Europe had begun to unravel, amidst the exertions of private enterprise and the decline of long-distance transportation costs (a benefit of steam propulsion on both land and sea). In many respects the creation of the modern global economy was an unintended result of the European pursuit of empire. The same can be said for the breakdown of Mephisto's trinity. Like the private trading companies of the Old Regime, it became a victim of its own success.

It was against this background that Americans sought to discover their place in the larger scheme of things, and how far a navy would be required to achieve it. Their perspective was unusual in European terms, but also revealing of how war at sea might be conceived quite differently from what Mephisto had in mind.

The new American Republic was aware of its military weakness, for which it sought to substitute its high value as a trading partner. As regards European political quarrels, the United States conceived itself as naturally neutral. Its leaders thought America's economic proficiency might somehow hold the world at bay while also inviting it to its door. American policy thus favoured high tariffs, but also neutral rights and freedom of the seas. Americans looked to oceanic glacis that surrounded them for their defense, and sought to make their way without the bother of either a large merchant fleet or a powerful navy, the latter being thought unnecessary absent the former.

Alfred Thayer Mahan's theoretical achievement was rooted in a desire to provide Americans with a more realistic appreciation of the importance of naval power to securing their country's economic future. Mahan read history as a record of God's providence, which was why he thought it was possible to derive scientific conclusions from it. If the Age of Sail had demonstrated the centrality of economic warfare as a naval concern, then the economic expansion and transformation of his own day could only mean that what had been true in the past was becoming even more true in the present. When he declared that there was nothing navies could do that 'at all compares with the protection and destruction of trade',[2] he was speaking a language that American official and public opinion already understood, albeit in terms that embraced an idealised vision of global order in which the marketplace, and not the battlefield, would have the last word. American policy at sea had always revolved around this basic proposition. The task of a navy was emphatically not to secure the indivisibly of war and trade, but as far as possible to keep them separate.

Mahan reconciled this requirement with possession of a powerful navy by way of a complex double move. First, he conceived America's interest at sea in terms of trade rather than shipping – trade being a familiar but helpfully elastic concept, whereas shipping was something of which the United States had little. He then located the military actions by which trade might be defended in a sphere of their own: ideally a great fleet action in which merchantmen and neutrals played no part. The net result was to marginalise trade as a direct object of naval warfare, while identifying it as the central concern of maritime strategy properly understood.

Mahan's vision was, in its native setting, a defensive one. Yet the international resonance that his work achieved shows that it was more than the idiosyncratic view of a sailor whose nation had been spared by Mother Nature from the difficult trade-offs that confronted European strategists. It also cohered well with prevailing military opinion. Mahan and Schlieffen were contemporaries, after all. Both were convinced that in modern war a decision must come early if it was to be worth having. Both believed that war waged against civilian

---

[2]   A.T. Mahan, 'The Possibilities of Anglo-American Reunion', *North American Review* 159 (New York, 1894), p. 561.

economic targets would just prolong the agony to no purpose, and might deprive the whole exercise of any useful result.

Mahan did not, strictly speaking, identify strategic excellence with the pursuit of a quick end to fighting. The historical examples that he studied had mostly been long wars, after all, in which the leverage afforded by command of the sea had only made itself felt with the passage of time. In contrast to Schlieffen, who conceived of war as a single continuous action, Mahan thought it would be enough if, in the wake of a decision at sea, seaborne trade (other than the enemy's) could be left to go about its business, even if the war itself were to drag on somewhere over the horizon. His thinking defined a middle path, between those who argued that the continued expansion and industrialisation of the world economy showed the way to perpetual peace, and those who believed that such conditions would become a recipe for perpetual strife. As so often, however, the middle path proved to be a winding one, narrowing as it went.

Two Maps

Anyone concerned with current naval affairs, in any country where the naval mission extends beyond law enforcement, will be familiar with two maps. One is a map of the world on which the deeply entangled routes of global seaborne trade are displayed. The other, smaller in scale, shows a stretch of coastline – in the United States it is usually the coast of China, but almost anywhere in Eurasia will do – from which multi-coloured phase lines extend out to sea. This map portrays the effective range of shore-based missiles and other weapons designed to hold warships at bay.

The first map portrays what Mahan called 'the Great Commons', the sea envisioned as a shared space, from whose orderly use all might profit. The second records the emergence of large tracts of the world's oceans as a kind of no man's land – contested space dividing adversaries, where neither possesses freedom of action. The confounding relationship between the economic and technological realities the two maps represent emerged as a central problem of maritime strategy in the first half of the twentieth century. It has grown more perplexing since.

Among the belligerents in the world wars it was Britain whose sailors thought most seriously about economic warfare. Around the turn of the century, when Britain's chief adversaries were presumed to be France and Russia, the problem presented itself in explicitly global terms: how to sustain far-flung operations against two rival empires whose economic connections were, in combination, almost as extensive as Britain's. That Britain's own commerce would require defence against two important rival navies was also obvious. But the manner in which these threats would present themselves was not.

The Crimean War, as Maartje Abbenhuis's chapter shows, offered ample evidence that modern economic relationships could impose themselves on navies in ways that confounded old habits. To this had been added a half-century of technological transformation. The Crimean War was fought by sailing navies. The steam-and-steel warships and merchantmen that subsequently displaced them were tied far more intimately to a shoreside infrastructure whose support was essential to extended operations at sea. Global trade had also continued to expand at an extraordinary rate. On its face this made the potential leverage of economic warfare higher than ever. Yet it was not clear that the new navies possessed the operational resilience required to attack the new economy, on whose financial system and supply chains they were themselves dependent.

These problems changed their form, though not their basic nature, in the years immediately preceding the outbreak of the First World War, as it became apparent that Britain's main challenger would be Germany. No one doubted that, in the event of war, Germany would seek to interfere with British trade if it could – though, as Matthew Seligmann's chapter shows, the ferocity of the German attack far outstripped what anyone imagined in advance. In any case, the most serious problem posed by German enmity was that its fleet was concentrated a few hundred miles from British shores. Behind it stood an army widely regarded as the best the world. The task facing the Royal Navy, with respect to economic warfare, was not merely how to secure British trade, while harrying that of the enemy. The real question was how such measures could be executed so as to contribute to the defeat of the German army.

Britain was the first maritime power to confront the problem posed by the two maps. Its naval leadership had concluded before war broke out that Germany's coastline was unapproachable, even to a range sufficient for observation. It is a fair question whether the destruction of Germany's High Seas fleet might somehow have shifted the balance so as to allow a traditional blockade to be enforced. As Bleddyn Bowen's account of sea power theory shows, the question when and how to seek battle at sea is always a paramount concern. But the same weapons that ruled out an observing flotilla also ruled out any direct descent upon the German fleet, and while that fleet existed operations in German coastal waters were impossible.

A blockade mounted further out to sea, at the western end of the Channel, and in the waters north of Scotland, was perfectly feasible in operational terms, and became the chosen strategy in the end. But a blockade hundreds of miles from enemy ports was bound to interfere with neutral trade, a problem whose risks, and escalatory potential, the British had long since learned to respect (on which see the chapters by John Hattendorf and Maartje Abbenhuis). There was, moreover, a good deal of doubt among naval planners whether such a measure, particularly if managed so as to preserve reasonable respect for neutral rights, could really produce useful results against an integrated industrial economy like Germany's.

A more decisive alternative presented itself once it was recognised that the Royal Navy's pre-eminence at sea might be reinforced by the Bank of England's equally commanding position in global financial markets. That economic warfare might be the business of bankers as well as sailors is readily recognisable as a harbinger of the future in which we now live. In the moment, however, the potential knock-on effects from bringing global credit to a standstill, as a way of bringing Germany to its knees, proved too unnerving to contemplate.[3] After the war was over few doubted that the economic privations of Britain's 'distant blockade' had taken a toll on German morale, particularly in the war's final year, after America's entry had rendered that blockade more nearly leak-proof. Yet the war's final year had been awfully slow in arriving, and the German army had remained capable of powerful offensive operations until very near the end.

As John Ferris's chapter on British prize courts illustrates, Britain went to considerable lengths to conduct the war at sea within a legal framework that would command the respect of neutral observers – a longstanding concern, as Anna Brinkman's chapter recalls. This was always a delicate matter, the more so because the range of social interests implicated by economic warfare had expanded in step with the world economy. Long-distance seaborne trade now included a wide range of mundane goods destined for mass consumption. Britain was determined that its blockade should encompass them, specifically including food, a practice that had already begun to raise the kind of humanitarian concerns that have only become more pressing in our own time. It was no less determined that its conduct be consistent with its claim to be fighting in defence of traditional norms of fair dealing and decency, even between adversaries at war.

Even a century ago, it was becoming difficult to reconcile such a claim with an attack on the sinews of civilian life. Britain's success in doing so was materially aided by Germany's decision to mount a counter-blockade employing submarines. Germany's conduct at sea put Britain's transgressions in the shade. Submarines must, by their nature, sink their targets rather than seize them. That such sinking should include killing the civilians on board was by no means inevitable. But to have taken the pains necessary to adhere to so-called 'cruiser rules', by which civilian lives might have been spared, would have reduced the U-boat campaign to a nuisance at worst, well below the war-winning measure Germany sought.

Germany's ability to stand off the Royal Navy created the sea space that made the submarine war possible, but the manner in which it was conducted was shaped by the same strategic pessimism that underlay the German war effort generally. In German eyes a war that ended in mutual exhaustion was tantamount to defeat, and avoiding such an outcome was reason enough to set claims of non-combatant immunity aside.[4] Britain had stayed its hand rather than risk overturning basic norms of international conduct or outraging neutral opinion.

---

[3]   N.A. Lambert, *Planning Armageddon: British Economic Warfare and the First World War* (Cambridge, MA: Harvard University Press, 2012).

[4]   On the legal and ethical dimensions of the German submarine campaign, see Isabel V. Hull, *A Scrap of Paper: Breaking and Making International Law during the Great War* (Ithaca, NY: Cornell University Press, 2014), pp. 240–75.

Germany did not, and it was the German campaign that pointed the way to the future. In its hands, economic warfare emerged as a modality of what would soon be called 'total war'.[5]

That Germany, a continental power par excellence, should have staked its future on a naval campaign against seaborne trade was itself evidence of how heavily concerns about economic vulnerability weighed in the strategic calculus of industrial societies. These concerns would be amplified by the development of war in the air, whose early advocates owed much to theorists of sea power. Giulio Douhet, an Italian evangelist of 'air power', declared that the air, like the sea, was something to be commanded. Those who did so would gain unique means of affecting the military result on land, not least by degrading the moral and economic reserves on which the war-making capacity of any nation rested.[6] Where powerful navies promised the gradual build-up of inexorable pressure, however, an air force promised swift, cataclysmic destruction. Navies would struggle to keep up.

It was in the skies above Japan that this new form of fighting achieved its purest expression. Japan was the only maritime power in the world wars that did not wage economic warfare at sea, at least not as distinct from waging war in general. As Greg Kennedy's chapter shows, Japan's early success insured that that there would be no such thing as neutral trade to worry about in the Western Pacific. Once the war broadened into global conflict, Japan's main adversaries were too rich and too far away for economic warfare to make sense. The Americans, for their part, set aside whatever scruples they might once have felt about the wholesale destruction of seaborne trade.[7] Their campaign against Japanese commerce became a model of remorseless efficiency, not least because the Japanese had neglected to make provision to protect their merchantmen. They were unable to improvise any means of doing so while in the throes of utter ruin.

Yet it was to America's airmen, rather than to its sailors, that credit for the ruin mainly attached. American planning for war against Japan reached back to the early years of the twentieth century. It had always envisioned a climactic

---

5   The coinage is owed to Erich Ludendorff, quartermaster general of the German general staff, and a strong advocate of the submarine war. See Ludendorff, *Der totale Krieg* (Munich: Albert Ebner, 1935).

6   See Giulio Douhet, *Il dominio dell'aria* (Rome: L'Amministrazione Della Guerra, 1921), published in English as *The Command of the Air*, trans. Dino Ferrari (New York: Coward-McCann, 1942).

7   The first general order issued to the United States Navy in the wake of Pearl Harbor was: 'execute against Japan unrestricted air and submarine warfare'; on the genesis of which, see Joel Eria Holwitt, *'Execute against Japan': The U.S. Decision to Conduct Unrestricted Submarine Warfare* (College Station: Texas A&M University Press, 2009). Woodrow Wilson's war message in 1917 had declared Germany's submarine campaign to be 'warfare against mankind'. See Woodrow Wilson, War Message, 65th Cong., 1st Sess. Senate Doc. No. 5, Serial No. 7264, Washington, DC, 1917, 3.

blockade, by which an island nation might be made to see reason.[8] When the time came, however, the real choice proved to be between direct invasion and aerial bombardment, both of which promised a more rapid and conclusive result, compared to slow strangulation from the sea.[9] The 'far-distant ships' that Mahan celebrated as guardians of the Great Commons had lost their place in the modern strategic imagination.[10] Economic war, total war – call it what you will; the only course that made sense was to get it over with.

Diplomacy without Gunboats

When the dust settled on the Second World War in Europe, 80 per cent of Germany's industrial plant was in working order. A third was less than five years old, compared with 10 per cent in 1939. At the same time, about 40 per cent of its people were without housing.[11] Both results have rightly been laid at the feet of allied air forces, but their significance for the post-war evolution of economic warfare in any form is apparent. In material terms, Germany ended the war a more economically advanced nation than it had been when it began. At the same time, the human cost of waging war on terms that defined the whole of the enemy's society as a legitimate target has come to seem inordinately high, if not actually counter-productive relative to the risk and effort involved. And while it is true that economic warfare waged from the sea does not normally knock people's houses down around them, the fact remains that, under modern conditions, the most immediate victims of economic warfare are liable to be those with the smallest personal margins of economic security, and little leverage over the actions of their governments.

Since the end of the world wars, the prospects for economic warfare have been limited by three inimical considerations: the practical difficulties of unteasing the fabric of modern economic relationships so as to apply strategically useful pressure to the enemy's economy without simultaneously harming your own or those of your friends; the increasing weight of humanitarian concern for the lives of innocents – founded upon a renewed conviction that there are in fact innocents in war; and the unlikelihood that the unrelenting economic mobilisation that defined industrialised warfare from the Marne to

---

[8]   On American naval planning against Japan, see Edward S. Miller, *War Plan Orange: The U.S. Strategy to Defeat Japan, 1897–1945* (Annapolis, MD: Naval Institute Press, 1991).

[9]   Richard B. Frank, *Downfall: The End of the Imperial Japanese Empire* (New York: Random House, 1999), pp. 117–63.

[10]   Captain A. T. Mahan, *The Influence of Sea Power Upon the French Revolution and Empire, 1793–1812* (Cambridge, MA: John Wilson and Son, 1892), vol. 2, p. 118.

[11]   Tony Judt, *Postwar: A History of Europe since 1945* (New York: Penguin Press, 2005), pp. 82–3.

Hiroshima will recur in a post-industrial world living in the shadow of nuclear weapons.

Advocates of economic warfare before 1914 appalled liberal opinion by their disregard for civilian lives, rights, and property, while simultaneously appealing to it by promising to cut off the vital resources that fuelled the carnage of combat on land.[12] The world wars seemed to resolve this tension by demonstrating that economic warfare under modern conditions could make war more destructive without making it more decisive. Yet economic warfare has never been as simple as seizing or destroying the wealth of the enemy. As the contributors to this volume have shown, it has always been a matter of understanding how different forms of economic pressure react upon broader elements of the international system. The pressure points have shifted in our time, but they are still there.

It is possible, for instance, to view the escalatory risks of economic warfare as a source of diplomatic leverage, however limited its effect may be on the enemy's capacity to make war. The Iran–Iraq war (1980–88) provides an example.[13] It began when Iraq seized some oil-rich Iranian territory opposite the Shatt al-Arab. Iraq aimed to keep some of what it had taken, trade some back in return for relief from Iranian-inspired subversion, and in so doing advance its claim to leadership among its Arab neighbours. After about a month of fighting, conditions seemed ripe for such an outcome, and Iraq proposed a ceasefire, the first of many similar efforts to reach a satisfactory end to what its autocratic president Saddam Hussein had regarded as a strictly limited-liability venture. All such overtures were rejected by Iran, whose revolutionary leadership judged the war to be an existential threat to their newly established Islamic Republic.

The war acquired a modest maritime dimension in 1981, when Iraq began attacking ships off Iranian ports at the northern end of the Persian Gulf, in retaliation for Iranian attacks on Iraqi oil infrastructure in the same area. In 1984, however, the war at sea acquired new significance when Iraq abandoned efforts to achieve a ceasefire and embarked on a more extensive and indiscriminate campaign against merchant vessels far removed from the scene of war on land, provided they were believed to be bound for Iran. Iran, having largely held its fire at sea up to then, responded in kind. By the end of the year the so-called 'Tanker War' had become a matter of widespread concern, not least in the United States,

---

[12] See Bernard Semmel, *Liberalism and Naval Strategy: Ideology, Interest and Sea Power during the Pax Britannica* (Boston: Allen & Unwin, 1986).

[13] The best general account of the war is Pierre Razoux, *The Iran-Iraq War*, trans. Nicholas Elliott (Cambridge, MA: Harvard University Press, 2015); for the war at sea, particularly in its political and legal dimensions, see Martin S. Navias and E.R. Hooton, *Tanker Wars: The Assault on Merchant Shipping During the Iran-Iraq Conflict, 1980–1988* (London: I.B. Tauris Publishers, 1996); and George K. Walker, *The Tanker War, 1980–88: Law and Policy* (Newport, RI: Naval War College, 2000).

which had proclaimed itself protector of the Gulf and its oil following the Soviet invasion of Afghanistan in 1979.

It is impossible to say precisely what balance of strategic calculation motivated the Iraqis to instigate the Tanker War. On its face it was scarcely a promising move, Iran being far better positioned to prey upon Gulf shipping. But that was almost certainly the point: Iraq, finding itself at war against an adversary stronger and less biddable than it had imagined, longed for American intervention; and it was indeed inconceivable that the United States would tolerate anything approximating a closure of the Gulf. Iran threatened as much, even while moderating its conduct so as to avoid provoking the Americans. By 1987, however, conditions had deteriorated to the point where neutral Gulf states were clamouring for protection. In May of that year the United States was finally pushed over the edge by an ostensibly inadvertent Iraqi attack on one of its own vessels, the USS *Stark*, on patrol off the Saudi coast. Even for American warships, the Gulf had become too dangerous to leave alone.

A week later the United States began offering protection to neutral shipping in the Gulf, some of which were 'reflagged' as American. The number of US warships there grew apace, as did clashes with Iranian vessels laying mines and harassing merchantmen in the shipping lanes the Americans were trying to protect. It was in the midst of one such encounter, in July 1988, that the USS *Vincennes* shot down an Iranian civilian airliner, mistaking it for an Iranian fighter – a calamity that finally persuaded the mullahs in Teheran to let go of a war that had helped to solidify their hold on power, but whose costs and risks were finally growing too exorbitant to bear.

The Tanker War is the most consequential example of economic warfare since 1945. It is instructive in a number of respects, most obviously as to the centrality of oil as the beating heart of the modern world economy. Oil's strategic significance was already evident in the 1930s, when Britain and France (and, separately, the United States) pondered whether to add an oil embargo to the list of economic sanctions applied against Italy, by way of persuading it to cease its aggression against Ethiopia. In the end, all three forbore to do so, fearing the escalatory consequences of such a move within an international environment that was already fraught with peril. That such concerns were far from groundless was demonstrated in lapidary fashion a few years later, when American efforts to restrict Japan's access to petroleum (or, more precisely, the dollars required to pay for it) was judged by the Japanese to be a de facto act of war, to which the attack on Pearl Harbor was the reply.

The latter catastrophe notwithstanding, both the Ethiopian and Japanese episodes share a fundamentally diplomatic character, and the Tanker War, too. They all aimed not to apply economic pressure in pursuit of military results but to alter political calculations in ways that were supposed to be conducive to negotiation. The same can be said of the oil embargo imposed by the leading members of OPEC during the October War of 1973 with a view to discouraging international support for Israel. Their action quadrupled oil prices around the world in a matter of months and established 'energy security' as a shibboleth of contemporary international relations; but it was not regarded as an act of war by anyone, including the Israelis.

In the whole course of the Tanker War only about 1 per cent of Gulf shipping ever came under any kind of attack. Some 200 mariners lost their lives. That such a militarily exiguous episode should stand out in the recent record of economic warfare is liable to be taken as evidence that such measures have disappeared from the repertoire of modern strategy. It is more accurate to say that they have been shifted from the realm of war to that of diplomacy, where they serve not to incite escalation, but to control it. Economic sanctions have become familiar tools of international politics because they afford means of coercive pressure, beyond what is possible in routine diplomatic relations, but below the threshold likely to incite a warlike response.[14] That such tasks now fall chiefly to bankers rather than sailors should not obscure the fact that their conceptual genealogy lies with the sea services of the past.

Sea services of the present do not, as a rule, contemplate attacking seaborne trade; which is not to say they would not do so in given circumstances. Yet it is broadly true that, as far as the world's major navies are concerned, economic warfare is a realm in which the defensive has triumphed. The United States Navy, which enjoys unprecedented preponderance over all its rivals, conceives itself to be a faithful guardian of Mahan's Great Commons. It often features the first of the two maps described earlier – the one showing global trade routes – in its recruitment and public relations advertising.

The extensive naval building programs that have been under way in South and East Asia since the turn of the century are also routinely portrayed as necessary to defend seaborne trade, particularly in the waters linking those regions to the oil producers of the Middle East. One need not doubt the sincerity of this motive to see that it begs the question who, exactly, might be envisioned as having the motive and capacity to interfere with such trade. And here the second map – the one showing the no man's land created by modern coastal defence systems – also comes to mind. The two maps overlap a good deal in the waters of South and East Asia, and it would be no more than prudence for the navies of that region to keep both of them in mind. You no longer need a navy to attack seaborne trade, but you probably do need one to protect it.

Yet the map is no more the territory at sea than on land. The true dimensions of the no man's land problem are far from certain. No more is the answer to the question of where the next attack on the world economy might come from. Should the trade-driven economies of rising Asia find themselves caught up in war at sea, there is no telling how well they would tolerate life behind an arsenal of

---

[14]   This is the understanding of 'gunboat diplomacy' advanced in the classic study by James Cable, *Gunboat Diplomacy 1919–1991: Political Applications of Limited Naval Force* (3rd ed.; New York: St Martin's Press, 1994).

anti-access weapons, even if those weapons succeeded in holding the enemy at bay. There is also no telling whether an adversary, confronting such a defense, will find the patience to sustain useful pressure against it without recourse to military methods that render the whole question of merely 'economic' warfare moot.

Nothing has happened since 1945 to add traction or hand-holds to the slippery slope that has come to connect economic warfare and total war. The most one can say with confidence is that a coherent defence of economic interests nowadays must accept that any nation, most especially when acting without the advice and support of allies and trading partners, can become the victim of its own actions in ways it does not foresee. Economic warfare has always been a two-edged sword. Nowadays, though, it is better to imagine a sword with no hilt. To wield it, you must be willing to grasp the blade.

Contributors

Maartje Abbenhuis is an associate professor in modern history at the University of Auckland, New Zealand. She specialises in the history of war, peace, neutrality, and internationalism, especially in the period 1815–1918. Her books include *The Art of Staying Neutral: The Netherlands in the First World War, 1914–1918* (2006); *An Age of Neutrals: Great Power Politics, 1815–1914* (2014); *The Hague Conferences in International Politics, 1898–1915* (2018); and *The First Age of Industrial Globalization: An International History, 1815–1918* (co-authored; 2019).

Bleddyn E. Bowen is a lecturer in international relations at the University of Leicester and convenes the Astropolitics Collective, an informal space politics research network in the UK. He is an expert in space policy and space-power theory and has published in several academic journals on sea power analogies to outer space and transatlantic space strategies, policies, and doctrines. He has a monograph forthcoming on space-power theory which is founded upon a collection of sea-power theories. Bowen's wider research interests include strategic theory and modern warfare.

Anna Brinkman is an AHRC post-doctoral research fellow at the University of Warwick. She received her PhD in war studies from King's College London in 2017 and MA in the history of warfare from King's College London in 2012. Her research focuses on Anglo-Spanish and Anglo-Dutch maritime history in the Atlantic during the long eighteenth century.

Stephen Conway is Professor of History at University College London. He is the author of *The War of American Independence* (1995); *The British Isles and the War of American Independence* (2000); *War, State, and Society in Mid-Eighteenth-Century Britain and Ireland* (2006); *Britain, Ireland, and Continental Europe in the Eighteenth Century: Similarities, Connections, Identities* (2011); *A Short History of the American Revolutionary War* (2013); and *Britannia's Auxiliaries: Continental Europeans and the British Empire, 1740–1800* (2017).

John Ferris is Professor of History at the University of Calgary, honorary professor in the Department of International Politics, Aberystwyth University and the Department of Politics and Law, Brunel University, and associate member, Nuffield College, Oxford. He has written widely on air, diplomatic, intelligence, international, military and naval history, and strategic studies. He is the author of the authorised history of GCHQ.

Louis Halewood is the Philip Nicholas Lecturer in Maritime History at the University of Plymouth. He was previously a Smith Richardson Foundation predoctoral fellow at International Security Studies, Yale University, and completed his DPhil at Merton College, Oxford, where he wrote a thesis on grand strategy, sea power, and the origins of the League of Nations between 1890 and 1919.

John B. Hattendorf is the Ernest J. King Professor Emeritus of Maritime History and senior mentor, John B. Hattendorf Centre for Maritime Historical Research at the US Naval War College in Newport, Rhode Island. He served as the E.J. King Professor for 32 years from 1984 to 2016. Additionally, he was chair of the college's maritime history department and director of the Naval War College Museum, 2003–16. A former officer in the US Navy, he earned his degrees in history from Kenyon College (AB, 1964), Brown University (AM, 1971), and the University of Oxford (DPhil., 1979; DLitt., 2016).

Greg Kennedy is Professor of Strategic Foreign Policy and the Director of the Corbett Centre for Maritime Policy Studies at the defence studies department, King's College London. He received his PhD from the University of Alberta in 1998. Before coming to England Greg taught at the Royal Military College of Canada, in Kingston, Ontario from 1993 to 2000. He is the author of the award-winning monograph, *Anglo-American Strategic Relations and the Far East, 1933–1939* (2002) and has published internationally on Anglo-American strategic relations from 1900 to 1945, contemporary UK security and defence policy, and maritime strategy and security.

Roger Knight left the National Maritime Museum when Deputy Director in 2000, after which he taught at the Greenwich Maritime Institute, University of Greenwich. He published a biography of Nelson, *The Pursuit of Victory*, in 2005, led a Leverhulme-funded project examining the victualling system of the navy at the time of the Napoleonic Wars, which in 2010 resulted in *Sustaining the Fleet*. In 2013, he published *Britain against Napoleon: The Organization of Victory*. He is now working on a book on the British convoy system, 1803–15, to be published in 2022.

Silvia Marzagalli is Professor of Early Modern History at the University Côte d'Azur in Nice and senior fellow of the Institut universitaire de France. Her research deals with merchant networks, the reconfiguration of shipping and trade in times of war, and consular information in eighteenth- and early nineteenth-century Atlantic and Mediterranean worlds. She is currently the P.I. of Portic, a project financed by the French Agence Nationale de la Recherche dealing with the online visualisation of data on maritime trade in France at the eve of the French Revolution.

Daniel Moran is a professor of international and military history in the Department of National Security Affairs at the Naval Postgraduate School in Monterey, California. He was educated at Yale and Stanford Universities, and has been a member of the Institute for Advanced Study at Princeton and a professor of strategy at the Naval War College in Newport, Rhode Island. He teaches and writes about the history of war and international relations in Europe and Asia. He is currently working on a historical atlas of the Second World War in Asia and the Pacific (1931–45).

David Morgan-Owen is Lecturer in Defence Studies at King's College London. His research focuses upon British military and naval history in the era of the First World War, with particular interests in notions of strategy and processes of strategy-making in the period 1870–1925. He is author of *The Fear of Invasion: Strategy, Politics, and British War Planning* (2017), which won the Templer Medal for best first book from the Society for Army Historical Research in 2018.

Erik Odegard studied maritime and colonial history at Leiden University. His dissertation focused on the career paths of Dutch colonial governors in the seventeenth century. During his PhD research, Erik was awarded the first research fellowship of the Dutch National Archives. He is currently employed by the Erasmus University in Rotterdam as a lecturer and a researcher at the *Mauritshuis* in The Hague. He was awarded the Dr Ernst Crone fellowship at the National Maritime Museum of the Netherlands in Amsterdam. Erik's research focuses on Dutch maritime and colonial history.

Matthew S. Seligmann is Professor of Naval History at Brunel University London. He is the author of *Spies in Uniform: British Military and Naval Intelligence on the Eve of the First World War* (2006); *Naval Intelligence from Germany: The Reports of the British Naval Attachés in Berlin, 1906–1914* (2007); *The Royal Navy and the German Threat, 1901–1914: Admiralty Plans to Protect British Trade in a War against Germany* (2012); *The Naval Route to the Abyss: The Anglo-German Naval Race, 1895–1914* (2015); and *Rum, Sodomy, Prayers and the Lash Revisited: Winston Churchill and Social Reform in the Royal Navy, 1900–1915* (2018).

Select Bibliography

Aaslestad, K.B. and J. Joor (eds), *Revisiting Napoleon's Continental System: Local, Regional, and European Experiences* (Basingstoke: Palgrave, 2014).

Abbenhuis, M., 'A Most Useful Tool of Diplomacy and Statecraft: Neutrality and the "Long" Nineteenth Century, 1815–1914', *International History Review* 35.1 (2013), pp. 1–22.

—— *An Age of Neutrals: Great Power Politics, 1815–1914* (Cambridge: Cambridge University Press, 2014).

—— *The Hague Conferences and International Politics, 1898–1915* (London: Bloomsbury, 2018).

Abbenhuis, M. and G. Morrell, *The First Age of Industrial Globalization: An International History, 1815–1914* (London: Bloomsbury, 2019).

Acemoglu, D., S. Johnson, and J. Robinson (eds), 'The Rise of Europe: Atlantic Trade, Institutional Change, and Economic Growth', *American Economic Review* 95.33 (2005), pp. 546–79.

Alimento, A. (ed.), *War Trade and Neutrality: Europe and the Mediterranean in the Seventeenth and Eighteenth Centuries* (Milan: Franco Angeli, 2011).

Allison, G., *Destined for War: Can America and China Escape Thucydides's Trap?* (London: Scribe, 2017).

Allison, R.J., *The Crescent Obscured: The United States and the Muslim World, 1776–1815* (2nd ed.; Chicago: University of Chicago Press, 2000).

Andersen, D.H. and P. Pourchasse, 'La navigation des flottes de l'Europe du Nord vers la Méditerranée (XVIIᵉ–XVIIIᵉ siècles)', in A. Bartolomei and S. Marzagalli (eds), 'La Méditerranée dans les circulations atlantiques au XVIIIᵉ siècle', *Revue d'histoire maritime* 13 (2011), pp. 21–44.

Anderson, J.L., 'Aspects of the Effect on the British Economy of the Wars against France, 1793–1815', *Australian Economic History Review* 12 (1972), pp. 1–20.

Anderson, O., *A Liberal State at War: English Politics and Economics during the Crimean War* (New York: St Martin's Press, 1967).

Andrew, C., *The Defence of the Realm: The Authorised History of MI5* (Toronto: Penguin Canada, 2009).

Arielli, N., G.A. Frei, and I. van Hulle, 'The Foreign Enlistment Act, International Law, and British Politics, 1819–2014', *International History Review* 38.4 (2016), pp. 636–56.

Armstrong, B.F. (ed.), *21st Century Mahan: Sound Military Conclusions for the Modern Era* (Annapolis, MD: Naval Institute Press, 2013).

Arthur, Brian, *How Britain Won the War of 1812: The Royal Navy's Blockades of the United States, 1812–1815* (Woodbridge: Boydell Press, 2011).

Avery, R.W., 'The Naval Protection of Britain's Maritime Trade, 1793–1802', DPhil. thesis, University of Oxford, 1983.

Baldwin, D.A., *Economic Statecraft* (Princeton, NJ: Princeton University Press, 1985).

Balzacq, T., P. Dombrowski, and S. Reich, 'Is Grand Strategy a Research Program? A Review Essay', *Security Studies* 40.1–2 (2017), pp. 295–324.

Bander, J., *Dutch Warships in the Age of Sail, 1600–1714: Design, Construction, Careers and Fates* (Barnsley: Seaforth Publishing, 2014).

Baugh, D., 'British Strategy during the First World War in the Context of Four Centuries', in D. Masterson (ed.), *Naval History: The Sixth Symposium of the US Naval Academy* (Wilmington, DE: Scholarly Resources, 1987), pp. 85–110.

—— 'Great Britain's "Blue-Water" Policy, 1689–1815', *International History Review* 10.1 (1988), pp. 33–58.

—— 'Maritime Strength and Atlantic Commerce: The Uses of "A Grand Maritime Empire"', in L. Stone (ed.), *An Imperial State at War: Britain from 1689 to 1815* (London: Psychology Press, 1994), pp. 185–223.

—— *The Global Seven Years War, 1754–1763: Britain and France in a Great Power Contest* (Abingdon: Routledge, 2011).

Belich, J., J. Darwin, M. Frenz, and C. Wickham (eds), *The Prospect of Global History* (Oxford: Oxford University Press, 2016).

Bell, A.C., *A History of the Blockade of Germany and of the Countries Associated with Her in the Great War, Austria–Hungary, Bulgaria and Turkey, 1914–1918* (London: HMSO, 1937).

Bell, C.M., '"Our Most Exposed Outpost": Hong Kong and the British Far Eastern Strategy, 1921–1941', *Journal of Military History* 60.3 (1996), pp. 61–88.

—— *The Royal Navy, Seapower and Strategy between the Wars* (London: Palgrave Macmillan, 2000).

—— 'The "Singapore Strategy" and the Deterrence of Japan: Winston Churchill, the Admiralty and the Dispatch of Force Z', *English Historical Review* 16.2 (2001), pp. 604–34.

Berghahn, V.R., *Der Tirpitz-Plan: Genesis und Verfall einer innenpolitischen Kreisenstrategie unter Wilhelm II* (Düsseldorf: Droste Verlag, 1971).

Best, G., *Humanity in Warfare: The Modern History of the International Law of Armed Conflicts* (London: Methuen, 1983).

Bickham, T., *The Weight of Vengeance: The United States, the British Empire, and the War of 1812* (New York: Oxford University Press, 2012).

Blakemore, R. and E. Murphy, *The British Civil Wars at Sea, 1638–1653* (Woodbridge: Boydell Press, 2018).

Blanchard, J.M.F. and N.M. Ripsman, 'A Political Theory of Economic Statecraft', *Foreign Policy Analysis* 4 (2008), p. 371.

Bönker, Dirk, *Militarism in a Global Age: Naval Ambitions in Germany and the United States before World War I* (Ithaca, NY: Cornell University Press, 2012).

Bourguignon, J.H., *Sir William Scott, Lord Stowell, Judge of the High Court of Admiralty, 1798–1828* (Cambridge: Cambridge University Press, 1987).

Bowen, B.E., 'From the Sea to Outer Space: The Command of Space as the Foundation of Spacepower Theory', *Journal of Strategic Studies* 42.2–3 (2019), pp. 532–56.

Bowen, H.V., 'The Contractor State, *c*.1650–1815', *International Journal of Maritime History* 25.1 (2013), pp. 239–74.

Boxer, C.R. (ed.), *The Journal of Maarten Harpertszoon Tromp Anno 1639* (Cambridge: Cambridge University Press, 1930).

—— *The Dutch in Brazil, 1624–1654* (Oxford: Clarendon Press, 1957).

Boyd, A., *The Royal Navy in Eastern Waters: Linchpin of Victory, 1935–1942* (London: Seaforth Publishing, 2017).

Boyle, F.A., *Foundations of World Order: The Legalist Approach to International Relations 1898–1922* (Durham, NC: Duke University Press, 1999).

Braisted, W.R., *The United States Navy in the Pacific, 1909–1922* (Austin: University of Texas Press, 1971).

Brakel, S. van, *De Hollandsche handelscompagnieën der zeventiende eeuw: Hun ontstaan – hunne inrichting* (The Hague: Martinus Nijhoff, 1908).

Brands, H., *What is Good Grand Strategy? Power and Purpose in American Statecraft from Harry S. Truman to George W. Bush* (Ithaca, NY: Cornell University Press, 2014).

Brands, H. and Z. Cooper, 'Getting Serious about Strategy in the South China Sea', *Naval War College Review* 71 (2018), pp. 13–21.

Brewer, J., *The Sinews of Power: War, Money and the English State, 1688–1783* (Cambridge, MA: Harvard University Press, 1990).

Broadberry, S. and M. Harrison (eds), *The Economics of World War I* (Cambridge: Cambridge University Press, 2009).

Bruijn, J.R., *The Dutch Navy of the Seventeenth and Eighteenth Centuries* (Columbia: University of South Carolina Press, 1993).

Buchet, C., *Marine, économie et société. Un exemple d'interaction: l'avitaillement de la Royal Navy durant la guerre de Sept Ans* (Paris: Honoré Champion, 1999).

Burk, K., *Old World, New World: Great Britain and America from the Beginning* (New York: Atlantic Monthly Press, 2007).

Butel, P., 'Réorientations du négoce français à la fin du XVIIIᵉ siècle, les Monneron et l'océan Indien', in Paul Butel and Louis M. Cullen (eds), *Négoce et industrie en France et en Irlande aux XVIIIᵉ et XIXᵉ siècles* (Paris: Editions du CNRS, 1980), pp. 65–73.

—— *Les négociants bordelais, l'Europe et les Îles au XVIIIᵉ siècle* (Paris: Aubier-Montaigne, 1996 [1974]).

Callwell, C.E., *Military Operations and Maritime Preponderance: Their Relation and Interdependence* (London: Blackwood and Sons, 1905).

Carless Davis, W.H.W., *History of the Blockade: Emergency Departments* (London: HMSO, 1920).

Castex, R., *Strategic Theories*, trans. Eugenai Kiesling (Annapolis, MD: Naval Institute Press, 1994).

Chadwick, E., *Traditional Neutrality Revisited: Law, Theory and Case Studies* (The Hague: Kluwer Law, 2002).

Chamberlain, J.P., *The Regime of International Rivers: Danube and the Rhine* (New York: Columbia University Press, 1923).

Chappell, J., 'Maritime Raiding, International Law and the Suppression of Piracy on the South China Coast 1842–1869', *International History Review* 40.3 (2018), pp. 473–92.

Clark, J.G., *La Rochelle and the Atlantic Economy during the Eighteenth Century* (Baltimore, MD: Johns Hopkins University Press, 1981).

Clarke, G., 'War, Trade and Food Supply', *National Review* 29 (1897).

Clauder, A., *American Commerce as Affected by the Wars of the French Revolution and Napoleon, 1793–1812* (Philadelphia: University of Pennsylvania, 1972 [1932]).

Clausewitz, C. von, *On War*, trans. Michael E. Howard and Peter Paret (Princeton, NJ: Princeton University Press, 1976).

Cohen, N., 'The Ministry of Economic Warfare and Britain's Conduct of Economic Warfare, 1939–1945', PhD thesis, King's College London, 2001.

Collins, G., 'A Maritime Oil Blockade against China – Tactically Tempting but Strategically Flawed', *Naval War College Review* 71 (2018), pp. 49–78.

Colville, Q. and J. Davey (eds), *A New Naval History* (Manchester: Manchester University Press, 2019).

Connolly, D., 'The Rise of the Chinese Navy: A Tirpitzian Perspective of Sea Power and International Relations', *Pacific Focus* 37.2 (2017), pp. 182–207.

Conway, S., *The British Isles and the War of American Independence* (Oxford: Oxford University Press, 2000).

—— 'Another Look at the Navigation Acts and the Coming of the American Revolution', in J. McAleer and C. Petley (eds), *The Royal Navy and the British Atlantic World, c.1750–1820* (London: Springer, 2016), pp. 77–96.

—— *Britannia's Auxiliaries: Continental Europeans and the British Empire, c.1740–1800* (Oxford: Oxford University Press, 2017).

—— 'Sea Power: The Royal Navy', in Alan Forrest (ed.), *The Cambridge History of the Napoleonic Wars*, 3 vols (Cambridge: Cambridge University Press, 2019).

Coogan, J.W., *The End of Neutrality: The United States, Britain and Maritime Rights 1899–1915* (Ithaca, NY: Cornell University Press, 1981).

Cook, A., *The Alabama Claims* (Ithaca, NY: Cornell University Press, 1975).

Coppolaro, L. and F. McKenzie (eds), *A Global History of Trade and Conflict since 1500* (Basingstoke: Palgrave, 2013).

Corbett, J.S., *England in the Seven Years War: A Study in Combined Strategy* (London: Longmans, Green and Company, 1907).

—— 'Staff Histories', in J.S. Corbett (ed.), *Naval and Military Essays* (Cambridge: Cambridge University Press, 1914).

—— *Naval Operations* (London: Longmans, Green and Company, 1921).

—— *Principles of Maritime Strategy* (Mineola, NY: Dover, 2004).

Cowman, I., *Dominion or Decline: Anglo-American Naval Relations in the Pacific, 1937–1941* (Oxford: Berg, 1996).

Cox, M., 'Hunger Games: Or How the Allied Blockade in the First World War Deprived German Children of Nutrition, and Allied Food Aid Subsequently Saved Them', *Economic History Review* 68.2 (2015), pp. 600–31.

Crosby, A.W., 'America, Russia, Hemp and Napoleon: A Study of Trade between the United States and Russia, 1783–1814', PhD thesis, Boston University, 1961.

Crouzet, F., *L'économie britannique et le blocus continental, 1806–1813*, 2 vols. (2nd ed.; Paris: Economica, 1987 [1958]).

Crowhurst, P., *The Defence of British Trade, 1689–1815* (Folkestone: Dawson, 1977).

—— *The French War on Trade: Privateering, 1793–1815* (Aldershot: Scolar Press, 1989).

Dancy, J.R., *The Myth of the Press Gang: Volunteers, Impressment, and the Naval Manpower Problem in the Late Eighteenth Century* (Woodbridge: Boydell & Brewer, 2015).

Daughan, G.C., *1812: The Navy's War* (New York: Basic Books, 2011).

Davey, J., *The Transformation of British Naval Strategy: Seapower and Supply in Northern Europe, 1808–1812* (Woodbridge: Boydell Press, 2012).

—— *In Nelson's Wake: The Navy and the Napoleonic Wars* (New Haven, CT: Yale University Press, 2015).

Davis, Lance E. and Stanley L. Engerman, *Naval Blockades in Peace and War: An Economic History since 1750* (Cambridge: Cambridge University Press, 2006).

Davis, R., *The Industrial Revolution and British Overseas Trade* (Leicester: Leicester University Press, 1979).

DeConde, A., *The Quasi War: The Politics and Diplomacy of the Undeclared War with France, 1797–1801* (New York: Scribner, 1966).

De Vries, J., 'The Limits of Globalization in the Early Modern World', *Economic History Review* 63.3 (2010), pp. 710–33.

Dehne, P., 'The Ministry of Blockade during the First World War and the Demise of Free Trade', *Twentieth Century British History* 27.3 (2016), pp. 333–56.

Deist, W., *Flottenpolitik und Flottenpropaganda: Das Nachrichtenbureau des Reichsmarineamtes, 1897–1914* (Stuttgart: Deutsche Verlags-Anstalt, 1976).

Dickerson, O., *The Navigation Acts and the American Revolution* (Philadelphia: University of Pennsylvania Press, 1951).

Dingman, R., *Power in the Pacific: The Origins of Naval Arms Limitation* (Chicago: University of Chicago Press, 1976).

Dobson, A.P., 'The Reagan Administration, Economic Warfare, and Starting to Close Down the Cold War', *Diplomatic History* 6.1 (2005), pp. 534–55.

Dowling, C., 'The Convoy System and the West Indian Trade 1803–1815', DPhil. thesis, University of Oxford, 1965.

Dülffer, J., 'Limitations on Naval Warfare and Germany's Future as a World Power: A German Debate, 1904–6', *War and Society* 3 (1985), pp. 23–43.

Earle, E.M., *Makers of Modern Strategy: Military Thought from Machiavelli to Hitler* (Princeton, NJ: Princeton University Press, 1943).

Elleman, B.A. and S.C.M. Paine (eds), *Naval Blockades and Seapower: Strategies and Counter-Strategies, 1805–2005* (London: Routledge, 2006).

Eloranta, J., E. Golson, P. Hedberg, and M.C. Moreira (eds), *Small and Medium Powers in Global History: Trade, Conflicts, and Neutrality from the 18th to the 20th Centuries* (London: Routledge, 2019).

Emmer, P., 'In Search of a System: The Atlantic Economy, 1500–1800', in Horst Pietschmann (ed.), *Atlantic History: History of the Atlantic System* (Göttingen: Vandenhoeck & Ruprecht, 2002), pp. 169–78.

Epkenhans, M., *Tirpitz: Architect of the German High Seas Fleet* (Washington, DC: Potomac Books, 2008).

Erickson, A.S., Lyle J. Goldstein, and Carnes Lord (eds), *China Goes to Sea: Maritime Transformation in Comparative Historical Perspective* (Annapolis, MD: Naval Institute Press, 2009).

Fabel, R.F.A., 'The Laws of War in the 1812 Conflict', *American Studies* 14.2 (1980), pp. 199–218.

Farrell, B.P. and S. Hunter (eds), *A Great Betrayal? The Fall of Singapore Revisited* (Singapore: Marshall Cavendish, 2010).

Fayle, C.E., 'The Employment of British Shipping', in C. Northcote Parkinson (ed.), *The Trade Winds: A Study of British Overseas Trade during the French Wars, 1793–1815* (London: George Allen & Unwin, 1948).

Ferris, J.R., 'The Origins of the Hunger Blockade: Irony, Intelligence and International Law, 1914–15', in M. Epkenhans and S. Huck, *Der Erste Weltkrieg zur See* (Munich: De Gruyter Oldenbourg, 2017), pp. 83–98.

Findlay, R. and K.H. O'Rourke, *Power and Plenty: Trade, War, and the World Economy in the Second Millennium* (Princeton, NJ: Princeton University Press, 2007).

Fleming, N.C., 'The Imperial Maritime League: British Navalism, Conflict and the Radical Right c.1907–1920', *War in History* 23.3 (2016), pp. 296–322.

Forssberg, A.M., M. Hallenberg, O. Husz, and J. Nordin (eds), *Organizing History: Studies in Honour of Jan Glete* (Lund: Nordic Academic Press, 2011).

Freedman, L., *The Official History of the Falklands Campaign*, 2 vols (London: Routledge, 2005).

Frei, G.A., 'Great Britain, Contraband and Future Maritime Conflict (1885–1916)', *Francia* 40 (2013), pp. 409–18.

Galani, K., *British Shipping in the Mediterranean during the Napoleonic Wars: The Untold Story of a Successful Adaptation* (Leiden: Brill, 2017).

Gee, J., *The Trade and Navigation of Great Britain Considered* (London: Sam. Buckley, 1767).

Gemzell, C.-A., *Organization, Conflict, and Innovation: A Study of German Naval Strategic Planning, 1888–1940* (Lund: Esselte Studium, 1973).

Glete, J., *Navies and Nations: Warships, Navies and State Building in Europe and America, 1500–1860*, 2 vols (Stockholm: Almqvist & Wiksell, 1993).

—— *War and the State in Early Modern Europe: Spain, the Dutch Republic and Sweden as Fiscal-Military States, 1500–1660* (London: Psychology Press, 2002).

—— 'Naval Power and Warfare 1815–2000', in J. Black (ed.), *War in the Modern World since 1815* (London: Routledge, 2003), pp. 217–36.

—— *Swedish Naval Administration, 1521–1721: Resource Flows and Organisational Capabilities* (Leiden: Brill, 2010).

Goldman, E.O., *Sunken Treaties: Naval Arms Control between the Wars* (University Park: Penn State Press, 1994).

Goldman, Z.K. and E. Rosenberg, *American Economic Power and the New Face of Financial Warfare* (Washington, DC: Center for a New American Security, 2015).

Goldrick, J. and J.B. Hattendorf (eds), *Mahan is Not Enough: The Proceedings of a Conference on the Works of Sir Julian Corbett and Admiral Sir Herbert Richmond* (Newport, RI: Naval War College Press, 1993).

Goldstein, E. and J. Maurer (eds), 'Special Issue on the Washington Conference, 1921–22: Naval Rivalry, East Asian Stability and the Road to Pearl Harbor', *Diplomacy & Statecraft* 4 (1993).

Golson, E.B., 'The Economics of Neutrality: Spain, Sweden, and Switzerland in the Second World War', PhD thesis, London School of Economics, 2011.

Gompert, D.C., *Sea Power and American Interests in the Western Pacific* (Santa Monica, CA: RAND Corporation, 2013).

Gore, J. (ed.), *Creevey's Life and Times: A Further Selection from the Correspondence of Thomas Creevey* (London: John Murray, 1934).

Grainger, J.D., *The First Pacific War: Britain and Russia, 1854–1856* (Woodbridge: Boydell Press, 2008).

Gray, C.S., *The Leverage of Sea Power: The Strategic Advantage of Navies in War* (New York: Free Press, 1992).

Greenfield, N.M., *The Battle of the St Lawrence: The Second World War in Canada* (Toronto: HarperCollins Canada, 2004).

Grenfell, R., *Main Fleet to Singapore* (London: Faber and Faber, 1951).

Gretton, P., *Maritime Strategy: A Study of British Defence Policy* (London: Cassell, 1965).

Guilmartin, J.F., Jr, *Galleons and Galleys* (London: Cassel and Co., 2002).

Hall, C.D., *Britain, America, and Arms Control* (New York: Palgrave Macmillan, 1987).

—— *British Strategy in the Napoleonic War, 1803–1815* (Manchester: Manchester University Press, 1992).

—— *Wellington's Navy: Sea Power and the Peninsular War, 1807–1814* (London: Chatham Publishing, 2004).

Halle, E. von, 'Die englische Seemachtpolitik und die Versorgung Großbritanniens in Kriegszeiten', *Marine Rundschau* 17 (1906), pp. 911–27 and 19 (1908), pp. 804–15.

Handel, M.I., *Masters of War: Classical Strategic Thought* (Abingdon: Routledge, 2001).

Harlaftis, G., 'The "Eastern Invasion": Greeks in Mediterranean Trade and Shipping in the Eighteenth and Early Nineteenth Centuries', in Maria Fusaro, Colin Heywood, and Mohamed-Salah Omri (eds), *Trade and Cultural Exchange in the Early Modern Mediterranean: Braudel's Maritime Legacy* (London: I.B. Tauris, 2010).

Harper, L., 'The Effects of the Navigation Acts on the Thirteen Colonies', in R. Morris (ed.), *The Era of the American Revolution* (New York: Columbia University Press, 1939), pp. 3–39.

Hattendorf, J.B., 'Maritime Conflict', in M. Howard, G.G. Andreopoulos, and M.R. Shulman (eds), *The Laws of War: Constraints on Warfare in the Western World* (New Haven, CT: Yale University Press, 1994), pp. 98–115.

—— 'The US Navy and the "Freedom of the Seas" 1775–1918', in R. Hobson and T. Kristiansen (eds), *Navies in Northern Waters, 1721–2000* (Portland, OR: Frank Cass, 2004), pp. 151–74.

—— 'The Third Allen Villiers Memorial Lecture: The War of 1812 in International Perspective', *Mariner's Mirror* 99.1 (2013), pp. 5–22.

Henderson, J., *Frigates, Sloops & Brigs* (Barnsley: Pen & Sword, 2005 [1970]).

Hepper, D.J., *British Warship Losses in the Age of Sail, 1650–1859* (Rotherfield: Jean Boudriot, 1994).

Hershey, A.S., *International Law and Diplomacy of the Russo-Japanese War* (Indianapolis, IN: Hollenbeck Press, 1906).

Herwig, H.H., 'The Failure of German Sea Power, 1914–1945: Mahan, Tirpitz, and Raeder Reconsidered', *International History Review* 10 (1988), pp. 68–105.

Heuser, B., *The Evolution of Strategy: Thinking War from Antiquity to the Present* (Cambridge: Cambridge University Press, 2010).

Hickey, D.R., *The War of 1812: A Forgotten Conflict*, bicentennial edition (Urbana: University of Illinois Press, 2012).

Higgins, Annalise, 'The Idea of Neutrality in British Newspapers at the Turn of the Twentieth Century, c.1898–1902', *New Zealand Journal of Research on Europe* 11 (2017), pp. 2–51.

Hill, R., *The Prizes of War: The Naval Prize System in the Napoleonic Wars, 1793–1815* (Stroud: Alan Sutton, 1998).

Hillman, H. and C. Gathmann, 'Overseas Trade and the Decline of Privateering', *Journal of Economic History* 71.3 (2011), pp. 730–61.

Hoboken, W.J. van, *Witte de With in Brazilie, 1648–1649* (Amsterdam: Noord-Hollandsche Uitgevers Maatschappij, 1955).

Holmes, J.R. and T. Yoshihara, *Chinese Naval Strategy in the 21st Century: The Turn to Mahan* (London: Routledge, 2012).

Hont, I., *Jealousy of Trade: International Competition and the Nation State in Historical Perspective* (Cambridge, MA: Harvard University Press, 2005).

Howard, M., 'The British Way in Warfare: A Reappraisal', in M. Howard (ed.), *The Causes of Wars and Other Essays* (Cambridge, MA: Harvard University Press, 1983).

Howland, D., 'Contraband and Private Property in the Age of Imperialism', *Journal of the History of International Law* 13 (2011), pp. 117–53.

Hufbauer, G.C., J.J. Schott, and K.A. Elliot, *Economic Sanctions Reconsidered: History and Current Policy* (2nd ed.; Washington, DC: Institute for International Economics, 1990).

Hughes, E (ed.), *The Private Correspondence of Admiral Lord Collingwood* (London: Navy Records Society, 1957).

James, W., *A Naval History of Great Britain*, 6 vols (London: Richard Bentley & Son, 1878).

—— *Naval Occurrences of the War of 1812* (London: Conway Maritime Press, 2004 [1817]).

John, L (ed.), *Despatches and Letters Relating to the Blockade of Brest, 1803–1805*, 2 vols (London: Navy Records Society, 1899–1902).

Keefer, S.A., '"An Obstacle, Though Not a Barrier": The Role of International Law in Security Planning during the *Pax Britannica*', *International History Review* 35.5 (2013), pp. 1031–51.

Kelly, P.J., *Tirpitz and the Imperial German Navy* (Bloomington: Indiana University Press, 2011).

Kennedy, G., '"To Throttle, Not Knock Out": The Role of Malta in the RN's Sea Denial, Interdiction and Naval Diplomacy Operations: 1939–43', in Ian Speller (ed.), *The Royal Navy and Maritime Power in the 20th Century* (London: Frank Cass, 2004).

—— 'Intelligence, Strategic Command and the Blockade: Britain's War Winning Strategy, 1914–1917', *Intelligence and National Security* 5.22 (2007), pp. 699–721.

—— 'Anglo-American Strategic Relations and the Blockade, 1914–1916', *Journal of Transatlantic Studies* 6.1 (2008), pp. 22–33.

—— 'The Royal Navy and Imperial Defence, 1919–1945', in G. Kennedy (ed.), *British Imperial Defence: The Old World Order, 1856–1956* (London: Routledge, 2008), pp. 133–52.

—— 'Strategy and Power: The Royal Navy, The Foreign Office and the Blockade, 1914–1917', *Journal of Defence Studies* 8.2 (2008), pp. 190–206.

—— (ed.), *Britain's War at Sea, 1914–1918: The War They Thought and the War They Fought* (Abingdon: Routledge, 2016).

—— 'The British Strategic Assessment of the United States as a Maritime Power, 1900–1917', in G. Kennedy (ed.), *Britain's War at Sea, 1914–1918: The War They Thought and the War They Fought* (Abingdon: Routledge, 2016), pp. 4–23.

—— 'Anglo-American Economic Warfare and the Deterrence of Japan, 1933–1941', in Anastasia Filippidou (ed.), *Concepts and Approaches to Deterrence* (Bristol: Bristol University Press, 2019).

Kennedy, P.M., *The Rise and Fall of British Naval Mastery* (London: Allen Lane, 1976).

—— *The Rise and Fall of the Great Powers: Economic Change and Military Conflict from 1500 to 2000* (New York: Vintage, 1986).

—— 'The Influence and the Limitations of Sea Power', *International History Review* 10.1 (1988), pp. 2–17.

—— 'Grand Strategy in War and Peace: Toward a Broader Definition', in P. Kennedy (ed.), *Grand Strategies in War and Peace* (New Haven, CT: Yale University Press, 1991).

Kert, Faye M., *Privateering: Patriots and Profits in the War of 1812* (Baltimore, MD: Johns Hopkins Universty Press, 2015).

King, Charles, *The Black Sea: A History* (Oxford, Oxford University Press, 2004).

Klooster, W., *Illicit Riches: Dutch Trade in the Caribbean, 1648–1795* (Leiden: KITLV Press, 1998).

Klooster, W. and G. Oostindie, *Realm between Empires: The Second Dutch Atlantic, 1690–1815* (Ithaca, NY: Cornell University Press, 2018).

Knight, R., *The Pursuit of Victory: The Life and Achievement of Horatio Nelson* (London: Allen Lane, 2005).

—— *Britain against Napoleon: The Organization of Victory, 1793–1815* (London: Allen Lane, 2013).

Knight, R. and M. Wilcox, *Sustaining the Fleet, 1793–1815: War, the British Navy and the Contractor State* (Woodridge: Boydell & Brewer, 2010).

Kramer, A., 'Blockade and Economic Warfare', in J.M. Winter (ed.), *The Cambridge History of the First World War*, vol. 2, *The State* (Cambridge: Cambridge University Press, 2014).

Kuethe, A. and K. Andrien, *The Spanish Atlantic World in the Eighteenth Century: War and the Bourbon Reforms, 1713–1796* (Cambridge: Cambridge University Press, 2014).

Kulsrud, C., *Maritime Neutrality to 1780: A History of the Main Principles Governing Neutrality and Belligerency to 1780* (Boston: Little, Brown, and Co., 1936).

Lacey, J (ed.), *Great Strategic Rivalries: From the Classical World to the Cold War* (Oxford: Oxford University Press, 2016).

Lambert, Andrew, 'Great Britain and Maritime Law from the Declaration of Paris to the Era of Total War', in R. Hobson and T. Kristiansen (eds), *Navies in Northern Waters, 1721–2000* (Portland, OR: Frank Cass, 2004), pp. 23–31.

—— 'The Naval War Course, Some Principles of Maritime Strategy and the Origins of "The British Way in Warfare"', in K. Neilson and G. Kennedy (eds), *The British Way in Warfare: Power and the International System, 1856–1956* (Farnham: Ashgate, 2010), pp. 219–56.

—— *The Challenge: Britain against America in the Naval War of 1812* (London: Faber and Faber, 2012).

—— *Seapower States: Maritime Culture, Continental Empires, and the Conflict that Made the Modern World* (New Haven, CT: Yale University Press, 2018).

Lambert, F., *The Barbary Wars: American Independence in the Atlantic World* (New York: Hill & Wang, 2005).

Lambert, N.A., 'Strategic Command and Control for Maneuver Warfare: Creation of the Royal Navy's "War Room" System, 1905–1915', *Journal of Military History* 69.2 (2005), pp. 361–410.

—— *Planning Armageddon: British Economic Warfare and the First World War* (Cambridge, MA: Harvard University Press, 2012).

Lambi, I.N., *The Navy and German Power Politics, 1862–1914* (Boston: Allen & Unwin, 1984).

Laughton, J.K., 'On Convoy', in T.A. Brassey (ed.), *The Naval Annual 1894* (Portsmouth: Brasseys, 1894).

Lawrence, T.J., 'Problems of Neutrality Connected with the Russo-Japanese War', *Royal United Services Institution Journal* 48.318 (1904), pp. 915–37.

Le Fevre, P. and R. Harding (eds), *British Admirals of the Napoleonic Wars: The Contemporaries of Nelson* (London: Chatham, 2005).

Lemnitzer, J., *Power, Law and the End of Privateering* (London: Palgrave Macmillan, 2014).

Lincoln, M., *Trading in War: London's Maritime World in the Age of Cook and Nelson* (New Haven, CT: Yale University Press, 2018).

Luke, M.H., *The Port of New York, 1800–1810: The Foreign Trade and Business Community* (New York: New York University, 1953).

Lynd, S. and D. Waldstreicher, 'Free Trade, Sovereignty, and Slavery: Toward an Economic Interpretation of American Independence', *William & Mary Quarterly*, 3rd series, 71 (2011), pp. 597–630.

McCarthy, M., '"A Delicate Question of a Political Nature": The *Corso Insurgente* and British Commercial Policy during the Spanish-American Wars of Independence, 1810–1824', *International Journal of Maritime History* 23.1 (2011), pp. 277–92.

McCranie, K.D., 'The War of 1812 in the Ongoing Napoleonic Wars: The Response of Britain's Royal Navy', *Journal of Military History* 76 (2012).

McCusker, John J., 'Worth a War? The Importance of Trade between British America and The Mediterranean', in S. Marzagalli, John J. McCusker, and J. Sofka (eds), *Rough Waters: American Involvement with the Mediterranean in the Eighteenth and Nineteenth Centuries* (St John's, Newfoundland: International Maritime Economic History Association, 2010).

McCusker, John J. and Russell R. Menard, *The Economy of British America, 1607–1789* (Chapel Hill: University of North Carolina Press, 1985; 2nd ed.; 1991).

McDonald, J.K., 'The Washington Conference and the Naval Balance of Power, 1921–22', in John B. Hattendorf and Robert S. Jordan (eds), *Maritime Strategy and the Balance of Power* (London: Springer, 1989).

Mackinder, H.J., 'Man-Power as a Measure of National and Imperial Strength', *National Review* (1905), pp. 136–43.

McMahon, C.J., 'Maritime Trade Warfare: A Strategy for the Twenty-First Century?' *Naval War College Review* (2017), pp. 15–38.

Mahan, A.T., *The Influence of Sea Power upon the French Revolution and Empire, 1793–1812*, 2 vols (London: Sampson Low, Marston and Co., 1893).

—— *Interest of America in Sea Power, Present and Future* (Port Washington, NY: Kennikat, 1897).

—— *Sea Power in its Relation to the War of 1812*, 2 vols (London: Sampson Low, Marston, 1905).

—— *The Influence of Sea Power upon History, 1660–1783* (Boston: Little Brown, 1918 [1890]).

Marcus, G., *A Naval History of England: The Age of Nelson* (Sheffield: Applebaum, 1971).

Marsden, A., 'The Blockade', in F.H. Hinsley (ed.), *British Foreign Policy under Sir Edward Grey* (Cambridge: Cambridge University Press, 1977), pp. 466–87.

Martel, W.C., *Grand Strategy in Theory and Practice: The Needs for an Effective American Foreign Policy* (Cambridge: Cambridge University Press, 2015).

Martin, T.G., *A Most Fortunate Ship: A Narrative History of Old Ironsides* (Annapolis, MD: Naval Institute Press, 1997).

Marzagalli, S., 'Establishing Transatlantic Trade Networks in Time of War: Bordeaux and the United States, 1793–1815', *Business History Review* 79.4 (2005), pp. 811–44.

—— 'Napoleon's Continental Blockade: An Effective Substitute to Naval Weakness?' in B.A. Elleman and S.C.M. Paine (eds), *Naval Blockades and Seapower: Strategies and Counter-Strategies, 1805–2005* (London: Routledge, 2006), pp. 23–33.

—— 'Limites et opportunités dans l'Atlantique français au XVIIIe siècle: le cas de la maison Gradis de Bordeaux', *Outre-Mers* 362–3 (2009), pp. 87–110.

—— *Bordeaux et les États-Unis, 1776–1815: politique et stratégies négociantes dans la genèse d'un réseau commercial* (Geneva: Droz, 2015).

—— 'La navigation américaine pendant les French Wars (1793–1815): une simple reconfiguration des circuits commerciaux par neutres interposés?' in Eric Schnakenbourg (ed.), *Neutres et neutralité dans l'espace atlantique durant le long XVIIIe siècle (1700–1820): une approche globale/Neutrals and Neutrality in the Atlantic World during the Long Eighteenth Century (1700–1820): A Global Approach* (Bécherel: Les Perseides, 2015).

—— '"However Illegal, Extraordinary or Almost Incredible Such Conduct Might Be": Americans and Neutrality Issues in the Mediterranean during the French Wars', *International Journal of Maritime History* 28.1 (2016), pp. 118–32.

—— 'Was Warfare Necessary for the Functioning of Eighteenth-Century Colonial Systems? Some Reflections on the Necessity of Cross-Imperial and Foreign Trade in the French Case', in Cátia A.P. Antunes and Amelia Polónia (eds), *Beyond Empires: Global, Self-Organizing, Cross-Imperial Networks, 1500–1800* (Leiden: Brill, 2016).

—— 'Le réseau consulaire des États-Unis en Méditerranée (1790–1815): logiques étatiques, logiques marchandes?' in Arnaud Bartolomei, Guillaume Calafat, and Jörg Ulbert (eds), *De l'utilité commerciale des consuls: l'institution consulaire et les marchands dans le monde méditerranéen (XVIIe–XIXe siècles)* (Rome and Madrid: Casa de Velázquez-EFR, 2017), pp. 295–307.

Marzagalli, S. and L. Müller (eds), '"In Apparent Disagreement with All Law of Nations in the World": Negotiating Neutrality for Shipping and Trade during the French Revolutionary Wars', *International Journal of Maritime History* 28.1 (2016), pp. 108–92.

Medlicott, W.N., *History of the Second World War: The Economic Blockade*, vol. 1 (London: HMSO, 1959).

Merli, F.J., *The Alabama, British Neutrality and the American Civil War* (Bloomington: Indiana University Press, 2004).

Miller, E.S., *War Plan Orange: The US Strategy to Defeat Japan, 1897–1945* (Annapolis, MD: Naval Institute Press, 1991).

Minchinton, W.E., *The Trade of Bristol in the Eighteenth Century* (Bristol: Bristol Record Society, 1957).

Mirski, S., 'Stranglehold: The Context, Conduct and Consequences of an American Naval Blockade of China', *Journal of Strategic Studies* 36.3 (2013), pp. 385–421.

Moran, D. and J. Russell, *Maritime Strategy and Global Order: Markets, Resources, Security* (Washington, DC: Georgetown University Press, 2016).

Morgan, K., *Bristol and the Atlantic Trade in the Eighteenth Century* (Cambridge: Cambridge University Press, 1993).

Morgan-Owen, D.G., 'War as it Might Have Been: British Sea Power and the First World War', *Journal of Military History* 83.4 (2019), pp. 1095–131.

Morriss, R., *The Royal Dockyards during the Revolutionary and Napoleonic Wars* (Leicester: Leicester University Press, 1983).

Moss, W.E., 'The End of the Crimean System: England, Russia and the Neutrality of the Black Sea, 1870–1', *Historical Journal* 4.2 (1961), pp. 164–90.

Müller, L., 'Swedish Merchant Shipping in Troubled Times: The French Revolutionary Wars and Sweden's Neutrality, 1793–1801', *International Journal of Maritime History* 28.1 (2016), pp. 147–64.

Mulligan, W., 'Mobs and Diplomats: The *Alabama* Affair and British Diplomacy, 1865–1872', in Markus Mösslang and Torsten Riotte (eds), *The Diplomats' World: A Cultural History of Diplomacy 1815–1914* (Oxford: Oxford University Press, 2008), pp. 105–32.

Murfett, M.H., 'Look Back in Anger: The Western Powers and the Washington Conference of 1921–1922', in B.J.C. McKercher (ed.), *Arms Limitation and Disarmament* (New York: Praeger, 1992), pp. 83–104.

Murray, S.L., 'Our Food Supply in Time of War and Imperial Defence', *Journal of the Royal United Services Institution* 45 (1901), pp. 656–729.

Neff, S., *The Rights and Duties of Neutrals: A General History* (Manchester: Manchester University Press, 2000).

—— *Justice among Nations: A History of International Law* (Cambridge, MA: Harvard University Press, 2014).

Negus, S., 'A Notorious Nest of Offence: Neutrals, Belligerents and Union Jails in Civil War Blockade Running', *Civil War History* 56.4 (2010), pp. 350–85.

Neilson, K., '"Unbroken Threat": Japan, Maritime Power and British Imperial Defence, 1920–32', in G. Kennedy (ed.), *British Naval Strategy East of Suez, 1900–2000: Influences and Actions* (London: Routledge, 2005).

Neilson, K. and G. Kennedy (eds), *The British Way in Warfare: Power and the International System, 1856–1956* (Farnham: Ashgate, 2010).

Newbolt, H., *Naval Operations* (London: Longmans, Green and Company, 1928).

Nish, I.H., *Alliance in Decline* (London: Athlone Press, 1972).

Nolan, C., *The Allure of Battle: A History of How Wars Have Been Won and Lost* (Oxford: Oxford University Press, 2017).

Norman, C.B., *The Corsairs of France* (London: Sampson Low, 1887).

Norris, W.J., *Chinese Economic Statecraft: Commercial Actors, Grand Strategy, and State Control* (Ithaca, NY: Cornell University Press, 2016).

North, Douglass C., 'The United States Balance of Payments, 1790–1860', *Trends in American Economy in the Nineteenth Century*, Studies in Income and Wealth 24 (Princeton, NJ: Princeton University Press, 1960), pp. 573–627.

O'Brian, P., 'European Economic Development: The Contribution of the Periphery', *Economic History Review* 35.1 (1982), pp. 1–18.

O'Brien, K., 'Global Warfare and Long-Term Economic Development, 1789–1939', *War in History* 3.4 (1996), pp. 437–50.

O'Brien, Patrick K., *The Economic Effects of the American Civil War* (Basingstoke: Palgrave Macmillan, 1988).

O'Brien, P.P., *How the War Was Won: Air–Sea Power and Allied Victory in World War II* (Cambridge: Cambridge University Press, 2015).

O'Malley, P., 'The Discipline of Violence: State, Capital and the Regulation of Naval Warfare', *Sociology* 22.2 (1988), pp. 253–70.

O'Shaughnessy, A., *An Empire Divided: The American Revolution and the British Caribbean* (Philadelphia: University of Pennsylvania Press, 2000).

Offer, A., *The First World War: An Agrarian Interpretation* (Oxford: Clarendon Press, 1989).

Osborne, E.W., *Britain's Economic Blockade of Germany, 1914–1919* (London: Frank Cass, 2004).

Overlack, P., 'German Commerce Warfare Planning for the Australia Station, 1900–1914', *War and Society* 14.1 (1996), pp. 17–48.

—— 'The Function of Commerce Warfare in an Anglo-German Conflict to 1914', *Journal of Strategic Studies* 20.4 (1997), pp. 91–114.

Overy, R., *Why the Allies Won* (London: Pimlico, 2006).

Paine, L., *The Sea and Civilization: A Maritime History of the World* (New York: Vintage, 2013).

Pares, R., *Colonial Blockade and Neutral Rights, 1739–1763* (Oxford: Clarendon Press, 1938).

Paret, Peter (ed.), *Makers of Modern Strategy from Machiavelli to the Nuclear Age* (Princeton, NJ: Princeton University Press, 1986).

Parrott, D., *The Business of War: Military Enterprise and Military Revolution in Early Modern Europe* (Cambridge: Cambridge University Press, 2012).

Parry, C., 'Foreign Policy and International Law', in F.H. Hinsley (ed.), *British Foreign Policy under Sir Edward Grey* (Cambridge: Cambridge University Press, 1977), pp. 89–112.

Parthesius, R., 'Dutch Ships in Tropical Waters: The Development of the Dutch East India Company (VOC) Shipping Network in Asia 1595–1660', PhD thesis, University of Amsterdam, 2007.

Pearce, A., *British Trade with Spanish America, 1763–1808* (Liverpool: Liverpool University Press, 2007).

Perkins, B., 'Sir William Scott and the *Essex*', *William and Mary Quarterly* 13 (1956), pp. 169–83.

Phillips, C.R., *Six Galleons for the King of Spain: Imperial Defense in the Early Sixteenth Century* (Baltimore, MD: Johns Hopkins University Press, 1986).

Pitkin, T., *A Statistical View of the Commerce of the United States* (Hartfort, CT: Charles Hosmer, 1816; repr. 1967; new ed.; New Haven, CT: Durrie & Peck, 1835).

Pomeranz, K., *The Great Divergence: China, Europe, and the Making of the Modern World Economy* (Princeton, NJ: Princeton University Press, 2000).

Pool, B., *Navy Board Contracts, 1660–1832: Contract Administration under the Navy Board* (London: Archon Books, 1966).

Pope, D., *The Battle of the River Plate: The Hunt for the German Pocket Battleship Graf Spee* (Ithaca, NY: McBooks Press, 2005).

Pourchasse, P., 'La guerre de la faim: l'approvisionnement de la République, le blocus britannique, et les bonnes affaires des neutres au cours des guerres révolutionnaires (1793–1795)', HdR thesis, Université de Bretagne Sud, 2013.

Ranft, B. (ed.), *Technical Change and British Naval Policy* (London: Hodder and Stoughton, 1977).

—— 'Restraints on War at Sea before 1945', in Michael Howard (ed.), *Restraints on War* (Oxford: Oxford University Press, 1979), pp. 39–56.

Ransom, R.L., 'British Policy and Colonial Growth: Some Implications of the Burdens of the Navigation Acts', *Journal of Economic History* 28.3 (1968), pp. 427–35.

Rath, A.C., *The Crimean War in Imperial Context, 1854–1856* (London: Palgrave Macmillan, 2015).

Redford, D., 'Collective Security and Internal Dissent: The Navy League's Attempts to Develop a New Policy towards British Naval Power between 1919 and the 1922 Washington Naval Treaty', *History* 96.321 (2011), pp. 48–67.

Reich, S. and P. Dombrowski, *The End of Grand Strategy: US Maritime Operations in the 21st Century* (Ithaca, NY: Cornell University Press, 2017).

Reinert, S., 'Rivalry: Greatness in Early Modern Political Economy', in P.J. Stern and C. Wennerlind (eds), *Mercantilism Reimagined: Political Economy in Early Modern Britain and its Empire* (Oxford: Oxford University Press, 2014).

Ricardo, J.L., *Anatomy of the Navigation Laws* (London: C. Gilpin, 1847).

Richardson, C. and P. Riden (eds), *Minutes of the Chesterfield Canal Company, 1771–1780* (Chesterfield: Derbyshire Record Society, 1996).

Richmond, H.W., *British Strategy, Military and Economic: A Historical Review and its Contemporary Lessons* (Cambridge: Cambridge University Press, 1941).

—— *Statesmen and Sea Power* (Oxford: Clarendon Press, 1946).

Riley, J.C., *The Seven Years War and the Old Regime in France: The Economic and Financial Toll* (Princeton, NJ: Princeton University Press, 1986).

Riste, O., *The Neutral's Ally: Norway's Relations with Belligerent Powers in the First World War* (Oslo: Universitetsforlaget, 1965).

Rodger, N.A.M., *The Wooden World: An Anatomy of the Georgian Navy* (London: Collins, 1988 [1986]).

—— *The Safeguard of the Sea: A Naval History of Britain*, vol. 1, *660–1649* (London: HarperCollins, 1997).

—— *The Command of the Ocean: A Naval History of Britain*, vol. 2, *1649–1815* (London: Allen Lane with the National Maritime Museum, 2004).

—— 'War as an Economic Activity in the "Long" Eighteenth Century', *International Journal of Maritime History* 22.2 (2010), pp. 1–18.

Rodger, N.A.M., J.R. Dancy, B. Darnell, and E. Wilson (eds), *Strategy and the Sea: Essays in Honour of John B. Hattendorf* (Woodbridge: Boydell & Brewer, 2016).

Rogers, N., 'British Impressment and its Discontents', *International Journal of Maritime History* 30 (2018), pp. 52–73.

Röhl, J.C.G., *Wilhelm II: Into the Abyss of War and Exile, 1900–1941* (Cambridge: Cambridge University Press, 2014).

Roksund, A., *The Jeune École: The Strategy of the Weak* (Leiden: Brill, 2007).

Roskill, S.W., *Naval Policy between the Wars*, vol. 1 (London: Walker, 1968).

—— *The Navy at War, 1939–1945* (London: Wordsworth, 1998).

Ruger, J., *Heligoland: Britain, Germany and the Struggle for the North Sea* (Oxford: Oxford University Press, 2017).

Ryan, A.N., 'The Defence of British Trade with the Baltic, 1808–1813', *English Historical Review* 74 (1959), pp. 443–66.

Sawers, Larry, 'The Navigation Acts Revisited', *Economic History Review* 45 (1992), pp. 262–84.

Schnakenbourg, C., 'Les sucreries de la Guadeloupe dans la seconde moitié du XVIIIᵉ siècle (1760–1790): contribution à l'étude de la crise de l'économie coloniale à la fin de l'ancien régime', Thèse d'état, Université de Paris II, 1972.

Schumpeter, E.B., *English Overseas Trade Statistics, 1697–1808* (Oxford: Clarendon Press, 1960).

Schurman, D.M., *Imperial Defence, 1868–1887*, ed. J. Beeler (London: Frank Cass, 2000).

Scott, H.M., *British Foreign Policy in the Age of the American Revolution* (Oxford: Clarendon Press, 1990).

Scott, J.B., *The Hague Peace Conferences of 1899 and 1907*, vol. 1 (Baltimore, MD: Johns Hopkins University Press, 1909).

Seligmann, M.S., *The Royal Navy and the German Threat, 1901–1914: Admiralty Plans to Protect British Trade in a War against Germany* (Oxford: Oxford University Press, 2012).

—— 'Germany's Ocean Greyhounds and the Royal Navy's First Battle Cruisers: An Historiographical Problem', *Diplomacy and Statecraft* 27 (2016), pp. 162–82.

—— 'Failing to Prepare for the Great War? The Absence of Grand Strategy in British War Planning before 1914', *War in History* 24.4 (2017), pp. 414–37.

Seligmann, M.S., Michael Epkenhans, and Frank Nägler (eds), *The Naval Route to the Abyss: The Anglo-German Naval Race, 1895–1914* (Farnham: Ashgate for the Navy Records Society, 2015).

Semmel, B., *Liberalism and Naval Strategy: Ideology, Interest and Sea Power during the* Pax Britannica (Boston: Allen & Unwin, 1986).

Sicking, L., *Neptune and the Netherlands: State, Economy, and War at Sea in the Renaissance* (Leiden: Brill, 2014).

Simpson, M. (ed.), *Anglo-American Naval Relations, 1917–1919* (Basingstoke: Navy Records Society, 1991).

Siney, M.C., *The Allied Blockade of Germany, 1914–1916* (Ann Arbor: University of Michigan Press, 1957).

—— 'British Official Histories of the Blockade of the Central Powers during the First World War', *American Historical Review* 68.2 (1963), pp. 392–401.

—— 'The Allied Blockade Committee and the Inter-Allied Trade Committees: The Machinery of Economic Warfare, 1917–1918', in K. Bourne and D.C. Watt (eds), *Studies in International History: Essays Presented to W. Norton Medlicott* (London: Archon Books, 1967), pp. 330–44.

Smith, R., 'Britain and the Strategy of the Economic Weapon in the War against Germany, 1914–1919', PhD thesis, Newcastle University, 2000.

Stagg, J.C.A., *The War of 1812: Conflict for a Continent* (Cambridge: Cambridge University Press, 2012).

Stapelbroek, K. (ed.), *Trade and War: The Neutrality of Commerce in the Inter-State System*, COLLeGIUM: Studies across Disciplines in the Humanities and Social Sciences, 10 (2011).

Stark, F.R., *Abolition of Privateering and the Declaration of Paris* (New York: Columbia University Press, 1987).

Starkey, D.J., *British Privateering Enterprise in the Eighteenth Century* (Exeter: University of Exeter Press, 1990).

Steffen, D., 'The Holtzendorff Memorandum of 22 December 1916 and Germany's Declaration of Unrestricted U-Boat Warfare', *Journal of Military History* 68 (2004), pp. 215–24.

Steinberg, J., *Yesterday's Deterrent: Tirpitz and the Birth of the German Battle Fleet* (New York: Macdonald, 1965).

Steiner, Z., *The Triumph of the Dark: European International History, 1933–1939* (Oxford: Oxford University Press, 2011).

Stephenson, D., K. Bourne, and D. Cameron Watt (eds), *British Documents on Foreign Affairs, Part II, Series H, the First World War, 1914–1918*, vol. 6, *Blockade and Economic Warfare, II, July 1915–January 1916* (Frederick, MD: University Publications of America, 1989).

Stockton, C.H., 'The Declaration of Paris', *American Journal of International Law* 14.3 (1920), pp. 356–68.

Stone, I.R., 'The Crimean War in the Arctic', *Polar Record* 21.135 (1985), pp. 577–81.

Strachan, H., *Carl von Clausewitz's On War: A Biography* (New York: Grove Press, 2008).

—— (ed.), *The Direction of War: Contemporary Strategy in Historical Perspective* (Cambridge: Cambridge University Press, 2013).

—— 'Strategy in Theory; Strategy in Practice', *Journal of Strategic Studies* 42.2 (2019), pp. 171–90.

Strachan, H. and A. Herberg-Rothe (eds), *Clausewitz in the Twenty-First Century* (Oxford: Oxford University Press, 2007).

Stradling, R.A., *The Armada of Flanders: Spanish Maritime Policy and European War, 1568–1668* (Cambridge: Cambridge University Press, 1992).

Sumida, J.T., *Inventing Grand Strategy and Teaching Command: The Classic Works of Alfred Thayer Mahan Reconsidered* (Baltimore, MD: Johns Hopkins University Press, 1999).

—— *Decoding Clausewitz: A New Approach to On War* (Lawrence: University Press of Kansas, 2008).

Sutcliffe, R.K., *British Expeditionary Warfare and the Defeat of Napoleon, 1793–1815* (Woodbridge: Boydell & Brewer, 2016).

Symonds, C.L., *Navalists and Antinavalists: The Naval Policy Debate in the United States, 1785–1827* (Newark: University of Delaware Press, 1980).

—— *World War II at Sea: A Global History* (New York: Oxford University Press, 2018).

Tarrade, J., *Le commerce colonial de la France à la fin de l'Ancien Régime*, 2 vols (Paris: Presses universitaires de France, 1972).

Thomas, R.P., 'A Quantitative Approach to the Study of the Effects of British Imperial Policy on Colonial Welfare: Some Preliminary Findings', *Journal of Economic History* 25 (1964), pp. 615–38.

—— 'British Imperial Policy and the Economic Interpretation of the American Revolution', *Journal of Economic History* 28 (1968), pp. 427–40.

Tooze, A., *The Wages of Destruction: The Making and Breaking of the Nazi Economy* (London: Allen Lane, 2006).

Tracy, Nicholas (ed.), *Sea Power and the Control of Trade: Belligerent Rights from the Russian War to the Beira Patrol, 1854–1970* (Aldershot: Ashgate and Navy Records Society, 2005).

Van der Burgh, A.H.H., *Inventaris van het archief van de Directie van de Levantse Handel en de Navigatie in de Middellandse Zee (1614) 1625–1826 (1828)* (The Hague: Nationaal Archief, 1882).

Verzijl, J.H.W., *International Law in Historical Perspective*, 10 (Leiden: A.W. Stijhoff, 1968).

Ville, S.P., *English Shipowning during the Industrial Revolution: Michael Henley and Son, London Shipowners 1770–1830* (Manchester: Manchester University Press, 1987).

Vincent, C.P., *The Politics of Hunger: The Allied Blockade of Germany, 1915–1919* (Athens: Ohio University Press, 1985).

Vliet, A.P. van, *Vissers en kapers: de zeevisserij vanuit het Maasmondgebied en de Duinkerker kapers (ca. 1580–1648)* (The Hague: Stichting Hollandse Historische Reeks, 1994).

Voelcker, T., *Admiral Saumarez versus Napoleon: The Baltic, 1807–12* (Woodbridge: Boydell Press, 2008).

Waddell, D.A.G., 'British Neutrality and Spanish-American Independence: The Problem of Foreign Enlistment', *Journal of Latin American Studies* 19.1 (1987), pp. 1–18.

Ward, J.R., *The Finance of Canal Building in Eighteenth-Century England* (Oxford: Oxford University Press, 1974).

Watson, A., *Ring of Steel: Germany and Austria–Hungary in World War I* (New York: Basic Books, 2014).

Weir, G.E., 'Tirpitz, Technology and U-Boat Building, 1897–1916', *International History Review* 6 (1984), pp. 174–90.

—— *Building the Kaiser's Navy: The Imperial Navy Office and German Industry in the von Tirpitz Era, 1890–1919* (Annapolis, MD: Naval Institute Press, 1992).

White, P.L. (ed.), *The Beekman Mercantile Papers, 1746–1799*, 3 vols (New York: New York Historical Society, 1956).

Widen, J.J., 'Julian Corbett and the Current British Maritime Doctrine', *Comparative Strategy* 28.2 (2009), pp. 170–85.

Wilkinson, C., *The British Navy and the State in the Eighteenth Century* (Woodbridge: Boydell Press, 2004).

Willis, S., *The Struggle for Sea Power: A Naval History of American Independence* (London: Atlantic Books, 2015).

Willmott, H.P., *Empires in the Balance: Japanese and Allied Pacific Strategies to April 1942* (Annapolis, MD: Naval Institute Press, 1982).

Winfield, R., *British Warships in the Age of Sail, 1793–1817* (London: Chatham Publishing, 2005).

Winter, J.M. (ed.), *War and Economic Development: Essays in Memory of David Joslin* (Cambridge: Cambridge University Press, 1975).

Witt, J.M., 'Smuggling and Blockade-Running during the Anglo-Danish War from 1807 to 1814', in K.B. Aaslestad and J. Joor (eds), *Revisiting Napoleon's Continental System: Local, Regional and European Experiences* (Basingstoke: Palgrave Macmillan, 2015).

Woodman, R., *Arctic Convoys, 1941–1945* (London: John Murray, 2004).

Wright, C. and C.E. Fayle, *A History of Lloyd's* (London: Macmillan, 1928).

Wright, Q., 'The Present Status of Neutrality', *American Journal of International Law* 34.3 (1940), pp. 391–415.

Wylie, J. C., *Military Strategy: A General Theory of Power Control*, ed. J.B. Hattendorf (Annapolis, MD: Naval Institute Press, 1989).

Index

Printed and bound by CPI Group (UK) Ltd, Croydon, CR0 4YY

27/10/2024

14580408-0004